Introduction to Semiconductor Physics and Devices

Mykhaylo Evstigneev

Introduction to Semiconductor Physics and Devices

 Springer

Mykhaylo Evstigneev
Department of Physics and Physical
Oceanography
Memorial University of Newfoundland
St. John's, NL, Canada

ISBN 978-3-031-08460-7   ISBN 978-3-031-08458-4   (eBook)
https://doi.org/10.1007/978-3-031-08458-4

This Springer imprint is published by the registered company Springer Nature Switzerland AG
The registered company address is: Gewerbestrasse 11, 6330 Cham, Switzerland

Preface

When I was assigned to teach physics of device materials at Memorial University of Newfoundland 7 years ago, I got excited. Who would not be happy to tell the students about such diverse areas as quantum mechanics, statistical physics, theory of solids, electrodynamics, and crystallography, not to mention some cool electronic applications, in a one-semester course? This opportunity was especially thrilling, because the majority of my audience consisted of second-year engineering students with little or no prior knowledge about most of these subjects. It is always great to be the first to tell young people about something tremendously important.

Finding a suitable text is the first challenge an instructor faces before teaching a course. This choice is usually made based on two factors: the knowledge the instructor intends to impart to the students and their background preparation level. There are many excellent books on semiconductor physics. But some of them implicitly assume the readers' familiarity with the topics mentioned above, while others dig into those topics too deeply and are more suitable for the advanced undergraduate or graduate physics students. Neither description applied to my audience.

Hence, I wrote a few pages on quantum mechanics and a few on statistical physics, which I shared with the students as a handout. As I was teaching this course in the following years, I added more material to those notes, which finally developed into this book.

The book contains three roughly equal parts, each of which can be covered in about one month's time. The first part, Fundamental Physics, gives the background information students will need to understand the rest of the text: quantum mechanics, crystal structure of solids, statistical mechanics, and band theory of solids. In the second part, this information is applied to explain the basic properties of semiconductors, namely the concentration of charge carriers in equilibrium, the interrelation between the carrier concentration and a non-uniform electric potential, the generation-recombination processes, and, finally, the response of the charge carriers to an electric field. The third part applies what the students have learned earlier in the book to explain some semiconductor devices: the Schottky diodes, the MOSFETs, the pn junction diodes, and some optoelectronic applications thereof.

Since my intention was to give only slightly more material than can be delivered within a one-semester course, many topics were left out. For example, the size-related effects in the semiconductor devices and the bipolar transistor are not discussed. An interested reader is referred to a number of excellent texts to fill those gaps, such as:

K. Brennan, The Physics of Semiconductors (Cambridge University Press, 1999)

J.P. Collinge and C.A. Collinge, Physics of Semiconductor Devices (Kluwer Academic Publishers, 2002)

J.A. del Alamo, Integrated Microelectronic Devices: Physics and Modeling (Pearson, 2018)

C. Hamaguchi, Basic Semiconductor Physics (Springer, 2010)

C. Hu, Modern Semiconductor Devices for Integrated Circuits (Pearson, 2009)

M. Grundmann, The Physics of Semiconductors (Springer, 2010)

J.P. McKelvey, Solid State and Semiconductor Physics (R.F. Krieger Publishing Company, 1966)

R.S. Muller and Th.I. Kamins, "Device Electronics for Integrated Circuits" (Wiley, 2002)

D. Neamen, Semiconductor Physics and Devices (McGraw-Hill, 2003)

C. Papadopoulos, Solid-State Electronic Devices (Springer, 2014)

R.F. Pieret, Semiconductor Device Fundamentals (Addison-Wesley Publishing Company, 1996)

D.K. Schroder, Semiconductor Material and Device Characterization (Wiley, 2006)

K. Seeger, Semiconductor Physics (Springer, 2004)

R.A. Smith, Semiconductors (Cambridge Univ. Press, 1978)

S.M. Sze and M.K. Lee, Semiconductor Devices (Wiley, 2012)

S.M. Sze and K.K. Ng, Physics of Semiconductor Devices (Wiley, 2007)

P.Y. Yu and M. Cardona, Fundamentlals of Semiconductors (Springer, 2010)

Reading a book like this is best done with pencil and paper. Ideally, a student should understand the material well enough to be able to explain it to another student. At the end of each chapter, the student will find problems that should, hopefully, strengthen their understanding of the chapter material. Each of the problem sections consists of three parts: solved problems, practice problems, and solutions. Students are encouraged to first try and tackle the problems by themselves and look into the solutions only if they are stuck.

Although I made every effort to make this book free of errors, I cannot guarantee that it is. I will greatly appreciate it if the readers inform me about any typo and mistake they find, or offer any other suggestions and criticism by email at mevstigneev@mun.ca.

I would like to thank my students and colleagues for their input. I am indebted to Dr. Michael Morrow for sharing his own materials with me when I got to teach this course for the first time. It has been a great pleasure to discuss pertinent subjects with Drs. Anatoliy Sachenko and Igor Sokolovskyi from the Institute of Semiconductor Physics in Kyiv, Ukraine. Last but not least, I thank Gleb Evstigneev for his suggestions that helped me improve the style of this text.

I dedicate this book to Ukraine, my brave home country.

St. John's, NL, Canada   Mykhaylo Evstigneev

Contents

Part I
Fundamental Physics

Chapter 1
Principles of Quantum Mechanics

1.1 Why Quantum Mechanics?

As far back as ca. 2700 years ago, some natural philosophers speculated that the building block of matter is a small particle they called an atom; its Greek name $\alpha\tau o\mu o\nu$ means "indivisible." Given its correctness, it comes as a surprise that atomism was a disputed idea for two and a half millennia.

Science is not all about brilliant ideas and deep insights into the nature's ways. It is also about healthy skepticism and reluctance to accept a particular theory before every competing theory is ruled out. People who were critical of atomism were not necessarily bad scientists; quite the opposite is true about many of them. The simple reason why it took so long for atoms to enter the mainstream of science is that seeing is believing. Atoms are too tiny to be seen even in the best optical microscope. One had to rely on the indirect evidence rather than direct observation of atoms.

For example, one can look at the barely visible small objects, such as pollen, coal dust, and similar particles suspended in water. Their irregular Brownian motion can be interpreted as caused by the unequal number of hits they receive from different sides at each instant of time. The hits are coming from atoms. Another example: chemists noticed that the masses of substances participating in chemical reactions are related to each other as integer numbers. This can be taken as an evidence of the minimal building blocks of matter that form chemical compounds by reacting with each other.

But with a bit of imagination, one can think of other (wrong) explanations of these facts based on the information that was available in the pre-nineteenth century science.

The big breakthrough was made by Joseph Thomson in 1897, who discovered that atoms are not as indivisible as previously believed. Thomson was interested in the flow of electric current between two metal plates placed against each other inside a vacuum tube. He showed that the current is carried by the negatively charged

M. Evstigneev, *Introduction to Semiconductor Physics and Devices*,
https://doi.org/10.1007/978-3-031-08458-4_1

particles (the electrons), whose charge-to-mass ratio could be measured by studying the deflection of an electron beam in electric and magnetic fields.

The next step was made in the 1910s by Geiger, Marsden, and Rutherford, who studied the transmission of alpha-particles through a thin metal foil. Their angular distribution signaled that alpha-particles underwent collisions with the positively charged nuclei thousands of times heavier than the electrons orbiting around it.

At a slightly earlier time, another line of research was actively being developed: the electromagnetism. It became established that an accelerated charge emits electromagnetic waves. Now, circular motion is a special type of an accelerated motion. An electron orbiting around the nucleus should emit electromagnetic waves, thereby losing energy and approaching the nucleus very quickly. Eventually, the electron should fall into the nucleus. An atom consisting of a heavy nucleus with light electrons orbiting around it cannot be stable.

One might have thought that this model was lacking some important piece of information that would explain the stability of atoms. But what was actually needed was an all-out revision of everything that physicists knew at that time. New physics had to be created, which we know as quantum mechanics. The first step toward its creation was made by Max Planck in 1900, who derived the thermal radiation spectrum of a body that absorbs all of the incident radiation.[1] The crucial step in his derivation was an assumption that the energy of an electromagnetic wave is directly proportional to its frequency. Making assumptions, especially as strong as this one, is dangerous. Planck had to take this desperate step simply because nothing else worked.

So, how does quantum mechanics explain the stability of atoms? At the most fundamental level, acceleration of an electron means that its state of motion changes; when this happens, an electron emits an electromagnetic wave. In quantum mechanics, the concept of the state of motion is replaced by that of a quantum state. Similar to the classical picture, it is only when an electron changes its quantum state that it emits an electromagnetic wave.

Unlike the classical state of motion, quantum states of an electron in an atom are not characterized by continuous variables such as electron's velocity and position. Rather, quantum states are described by a not too large set of discrete parameters, called quantum numbers. To each quantum state corresponds its own energy. When an electron goes from a higher-energy quantum state to a lower-energy one, it emits an electromagnetic wave. Out of all electron states in an atom, there is a special one with the lowest energy. It is called the ground state. An electron in the ground state cannot lose energy anymore; the only way out of the ground state is up to a higher-energy state. For this reason, an electron cannot fall into the nucleus.

This physical picture allows to make an important statement that atoms radiate light only of well-defined discrete energies that correspond to the energy differences between different quantum states. This is indeed what is going on in real life, as was established by the spectrometrists well before the advent of quantum mechanics.

---

[1] This derivation is presented in Appendix A.5.

Quantum mechanics has a huge predictive power. Not only does it explain quantitatively the stability of atoms and their radiation spectra, but also it tells us why and how atoms interact with each other to form small molecules or large crystalline structures. It explains the properties of electrons in solids, in particular, in semiconductors that form the basis of electronic devices. At present, it revolutionizes the whole computer industry by introducing quantum computers that operate based on its rules.

It is impossible to understand anything in modern physics without a solid knowledge of quantum theory. The purpose of this chapter is to give the reader just enough background information in quantum mechanics to be able to understand the rest of the book.

1.2   Wave–Particle Duality

There is an obvious difference between waves and particles. A particle is a localized point-like object with well-defined mass, m, position, $\vec{r}$, and velocity, $\vec{v}$, and also energy, E, and momentum, $\vec{p}$. A free particle moves with a constant velocity from the initial position $\vec{r}_0$. At a given time $t > 0$, its position is $\vec{r} = \vec{v}t + \vec{r}_0$. Its kinetic energy is $E = mv^2/2$, and its momentum is $\vec{p} = m\vec{v}$.

A wave, on the other hand, is a time-dependent perturbation that propagates in space. This perturbation may be, e.g.,[2] pressure and density variations in a sound wave, liquid level oscillations in a surface wave, or electric and magnetic fields in an electromagnetic wave. A plane wave is completely delocalized in space. It is described by the wave function, which is proportional, e.g., to $\sin(\vec{k} \cdot \vec{r} - \omega t)$, where $\vec{k}$ is the wave vector and ω is the angular frequency of the wave, related to its frequency, f, as $\omega = 2\pi f$. If one fixes the position $\vec{r}$, then the wave will repeat itself with the periodicity $T = 1/f = 2\pi/\omega$, which is the period of the wave. If one takes a snapshot of the wave at a fixed time t, one will find that the wave repeats itself in space with the space period λ, the wavelength. The magnitude of the wave vector is related to the wavelength, λ, by $\lambda = 2\pi/|\vec{k}|$. The direction of $\vec{k}$ is the direction of wave propagation. The velocity of the wave is the velocity with which a particular peak of the wave function moves. It is obtained from the condition of constant phase: $\vec{k} \cdot \vec{r} - \omega t = \text{const}$, giving $\vec{r} = \vec{v}t + \vec{r}_0$ with $\vec{v} = \vec{k}\frac{\omega}{k^2} = \frac{\omega}{k}\frac{\vec{k}}{k} = \frac{\omega\lambda}{2\pi}\frac{\vec{k}}{k} = \frac{\lambda}{T}\frac{\vec{k}}{k}$.

Before the twentieth century, it was believed that the distinction between particles and waves was clear-cut. Atoms and electrons were thought to be particles, whereas light was considered to be a wave. The situation changed when it was discovered that sometimes particles behave like waves, and waves behave like particles. This is called the wave–particle duality.

---

[2] Some Latin abbreviations commonly used in the scientific literature are: e.g. = exempli gratia (for example), cf. = confer (compare), i.e. = id est (that is), etc. = et cetera (and so on), i.a. = inter alia (among others), viz. = videlicet (namely), vs. = versus (against), etc.

The first indication that waves displayed particle properties was discovered by Max Planck (1900), who tried to explain thermal radiation of bodies that do not reflect incident light but absorb it completely. Even though their absorption coefficient is 1 and reflection coefficient is 0, these so-called black bodies do radiate light when heated; the higher their temperature, the more light they produce. In order to understand the experimentally observed spectrum, Planck had to postulate that light propagates in elementary quanta, which were later called photons. Each photon carries the energy proportional to the frequency,

$$E = hf = \hbar\omega$$

with the proportionality constants being

$$h = 6.6260704 \ldots \cdot 10^{-34} \, \text{J} \cdot \text{s} = 4.13566766 \ldots \cdot 10^{-15} \, \text{eV} \cdot \text{s};$$

$$\hbar = 1.0545718 \ldots \cdot 10^{-34} \, \text{J} \cdot \text{s} = 6.58211951 \ldots \cdot 10^{-16} \, \text{eV} \cdot \text{s}.$$

The constant h is called Planck's constant, and $\hbar = h/(2\pi)$ is known as the reduced Planck constant. The values of h and $\hbar$ can be established by fitting the experimental black-body radiation spectrum with Planck's theoretical formula.

An independent test of the theory, which also allows one to measure Planck's constant, is the photoelectric effect, that is, light-induced emission of electrons by metal surfaces. As was found experimentally, the energy of the electrons produced is proportional to the frequency of the incident light with the proportionality constant h. This finding was interpreted by Albert Einstein (1905) as an indication of the fact that light quanta give their energy hf to the electrons in the metal, thus enabling them to leave the metal.

About two decades later, Arthur Compton (1923) studied the scattering of X-rays by electrons. He found that the wavelength of the scattered X-rays was shifted with respect to the incident X-ray wavelength. In order to explain these findings, Compton postulated that light quanta carry not only energy, but also momentum, which is inversely proportional to the wavelength

$$p = \frac{h}{\lambda},$$

with the proportionality coefficient once again being Planck's constant.

A year later, Louis de Broglie (1924) proposed that this momentum–wavelength relation is more general and applies not only to photons, but to all particles as well. The particle waves are characterized by the wavelength

$$\lambda = \frac{h}{p}.$$

The fundamental difference between waves and particles is that waves are capable of exhibiting the phenomenon of interference, where two coherent waves either magnify each other constructively or suppress each other destructively. In 1925, Clinton Davisson and Lester Germer observed the interference of electrons reflected by different crystalline planes of a nickel sample. The results of their experiment indicated that the wavelength of a particle is indeed given by de Broglie's formula. Later, wave properties of atoms and other particles were also demonstrated experimentally.

What photons, electrons, and atoms have in common is that these waves can propagate in vacuum; they do not require a medium to carry them. One may wonder about other waves that do require such a medium: do they behave like particles? It turns out that they do. For example, sound propagates in discrete chunks, or quanta, called phonons. The quantum nature of sound exhibits itself in the specific temperature dependence of the heat capacity of solids. There are all sorts of elementary excitations in solids that sometimes behave as particles and sometimes as waves, such as waves quanta of charge density (plasmons) or magnetization (magnons). The wave–particle duality is hard to understand intuitively, but it is supported by a plenty of experimental evidence.

1.3   Wavelength of a Free Particle in Terms of Its Energy

It is often more convenient to express the de Broglie wavelength of a free particle in terms of its energy rather than its momentum. Such an expression is obtainable from the particle's energy–momentum relation, $E(p)$, which can be inverted to obtain the magnitude of momentum as a function of energy. Hence,

$$\lambda = \frac{h}{p(E)}.$$

The kinetic energy of a massive particle is

$$E(p) = \frac{p^2}{2m} \quad \Rightarrow \quad p = \sqrt{2mE} \quad \Rightarrow \quad \lambda = \frac{h}{\sqrt{2mE}} = \sqrt{\frac{h^2}{2mE}}.$$

In particular, the combination of Planck's constant and electron mass, m_e, under the square root has the numerical value

$$\frac{h^2}{2m_e} = 150.412\ldots \text{ eV} \cdot \text{Å}^2,$$

so the wavelength (in Angstroms) of a free electron is related to its energy (measured in electron-volts) as

$$\lambda = \sqrt{\frac{150.4\,\text{eV}}{E}}\ \text{Å}.$$

The wavelength of any other particle of mass m can be quickly found as

$$\lambda = \sqrt{\frac{150.4}{E}\frac{m_e}{m}}.$$

As for the massless photons, the energy–momentum relation is

$$E = hf = cp,$$

where c is the speed of light in vacuum. A photon's de Broglie wavelength is then just its wavelength:

$$\lambda = \frac{c}{f} = \frac{h}{p} = \frac{hc}{hf} = \frac{hc}{E}.$$

The product of Planck's constant and the speed of light has the numerical value

$$hc = 1.23985\ldots\,\text{eV}\cdot\mu\text{m},$$

so, numerically, the relation between photon wavelength and energy is

$$\lambda = \frac{1.24\,\text{eV}\cdot\mu\text{m}}{hf}.$$

1.4   Energy Quantization

An important conclusion that can be drawn from the wave nature of particles is that a particle in a confined geometry cannot have an arbitrary energy. Rather, its energy must be quantized. This statement is demonstrated by two examples.

Example 1 A particle between two walls. Consider a particle of mass m, which bounces between two parallel walls separated by the distance d. Every time the particle hits a wall, it gets reflected. If the particle is viewed as a wave, this must be a standing wave. Hence, the distance between the walls must equal an integer multiple of half-wavelength: $d = n\lambda/2$, where n is an integer, see Fig. 1.1. We can combine this with de Broglie's formula to establish the allowed values of momentum $p = nh/(2d)$ with $n = 1, 2, 3, \ldots$. It then follows that the energy of the particle confined between two walls assumes quantized values of

$$E = \frac{p^2}{2m} = n^2\frac{h^2}{8md^2}.$$

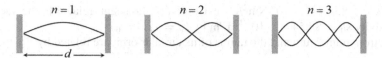

Fig. 1.1 The wave patterns of a particle between two walls at the three lowest energies

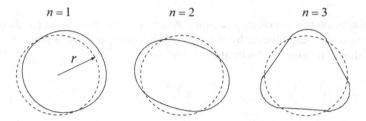

Fig. 1.2 The wave patterns of a particle traveling on a circle at the three lowest energies

Example 2 A particle on a circle. Suppose now that a particle can move on a circular path of radius r. Then, the wave associated with the particle must have a wavelength such that the circular path length is its integer multiple; otherwise, the wave will interfere with itself destructively. This implies the quantization condition $2\pi r = n\lambda = nh/p$, $n = 1, 2, 3, \ldots$, see Fig. 1.2. We obtain the possible momentum values $p = nh/(2\pi r)$ and the allowed energies

$$E = \frac{p^2}{2m} = n^2 \frac{h^2}{8\pi^2 r^2 m}.$$

1.5 Radiation Spectrum of Hydrogen

A hydrogen atom consists of a heavy proton and a single electron that orbits around it. Electron mass and charge are

$$m_e = 9.109\ldots \cdot 10^{-31}\,\text{kg}, \quad q_e = -e = -1.602\ldots \cdot 10^{-19}\,\text{C}.$$

The magnitude of the electron charge, $|q_e| = e$, is called the elementary charge.

Because a proton is 1836 times heavier than an electron, it can be considered as stationary. The total energy of the electron is the sum of its kinetic energy and the potential energy of its electrostatic interaction with the nucleus:

$$E = \frac{p^2}{2m_e} - k_e \frac{e^2}{r},$$

where $k_e = 8.987 \ldots \cdot 10^9 \, \text{Nm}^2/\text{C}^2$ is Coulomb's constant, related to the vacuum permittivity $\varepsilon_0 = 8.854 \ldots \cdot 10^{-12} \, \text{F/m}$ by $k_e = 1/(4\pi\varepsilon_0)$.

Electron momentum and the radius of its circular orbit are related by

$$2\pi r = n\lambda = n\frac{h}{p} \quad \Rightarrow \quad p = \frac{nh}{2\pi r} = \frac{n\hbar}{r},$$

as explained in the previous example. A further relation between these two parameters can be established by noting that the centripetal force that acts on the electron should be identified with the Coulomb force:

$$\frac{p^2}{m_e r} = k_e \frac{e^2}{r^2}.$$

The last two equations can be used to express p and r in terms of the quantum number n. For instance, dividing the second equation by the first one, we find p; after substituting this p into the first equation, we find r. The result is

$$p_n = \frac{k_e m_e e^2}{n\hbar}, \quad r_n = \frac{n^2 \hbar^2}{k_e m_e e^2}.$$

The value of $r_B \equiv r_1 = \frac{\hbar^2}{k_e m_e e^2} = 0.529 \, \text{Å}$ is called Bohr radius. Substitution of these expressions into the electron energy formula gives the quantized energy values:

$$E = -\frac{1 \, \text{Ry}}{n^2},$$

where

$$1 \, \text{Ry} = \frac{k_e^2 m_e e^4}{2\hbar^2} \approx 13.606 \, \text{eV}$$

is the so-called Rydberg's constant.

When the electron performs a transition from a higher-energy level characterized by a quantum number m to a lower level with the quantum number $n < m$, it emits a photon of frequency

$$f = \frac{E_m - E_n}{h} = \frac{1 \, \text{Ry}}{h}\left(\frac{1}{n^2} - \frac{1}{m^2}\right) \approx (3.290 \cdot 10^{15} \, \text{Hz}) \cdot \left(\frac{1}{n^2} - \frac{1}{m^2}\right).$$

The radiation spectrum of hydrogen indeed consists of spectral lines, whose frequencies are described by this relation to a very high accuracy. This relation was established experimentally by Johannes Rydberg in 1888 and was explained using the above reasoning by Niels Bohr in 1913.

The last formula for the frequency of a photon suggests that also the electron wave in the nth quantum state is endowed with the frequency

$$f_n = E_n/h, \quad \omega_n = 2\pi f_n = E_n/\hbar.$$

1.6   The Wave Function

The success of Bohr's model of hydrogen energy levels indicates that an electron in an atom really behaves as a wave. However, the version of quantum mechanics based on de Broglie's wavelength expression is of very limited use. It can be applied only to very simple problems and describe systems with just one particle. It is of little help when the potential depends on more than one coordinate, when it changes along the path of the wave, or when the system has many particles that interact with each other, as they do in real life.

To treat a broader spectrum of problems, an approach based on the wave equation is used. At the heart of this approach is the wave function of the particle, ψ, which depends both on space coordinate and time. It describes a certain oscillatory process that propagates in space (i.e., a wave) and contains information about the particle's state of motion.

Let us return to the electron moving on a circle of radius r around a stationary proton. Let us measure the position x along the circular orbit, see Fig. 1.3. The wave function of an electron depends on x and t. It should be a periodic function of x with the periodicity $\lambda = h/p$, and it should repeat itself every time the electron makes one full rotation. Hence,

$$\psi(x + \lambda, t) = \psi(x, t), \quad \psi(x + 2\pi r, t) = \psi(x, t).$$

It then follows that $2\pi r = n\lambda$, as stated in the previous section.

The possible natural candidates for the wave function are the trigonometric functions:

Fig. 1.3 The (real or imaginary part of the) wave function of an electron moving on a circular orbit

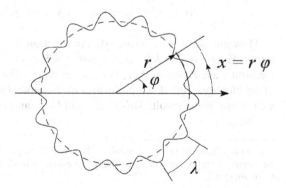

$$\psi(x, t) \propto \sin(kx - \omega t) \quad \text{or} \quad \psi(x, t) \propto \cos(kx - \omega t),$$

where $k = 2\pi/\lambda$ is the wave number and $\omega = E/\hbar$ is the angular frequency of the electron wave of energy E. However, there are insurmountable interpretation difficulties associated with these choices. We still do not quite know how $\psi(x, t)$ is to be interpreted, but what we do know is that the absence of the particle means that its wave function is zero. The trigonometric functions above do turn to zero when their argument is an integer or a half-integer multiple of π. This means that there are some points in space, where the electron cannot be found, see Fig. 1.3. But there is nothing special about any point on the circular path of the electron wave. They are all identical. For this reason, neither candidate can be accepted as a wave function of an electron moving on a circle.

There is a simple remedy to this problem. It is based on the trigonometric identity $\sin^2(\theta) + \cos^2(\theta) = 1$. Namely, we can treat the wave function as a mathematical object with two components,

$$\psi(x, t) = (\psi_1(x, t), \psi_2(x, t)),$$

$$\psi_1(x, t) \propto \cos(kx - \omega t), \quad \psi_2(x, t) \propto \sin(kx - \omega t),$$

and relate the presence of an electron to the magnitude squared of the wave function:

$$|\psi(x, t)|^2 = \psi_1^2(x, t) + \psi_2^2(x, t) = \text{const.}$$

It is possible, in principle, to further develop quantum mechanics using the language of a two-component wave function. However, it is more convenient to treat it as a complex number,[3] whose real and imaginary parts are ψ_1 and ψ_2:

$$\psi = \psi_1 + i\psi_2, \quad \psi_1 = \text{Re}\psi, \quad \psi_2 = \text{Im}\psi, \quad i^2 = -1.$$

This type of description is standard. It is equivalent to the two-component one. The "length squared" of ψ in the language of complex numbers is called the modulus squared:

$$|\psi|^2 = \psi^*\psi = (\psi_1 - i\psi_2)(\psi_1 + i\psi_2) = \psi_1^2 + \psi_2^2.$$

How should it be interpreted? An electron is a particle described in the wave terms. The wave is a delocalized entity; hence, an electron wave is delocalized. The electron wave function $\psi(x, t)$ does not carry information about the exact location of an electron. But if we perform a measurement of the electron's position x, we will obtain some result. This result will be a random number uniformly distributed

---

[3] A crash course in complex numbers is given in Appendix A.1. Those readers who did not encounter complex numbers previously are asked to read this appendix before proceeding with the main text.

along the circle of length $L = 2\pi r$. The probability that a measurement of the electron's position will yield a result anywhere between the positions x_1 and x_2 with $0 < x_1 < x_2 < L$ is, then,

$$P(x_1 < x < x_2) = \frac{x_2 - x_1}{L}.$$

This probability must be related to the modulus squared of the wave function. We write

$$P(x_1 < x < x_2) = (x_2 - x_1)|\psi(x, t)|^2 \text{ with } |\psi(x, t)|^2 = \frac{1}{L}.$$

We interpret $|\psi(x, t)|^2 \, dx$ as the probability to find an electron between two infinitely close points $x_1 = x$ and $x_2 = x + dx$:

$$dP(x) = |\psi(x, t)|^2 \, dx = \frac{dx}{L}.$$

This means that $|\psi(x, t)|^2$ has the physical meaning of the probability density. Note that the wave function is normalized to 1:

$$\int_0^L dP(x) = \int_0^L dx \, |\psi(x, t)|^2 = 1,$$

meaning that the probability to find an electron anywhere is unity.

Coming to the three-dimensional space $\vec{r}$ instead of one-dimensional space x, the wave function is complex-valued. It is expressed as a sum of its real and imaginary parts:

$$\psi(\vec{r}, t) = \text{Re}\psi(\vec{r}, t) + i \,\text{Im}\psi(\vec{r}, t).$$

The probability to find the particle in a small volume $d^3r = dx dy dz$ around the point $\vec{r}$ at the moment of time t is $d^3r \, |\psi(\vec{r}, t)|^2$.

Suppose a particle is placed in a box of volume V. Because the total probability of finding the particle anywhere in this box is equal to one, the wave function must satisfy the normalization condition:

$$\int_V dx \, dy \, dz \, |\psi(\vec{r}, t)|^2 = 1.$$

Occasionally, one has to consider a particle moving on a two-dimensional surface, say, in the (x, y)-plane, or performing a one-dimensional motion, say, along the x-axis. Then, the wave function would depend, in the former case, on the coordinates x, y and time t, and in the latter case, on the coordinate x and time t. The normalization condition will then be formulated as either a two-dimensional integral, $\int dx\,dy\,|\psi(x, y, t)|^2 = 1$, or a line integral, $\int dx\,|\psi(x, t)|^2 = 1$, respectively.

1.7 The Wave Function of a Free Particle

The wave function of a free particle has the form of a plane wave,

$$\psi(\vec{r}, t) = Ae^{i(\vec{k}\cdot\vec{r}-\omega t)} = A\left(\cos(\vec{k}\cdot\vec{r} - \omega t) + i\sin(\vec{k}\cdot\vec{r} - \omega t)\right),$$

characterized by the wave vector $\vec{k}$ and angular frequency ω. The prefactor A is to be determined from the normalization condition.

The square of the wave function is independent of the position,

$$|\psi(\vec{r}, t)|^2 = A^2,$$

meaning that the probability density to find the particle is the same for any point of space. Here, we encounter a challenge: the integral of the probability density over the whole space of infinite volume is infinite.

The standard way out of this difficulty is to consider the particle inside a cube of side length L and finite volume L^3 with the periodic boundary conditions expressed as

$$\psi(\vec{r} + \vec{n}L, t) = \psi(\vec{r}, t),$$

where $\vec{n}$ is a vector, all three of whose components are integers. This means that whenever the wave reaches one side of the cube, it reappears on the opposite side with the same value of the wave function.

With this choice of the boundary condition, integration $|\psi(\vec{r}, t)|^2$ is to be performed over a finite volume L^3, giving

$$\int_{L^3} d^3r\,|\psi(\vec{r}, t)|^2 = A^2L^3 = 1,$$

and thus

$$A = \frac{1}{L^{3/2}}.$$

The plane wave is therefore described by the wave function

$$\psi(\vec{r}, t) = \frac{1}{L^{3/2}}e^{i(\vec{k}\cdot\vec{r}-\omega t)}.$$

Periodic boundary conditions imply that the wave vector can have only discrete values,

$$\vec{k} = \frac{2\pi}{L}\vec{n}.$$

At the end of the calculations, the limit $L \to \infty$ is taken. This makes the wave vector $\vec{k}$ and momentum $\vec{p} = \hbar\vec{k}$ continuous because the difference between two successive $\vec{p}$-values goes to zero.

1.8 Schrödinger's Equation

1.8.1 Time-Dependent Schrödinger's Equation

The de Broglie expression for the wavelength $\lambda = h/p$ and the expression for the magnitude of the wave number, $k = 2\pi/\lambda$, imply that momentum is $\vec{p} = \hbar\vec{k}/(2\pi) = \hbar\vec{k}$. Similarly, the energy is related to the angular frequency as $E = \hbar\omega$. Then, the wave function of a free particle can be expressed in terms of momentum and energy as

$$\psi(\vec{r}, t) = \frac{1}{L^{3/2}}e^{i(\vec{p}\cdot\vec{r}-Et)/\hbar} = \frac{1}{L^{3/2}}e^{i(p_x x + p_y y + p_z z - Et)/\hbar}.$$

Note that energy can be recovered from this expression as the time derivative:

$$E\psi(\vec{r}, t) = i\hbar\frac{\partial\psi}{\partial t}.$$

Likewise, the momentum can be recovered by differentiation with respect to space:

$$p_x\psi(\vec{r}, t) = -i\hbar\frac{\partial\psi}{\partial x}, \quad p_y\psi(\vec{r}, t) = -i\hbar\frac{\partial\psi}{\partial y}, \quad p_z\psi(\vec{r}, t) = -i\hbar\frac{\partial\psi}{\partial z}.$$

The three spatial derivatives can be united into a single vector operator, called the gradient,

$$\vec{\nabla} = \vec{e}_x\frac{\partial}{\partial x} + \vec{e}_y\frac{\partial}{\partial y} + \vec{e}_z\frac{\partial}{\partial z},$$

in terms of which the momentum vector is obtained as

$$\vec{p}\psi(\vec{r}, t) = -i\hbar\vec{\nabla}\psi(\vec{r}, t).$$

We see that energy and momentum can be identified with the operators

$$\hat{E} = i\hbar\frac{\partial}{\partial t}, \quad \hat{\vec{p}} = -i\hbar\vec{\nabla},$$

which we denote with a hat in order to distinguish from the energy and momentum values E and $\vec{p}$. When these operators act on the wave function of a free particle, its energy and momentum values are obtained:

$$\hat{E}\,\psi(\vec{r},t) = E\,\psi(\vec{r},t), \quad \hat{\vec{p}}\,\psi(\vec{r},t) = \vec{p}\,\psi(\vec{r},t).$$

Acting on both sides of the second expression with $\hat{\vec{p}} = -i\hbar\vec{\nabla}$ again, we obtain the square of the momentum:

$$\hat{\vec{p}}^2\psi = \vec{p}^{\,2}\psi = -\hbar^2\vec{\nabla}^2\psi,$$

where

$$\vec{\nabla}^2 = \frac{\partial^2}{\partial x^2} + \frac{\partial^2}{\partial y^2} + \frac{\partial^2}{\partial z^2}$$

is called the Laplacian operator.

The relation between the kinetic energy and the momentum of a free particle is $E = \frac{\vec{p}^{\,2}}{2m}$. This same relation can be written using the derivatives:

$$i\hbar\frac{\partial\psi}{\partial t} = -\frac{\hbar^2}{2m}\vec{\nabla}^2\psi.$$

This is Schrödinger's equation for a free particle. Even though we have obtained it assuming the plane-wave form of $\psi(\vec{r},t)$, it is valid for any type of wave function in free space.

If the particle finds itself in some potential $V(\vec{r})$, then the energy expression should include its potential energy in addition to the kinetic energy,

$$E\psi = \frac{\vec{p}^{\,2}}{2m}\psi + V(\vec{r})\psi,$$

and expressing energy and momentum components in terms of the partial derivatives, we finally obtain the time-dependent Schrödinger equation:

$$i\hbar\frac{\partial\psi(\vec{r},t)}{\partial t} = -\frac{\hbar^2}{2m}\vec{\nabla}^2\psi(\vec{r},t) + V(\vec{r})\psi(\vec{r},t).$$

1.8.2 Time-Independent Schrödinger's Equation

To find the possible values of energy that a particle can have, the solution of the time-dependent Scrhödinger equation is written in the form

$$\psi(\vec{r}, t) = \Psi(\vec{r})e^{-iEt/\hbar}$$

with time-independent $\Psi(\vec{r})$. Substitution of this decomposition into the time-dependent Schrödinger equation gives the time-independent version of this equation for a single particle:

$$E\Psi(\vec{r}) = -\frac{\hbar^2}{2m}\vec{\nabla}^2\Psi(\vec{r}) + V(\vec{r})\psi(\vec{r}).$$

The solution of this equation with the boundary conditions appropriate for a given problem at hand yields the possible energies E_α that the particle can have, and the corresponding stationary wave functions, $\Psi_\alpha(\vec{r})$. Both depend on a set of quantum numbers

$$\alpha = (\alpha_1, \alpha_2, \ldots),$$

which label the specific quantum state of the particle.

The quantum state with the lowest energy found in the energy spectrum E_α is called the ground state.

1.9 Probabilistic Interpretation and the Collapse of the Wave Function

The states of a particle moving in the positive and negative x-directions are described by the plane waves:

$$\psi_\rightarrow(x, t) = \frac{1}{\sqrt{L}}e^{i(px-Et)/\hbar}, \quad \psi_\leftarrow(x, t) = \frac{1}{\sqrt{L}}e^{i(-px-Et)/\hbar}$$

with $p = \frac{nh}{L} > 0$ and $E = p^2/(2m)$.

Both wave functions, $\psi_\rightarrow(x, t)$ and $\psi_\leftarrow(x, t)$, satisfy the time-independent Schrödinger equation for a free particle. Any linear combination of these two wave functions satisfies the same Schrödinger equation and therefore describes a possible quantum state of the particle. For example, consider the sum

$$\psi_{sw}(x, t) = \frac{1}{\sqrt{2}}(\psi_\rightarrow(x, t) + \psi_\leftarrow(x, t)),$$

where the factor $1/\sqrt{2}$ comes from normalization of $\psi_{sw}(x, t)$ to 1:

$$\int_0^L dx \, |\psi_{sw}(x, t)|^2 = \int_0^L dx \, \psi_{sw}^*(x, t) \, \psi_{sw}(x, t)$$

$$= \frac{1}{2} \left(\int_0^L dx \, |\psi_\rightarrow|^2 + \int_0^L dx \, |\psi_\leftarrow|^2 \right) = \frac{1}{2}(1 + 1) = 1.$$

Here, we used the fact that

$$\int_0^L dx \, \psi_\leftarrow^*(x, t) \psi_\rightarrow(x, t) = \frac{1}{L} \int_0^L dx \, e^{i2px/\hbar} = \frac{\hbar}{i2pL} e^{i2px/\hbar} \Big|_{x=0}^L$$

$$= \frac{\hbar}{i2pL} e^{i2\pi nx/L} \Big|_{x=0}^L = \frac{\hbar}{i2pL} (e^{i2\pi n} - 1) = 0$$

and similar with $\int_0^L dx \, \psi_\rightarrow^* \psi_\leftarrow = 0$. Using the expression $\frac{e^{i\alpha} + e^{-i\alpha}}{2} = \cos \alpha$, we find

$$\psi_{sw}(x, t) = \sqrt{\frac{2}{L}} \cos \left(\frac{px}{\hbar} \right) e^{-iEt/\hbar},$$

i.e., $\psi_{sw}(x, t)$ describes a standing wave.

Is the particle in the standing wave state moving to the right or to the left? To understand this, we apply the rule, according to which the momentum of a particle is to be found by differentiating its wave function with respect to the spatial coordinate. For the one-dimensional motion, this rule reads:

$$\hat{p}\psi(x, t) = -i\hbar \frac{\partial \psi(x, t)}{\partial x} = p\psi(x, t).$$

The problem with the standing wave is that the derivative

$$-i\hbar \frac{\partial \psi_{sw}(x, t)}{\partial x} = -\sqrt{\frac{2}{L}} \sin \left(\frac{px}{\hbar} \right) e^{-iEt/\hbar}$$

is not given by a product of momentum p and the wave function $\psi_{sw}(x, t)$. The reason is that the "momentum rule" applies to free particles only, whereas a standing wave appears if the particle is not free but is confined between two walls. It is moving to the right 50% of the time and to the left 50% of the time. In other words, if one measured the direction of its motion, one would get either the result "to the right" or "to the left" with the probability of 50%. Note that the probability of 50% is the square of the factor $1/\sqrt{2}$ that multiplies both the right-propagating, $\psi_\rightarrow(x, t)$, and the left-propagating, $\psi_\leftarrow(x, t)$, components of the standing wave $\psi_{sw}(x, t)$ above.

One can form an arbitrary combination of the right-propagating and the left-propagating plane waves:

$$\psi(x,t) = c_\rightarrow \psi_\rightarrow(x,t) + c_\leftarrow \psi_\leftarrow(x,t)$$

with the coefficients $c_\rightarrow$ and $c_\leftarrow$ such that

$$|c_\rightarrow|^2 + |c_\leftarrow|^2 = 1.$$

For example, a more general expression for the standing wave $\psi_{sw}(x,t)$ is

$$\psi_{sw}(x,t) = \frac{e^{i\theta_\rightarrow}}{\sqrt{2}}\psi_\rightarrow(x,t) + \frac{e^{i\theta_\leftarrow}}{\sqrt{2}}\psi_\leftarrow(x,t)$$

with some real-valued phases $\theta_\rightarrow$ and $\theta_\leftarrow$. The probabilities of particular outcomes of a measurement of the particle's direction of motion for this more general standing wave are

$$P_\rightarrow = |c_\rightarrow|^2, \quad P_\leftarrow = |c_\leftarrow|^2.$$

Both are equal 1/2.

Let us say, we measured the direction of propagation to be to the right. Is it possible for another measurement, performed immediately after the first one, to yield the result "to the left" with any probability? The answer is no. If two measurements performed immediately one after the other gave different momenta, p and $-p$, this would be at odds with the law of conservation of momentum. A particle cannot change its direction of motion, unless acted upon by a force. Same is true about the wave associated with the particle. In the wave function language, this means that a measurement induces a collapse of the original wave function $\psi_{sw}(x,t)$ to either $\psi_\rightarrow(x,t)$ or $\psi_\leftarrow(x,t)$ with the probabilities $|c_\rightarrow|^2$ and $|c_\leftarrow|^2$. The second measurement will yield the same momentum as the first one, provided that the two measurements are performed immediately one after the other.

If the second measurement is performed after a finite time after the first one, then its result really depends on the external forces that may change the particle's momentum. If such forces are present, e.g., because of the walls between which the particle is confined, the wave function of the particle will evolve from the plane wave back to the standing wave.

As a further generalization of these statements, consider a wave function, which is a linear combination of many plane waves with momenta $p_1, p_2, \ldots$:

$$\psi(x,t) = \sum_n \frac{c_n}{\sqrt{L}}e^{i(p_n x - E_n t)/\hbar},$$

where the momenta are

$$p_n = nh/L, \quad n = 0, \pm 1, \pm 2, \ldots,$$

the coefficients c_n are such that

$$\sum_n |c_n|^2 = |c_1|^2 + |c_2|^2 + \ldots = 1,$$

and the energies are $E_n = p_n^2/(2m)$. This time, the plane-wave contributions to $\psi(x, t)$ differ not only in the direction, but also in the magnitude of their individual momenta p_n. The parameters $|c_1|^2, |c_2|^2, \ldots$ are the probabilities that a measurement of the particle's momentum will yield the results $p_1, p_2, \ldots$, respectively. After a measurement that resulted in a particular momentum value p_n, the particle's wave function above collapses to the single-momentum wave function $\frac{1}{\sqrt{L}}e^{i(p_n x - E_n t)/\hbar}$. If there are no external forces acting on the particle, the particle will remain in that state indefinitely. All subsequent momentum measurements will yield the same result $p = p_n$ with probability 1. Otherwise, the wave function will evolve from the plane-wave state with the momentum p_n to a more complicated sum of plane waves with different momenta.

Now, suppose that we perform a momentum measurement many times, and after each measurement, we wait long enough for the wave function to evolve to the original state given by the sum over the plane waves above. Every time we measure the momentum, we will get a different number. But the average momentum resulting from a large number of such measurements is a well-defined value:

$$\langle p \rangle = \sum_n |c_n|^2 p_n,$$

as $|c_n|^2$ is the probability to get the momentum value p_n in a single measurement. The same average momentum can be obtained with the help of the momentum operator $\hat{p} = -i\hbar\partial/\partial x$, see Sect. 1.8.1. Namely,

$$\langle p \rangle = -i\hbar \int_0^L dx \, \psi^*(x, t) \frac{\partial \psi(x, t)}{\partial x}.$$

Indeed, multiplying

$$\psi^*(x, t) = \sum_n \frac{c_n^*}{\sqrt{L}} e^{-i(p_n x - E_n t)/\hbar} \quad \text{and} \quad \frac{\partial \psi(x, t)}{\partial x} = \sum_n \frac{c_n}{\sqrt{L}} \frac{i p_n}{\hbar} e^{i(p_n x - E_n t)/\hbar}$$

together, we obtain

$$\psi^*(x, t) \frac{\partial \psi(x, t)}{\partial x} = \sum_{n,n'} \frac{c_n^* c_{n'}}{L} \frac{i p_{n'}}{\hbar} e^{i 2\pi (n'-n)x/L} e^{-i(E_{n'} - E_n)t/\hbar}.$$

After integration over x from 0 to L, only the terms with $n' = n$ survive in the sum. Multiplication of the result with $-i\hbar$ gives $\langle p \rangle$.

The above expression for the average momentum can be generalized to the three-dimensional space, in which the average momentum is

$$\langle \vec{p} \rangle = -i\hbar \int d^3r \, \psi^*(\vec{r}, t) \, \vec{\nabla} \psi(\vec{r}, t).$$

1.10 Measurable and Unmeasurable in Quantum Mechanics

The complex-valued wave function can be expressed in an equivalent way, viz., in terms of its modulus $|\psi(\vec{r}, t)|$ and phase $\theta(\vec{r}, t)|$:

$$\psi(\vec{r}, t) = |\psi(\vec{r}, t)| e^{i\theta(\vec{r},t)},$$

$$|\psi| = \sqrt{(\mathrm{Re}\psi)^2 + (\mathrm{Im}\psi)^2} = \sqrt{\psi\psi^*}, \quad \theta = \tan^{-1}\frac{\mathrm{Im}\psi}{\mathrm{Re}\psi}.$$

The arguments $\vec{r}$, t are not written explicitly in the second line for the sake of brevity. With Euler's formula, $e^{i\theta} = \cos(\theta) + i\sin(\theta)$, the real and imaginary parts of the wave function are

$$\mathrm{Re}\psi = |\psi|\cos(\theta), \quad \mathrm{Im}\psi = |\psi|\sin(\theta).$$

Out of the two parameters, $|\psi|$ and θ, it is the former, $|\psi(\vec{r}, t)|$, that has a tangible physical meaning of the square root of the probability density to find a particle. Because $|e^{i\theta(\vec{r},t)}|^2 = 1$ for any real-valued phase $\theta(\vec{r}, t)$, the probability density does not depend on the phase. Only variations of the phase $\theta(\vec{r}, t)$ in time and space lead to observable physical effects, but the phase itself cannot be measured in any experiment. Hence, the wave function $\psi(\vec{r}, t)$ is also not measurable, only its magnitude is.

In other words, if two wave functions have different phases but the same magnitude,

$$\psi_1 = |\psi| e^{i\theta_1}, \quad \psi_2 = |\psi| e^{i\theta_2}, \quad \theta_1(\vec{r}, t) \neq \theta_2(\vec{r}, t),$$

then they are mathematically distinct, but physically equivalent. There is absolutely no way to determine experimentally which wave function is the correct one.

This statement immediately encounters two objections. First, it is too self-contradictory to be acceptable. An even more serious second objection is that this statement is not consistent with the time-dependent Schrödinger equation. Namely, if two wave functions

$$\psi_1(\vec{r}, t) = \psi(\vec{r}, t) \text{ and } \psi_2(\vec{r}, t) == \psi(\vec{r}, t) e^{i\theta(\vec{r},t)}$$

are physically equivalent, then they should satisfy the same Schrödinger equation. In particular, we should have

$$i\hbar \frac{\partial \psi_2(\vec{r}, t)}{\partial t} = -\frac{\hbar^2}{2m} \vec{\nabla}^2 \psi_2(\vec{r}, t) + U(\vec{r}) \, \psi_2(\vec{r}, t).$$

But a simple substitution of $\psi_2(\vec{r}, t) = \psi(\vec{r}, t) \, e^{i\theta(\vec{r}, t)}$ results in an equation for $\psi(\vec{r}, t)$ that would look similar to the already familiar Schrödinger equation if not for a few extra terms with the time and spatial derivatives of $\theta(\vec{r}, t)$.

To address both objections, we note that the statement is not more self-contradictory than the assertion that the energy is defined up to an additive constant: it is the energy variations that are measurable in physical processes. Consider the electric field $\vec{\mathcal{E}}$ as an example. It is derived from the electric potential $\phi(\vec{r})$ as its spatial derivative:

$$\vec{\mathcal{E}}(\vec{r}) = -\vec{\nabla}\phi(\vec{r}).$$

The energy of a charge q in an electric field is $q\phi(\vec{r})$. Adding an arbitrary constant ϕ_0 to $\phi(\vec{r})$ changes the energy by $q\phi_0$, but it does not change the electric field $-\vec{\nabla}(\phi(\vec{r}) + \phi_0) = -\vec{\nabla}\phi(\vec{r}) = \vec{\mathcal{E}}(\vec{r})$. Because it is the force $q\vec{\mathcal{E}}$ that determines the motion of the charge, the charge dynamics is independent of the constant ϕ_0.

More generally, the dynamics of a charge q is determined by the electric and the magnetic fields, $\vec{\mathcal{E}}(\vec{r}, t)$ and $\vec{B}(\vec{r}, t)$, which are derivable from the scalar potential $\phi(\vec{r}, t)$ and the vector potential $\vec{A}(\vec{r}, t)$ as

$$\vec{\mathcal{E}} = -\vec{\nabla}\phi(\vec{r}, t) - \frac{\partial \vec{A}(\vec{r}, t)}{\partial t}, \quad \vec{B}(\vec{r}, t) = \vec{\nabla} \times \vec{A}(\vec{r}, t).$$

A simultaneous change of the scalar and vector potentials to

$$\phi(\vec{r}, t) \rightarrow \phi(\vec{r}, t) - \frac{\partial g(\vec{r}, t)}{\partial t}; \quad \vec{A}(\vec{r}, t) \rightarrow \vec{A}(\vec{r}, t) + \vec{\nabla}g(\vec{r}, t)$$

does not change the electric and magnetic fields, no matter how the scalar function $g(\vec{r}, t)$ is chosen. This property of the electromagnetic field is called the gauge invariance. The transformation of the scalar and vector potentials above is called the gauge transformation.

Now, when we say that the particle finds itself in a potential $U(\vec{r})$, what we really mean is that the particle is charged and the potential is generated by an electric field:

$$U(\vec{r}) = q\phi(\vec{r}).$$

To be fully consistent with electrodynamics, Schrödinger's equation must involve not only the electric, but also the vector potential. It turns out that a multiplication of the wave function by a phase factor $e^{i\theta(\vec{r}, t)}$ and a gauge transformation of the potentials $\phi(\vec{r}, t)$ and $\vec{A}(\vec{r}, t)$ with a suitably chosen gauge function $g(\vec{r}, t)$ indeed leave Schrödinger's equation unchanged.

We will not go into the details of this derivation. Likewise, we will not include the vector potential into consideration. But the two statements made in this section, namely, that the phase of the wave function is unmeasurable and that two wave functions that differ by phase only are physically the same, should be borne in mind.

1.11   Electron States in a Hydrogen Atom

An electron in a hydrogen atom finds itself in a potential

$$V(\vec{r}) = -k_e \frac{e^2}{r}.$$

It turns out that with this potential, the time-independent Schrödinger equation can be solved exactly. The resulting stationary wave functions turn out to depend on three quantum numbers, usually denoted as n, l, and m.

The principal quantum number, n, determines the energy of the electron, which turns out to be exactly the same as what we obtained in Sect. 1.5: $E_n = -1\,\mathrm{Ry}/n^2$. The principal quantum number assumes positive integer values:

$$n = 1, 2, 3, \ldots.$$

The orbital quantum number, l, determines the angular momentum of the electron orbiting around the nucleus. It is given by

$$L = \hbar\sqrt{l(l + 1)}.$$

For a given principal quantum number n, the orbital quantum number can assume n different values:

$$l = 0, 1, \ldots, n - 1.$$

This result differs from what we obtained within the simplistic treatment of Sect. 1.5. There, we obtained the angular momentum values $pr = n\hbar$, which are incorrect.

The magnetic quantum number, m (not to be confused with the electron mass, m_e), determines the projection of the orbital angular momentum on some selected coordinate axis, say, the z-axis:

$$L_z = \hbar m_z.$$

For a given orbital quantum number, the magnetic quantum number assumes $2l + 1$ different values:

$$m = l, l + 1, \ldots, l - 1, l.$$

1.12 Spin

In addition to the quantum numbers that characterize its motion in space, an electron also possesses its own internal degree of freedom, which is not related to the electron's orbital motion. Associated with it is the electron's intrinsic angular momentum called spin. The spin of a particle is an intrinsic property that does not change in time. It was explained by Paul Dirac in 1928 by combining the rules of quantum mechanics with special relativity.

Spin angular momentum is characterized by an additional quantum number, s, which can assume only two values: $s = \frac{1}{2}$ and $s = -\frac{1}{2}$. The projection of the electron spin angular momentum on some selected axes is $S_z = \hbar s$. Associated with spin angular momentum is the magnetic moment of an electron, $\vec{m}$, which is important in the presence of a magnetic field, $\vec{B}$. It manifests itself as an additional energy $-\vec{m} \cdot \vec{B}$.

Other particles may have other spin values, either integer or half-integer. For instance, photons have spin $s = 1$; some particles of the meson family have spin $s = 0$ and others have spin $s = 1$; particles of the baryon family may have either spin $1/2$ (e.g., protons and neutrons) or $3/2$; gravitons have spin 2; Higgs bosons have spin 0, etc.

1.13 Degeneracy

One should not confuse the concepts of energy level and quantum state. An energy level is a possible energy value that the system of interest can have according to the time-independent Schrödinger equation. The quantum state is specified by a set of quantum numbers α that label the system's wave function $\Psi_\alpha(\vec{r})$. Different quantum states may have different energies, or they may belong to the same energy level.

The situation when the same energy level accommodates more than one quantum state is called degeneracy.

The degree of degeneracy of a particular energy level is the number of different quantum states that belong to this energy level. In other words, it is the number of different sets of quantum numbers α such that E_α is the same.

Consider a hydrogen atom as an example. Quantum states of an electron in a hydrogen atom are labeled by the hydrogen quantum numbers

$$\alpha = (n, l, m, s).$$

The energy does not depend on l, m, and s. The degree of degeneracy of the energy level characterized by the principle quantum number n will be denoted as g_n. It is found as the number of possible combinations of $l = 0, \ldots, n-1$, $m = -l, \ldots, l$, and $s = 1/2, -1/2$ that correspond to the same n-value:

$$g_n = 2 \sum_{l=0}^{n-1} \sum_{m=-l}^{l} 1.$$

The factor 2 comes from two possible spin values. The double sum runs over all possible values of l and m; every time either of this quantum number changes, we should increase the result by 1. Using the fact that

$$\sum_{m=-l}^{l} 1 = 2l + 1,$$

we find, using the formula for arithmetic series, that

$$g_n = 2 \sum_{l=0}^{n-1} (2l + 1) = 2n^2.$$

The lowest energy level, $n = 1$, consists of 2 quantum states with $l = 0$, $m = 0$, and $s_z = \pm 1/2$. This means that the ground state energy level of hydrogen is double degenerate. The next energy level, $n = 2$, has the degeneracy $g_2 = 8$, i.e., 8 states have the energy E_2. These states are characterized by the following quantum numbers: $l = 0$, $m = 0$, and $s_z = \pm 1/2$ (2 states), and $l = 1$, $m = -1, 0, 1$, and $s_z = \pm 1/2$ (6 states).

1.14 Indistinguishability of Quantum Particles

Consider the situation when we have more than one particle of the same kind, e.g., N electrons. In classical physics, each particle follows its own trajectory, and so it is possible, in principle, to follow the motion of any particle and to be certain that we are looking at the same particle on the trajectory. But in quantum physics, the concept of trajectory is meaningless because it is impossible to define the trajectory of a wave distributed over the whole of space. Impossibility to follow the motion of an individual particle implies the impossibility to label quantum particles, i.e., to distinguish them from one another.

The state of N particles is described by an N-particle wave function, which depends on N coordinates and time, $\psi(\vec{r}_1, \ldots, \vec{r}_N, t)$. It must be normalized to 1:

$$\int d^3 r_1 \ldots d^3 r_N |\psi(\vec{r}_1, \ldots, \vec{r}_N, t)|^2 = 1.$$

It evolves in time according to the time-dependent Schrödinger equation:

$$i\hbar \frac{\partial \psi(\vec{r}_1, \ldots, \vec{r}_N, t)}{\partial t} = -\frac{\hbar^2}{2m} \left(\vec{\nabla}_1^2 \psi + \ldots + \vec{\nabla}_N^2 \psi \right) + V(\vec{r}_1, \ldots, \vec{r}_N)\psi,$$

where $\vec{\nabla}_n^2 = \frac{\partial^2}{\partial x_n^2} + \frac{\partial^2}{\partial y_n^2} + \frac{\partial^2}{\partial z_n^2}$, $n = 1, \ldots, N$, and the N-particle potential energy $V(\vec{r}_1, \ldots, \vec{r}_N)$ includes an external potential and the potential energy of interaction between the particles. So far, the difference between the many-particle state and a single-particle state is just in the number of arguments and Laplacian operators.

Mathematically, indistinguishability of quantum particles means that the wave function $\psi(\vec{r}_1, \vec{r}_2, \ldots, \vec{r}_N, t)$ and the one with an arbitrary permutation of the coordinates $\vec{r}_1, \vec{r}_2, \ldots, \vec{r}_N$ are physically equivalent. The wave function itself is unmeasurable, only its modulus is. Then, an interchange of any two coordinates gives a wave function that has the same modulus as the original one:

$$|\psi(\ldots, \vec{r}_i, \ldots, \vec{r}_j, \ldots, t)|^2 = |\psi(\ldots, \vec{r}_j, \ldots, \vec{r}_i, \ldots, t)|^2$$

for any pair of indices i and j. But this is only possible if the two wave functions differ by a multiplicative constant of unit modulus:

$$\psi(\ldots, \vec{r}_j, \ldots, \vec{r}_i, \ldots, t) = c\psi(\ldots, \vec{r}_i, \ldots, \vec{r}_j, \ldots, t), \quad |c|^2 = 1.$$

From this, we can conclude that the constant is of the form $c = e^{i\theta}$, where θ is an arbitrary real number, implying infinitely many possible values of c. But actually, the spectrum of the possible c-values is much narrower.

Indeed, performing an interchange of the coordinates $\vec{r}_i$ and $\vec{r}_j$ again, we recover the original wave function:

$$\psi(\ldots, \vec{r}_i, \ldots, \vec{r}_j, \ldots, t) = c\psi(\ldots, \vec{r}_j, \ldots, \vec{r}_i, \ldots, t)$$
$$= c^2 \psi(\ldots, \vec{r}_i, \ldots, \vec{r}_j, \ldots, t).$$

This means that the prefactor squared is one, $c^2 = 1$, and therefore, the prefactor can only have two values:

$$c = \pm 1.$$

1.15 Spin-Statistics Theorem

There is an important result, called the spin-statistics theorem (Pauli, 1940):

Particles with integer spin (i.e., $s = 0, 1, 2, \ldots$) follow a many-particle wave function that remains the same if any two coordinates are interchanged, e.g., for two particles $\psi(\vec{r}_1, \vec{r}_2, t) = \psi(\vec{r}_2, \vec{r}_1, t)$. Such particles are called bosons.

Particles with half-integer spin (i.e., $s = 1/2, 3/2, 5/2, \ldots$) follow a many-particle wave function that changes sign if any two coordinates are interchanged, e.g., for two particles $\psi(\vec{r}_1, \vec{r}_2, t) = -\psi(\vec{r}_2, \vec{r}_1, t)$. Such particles are called fermions.

The particles that are most relevant for the purposes of this book are electrons. Their spin is 1/2, and therefore, they are fermions.

1.16 Pauli's Exclusion Principle

Electrons are spin-1/2 fermions with an antisymmetric wave function. Now, suppose you have a system of two electrons, and assume that interaction between them is negligibly small. Both electrons find themselves in the same potential produced, say, by an atomic nucleus. The energy levels of the electrons can be found by solving the time-independent Schrödinger equation, and the solution will be characterized by some quantum numbers, which we denote collectively as α and β. That is, the energy of one electron is E_α, its stationary wave function is $\Psi_\alpha(\vec{r}_1)$, the energy of the other electron is E_β, and its stationary wave function is $\Psi_\beta(\vec{r}_2)$.

The net wave function of both electrons must be obtained as some combination of $\Psi_\alpha(\vec{r}_1)$ and $\Psi_\beta(\vec{r}_2)$. A simple product, $\Psi(\vec{r}_1, \vec{r}_2) = \Psi_\alpha(\vec{r}_1)\Psi_\beta(\vec{r}_2)$, will respect the time-independent Schrödinger equation for two electrons. However, it is not an acceptable two-electron wave function because it does not change sign upon the interchange of $\vec{r}_1$ and $\vec{r}_2$. An expression that does have this property is

$$\Psi_{\alpha\beta}(\vec{r}_1, \vec{r}_2) = \frac{1}{\sqrt{2}}\left(\Psi_\alpha(\vec{r}_1)\Psi_\beta(\vec{r}_2) - \Psi_\alpha(\vec{r}_2)\Psi_\beta(\vec{r}_1)\right) = -\Psi_{\alpha\beta}(\vec{r}_2, \vec{r}_1).$$

The prefactor $1/\sqrt{2}$ is necessary for the correct normalization to one.

Now, if α and β are the same, we have

$$\Psi_{\alpha\alpha}(\vec{r}_1, \vec{r}_2) = \frac{1}{\sqrt{2}}\left(\Psi_\alpha(\vec{r}_1)\Psi_\alpha(\vec{r}_2) - \Psi_\alpha(\vec{r}_2)\Psi_\alpha(\vec{r}_1)\right) = 0.$$

The wave function of two fermions in the same quantum state is zero. This means that in a system of more than one fermion (e.g., electrons), only one fermion can occupy the same quantum state. In other words, a single-particle quantum state may be either occupied by only one fermion or empty.

1.17 Problems

1.17.1 Solved Problems

Problem 1 A photon with wavelength 300 nm kicks an electron out of a metal. The energy needed to remove an electron from the metal is $w = 3.5\,\text{eV}$. Determine de Broglie wavelength of this electron.

Problem 2 An electron is moving in the xy-plane in the presence of a magnetic field $\vec{B}$ pointing in the z-direction. What are the possible values of the electron's kinetic energy and magnetic moment?

Problem 3 The wave function of an electron is given by

$$\psi(x, t) = A \, \sin(kx) \, e^{-b|x|} \, e^{i\omega t}$$

with $k = 1\,\text{Å}^{-1}$ and $b = 2\,\text{Å}^{-1}$.

(a) Find the normalization constant A. Hint: Use Euler's formula to evaluate the integral.
(b) Determine the locations where the electron is most likely to be found.

Problem 4 An electron is confined to the x-axis, where it is placed inside an infinite potential well of width L:

$$V(x) = \begin{cases} 0 \text{ for } 0 < x < L \\ \infty \text{ for } x \leq 0 \text{ and } x > L. \end{cases}$$

By solving the time-independent Schrödingers equation

$$-\frac{\hbar^2}{2m_e}\frac{d^2\Psi(x)}{dx^2} = E\Psi(x) \text{ for } 0 < x < L$$

with the boundary conditions $\Psi(x = 0) = \Psi(x = L) = 0$ and normalization $\int_0^L dx \, |\Psi(x)|^2 = 1$, find all possible wave functions $\Psi(x)$ and energies E that the electron can have.

Problem 5 The wave function of a free electron moving in the x-direction is given by

$$\psi(x, t) = \frac{1}{\sqrt{L}} \left(\frac{\sqrt{3}}{2} e^{i(x/(0.2\,\text{nm})-\omega_1 t)} - \frac{1-i}{2\sqrt{2}} e^{i(-x/(1\,\text{nm})-\omega_2 t)} \right)$$

with the periodic boundary conditions $\psi(x + L, t) = \psi(x, t)$.

(a) What are the angular frequencies ω_1 and ω_2?
(b) What possible momentum values of this electron can be measured and what are the probabilities to measure those momentum values?
(c) If many electrons are prepared in the state described by the wave function above, what is their average momentum?

Problem 6 The quantum states of an isotropic two-dimensional simple harmonic oscillator of mass m and spring constant k depend on 2 quantum numbers, n_1 and n_2, which can take on non-negative integer values $0, 1, 2, 3, \ldots$. The energy is

$$E_{n_1 n_2} = \hbar\omega(n_1 + n_2 + 1)$$

with the natural frequency $\omega = \sqrt{k/m}$.

(a) Find the degeneracy g_n of the nth energy level with energy $\hbar\omega(n + 1)$.
(b) The quantum states of an isotropic three-dimensional simple harmonic oscillator of mass m and spring constant k depend on 3 quantum numbers, n_1, n_2, and n_3, which can take on non-negative integer values $0, 1, 2, 3, \ldots$. The energy is

$$E_{n_1 n_2 n_3} = \hbar\omega(n_1 + n_2 + n_3 + 3/2).$$

Find the degeneracy g_n of the nth energy level with energy $\hbar\omega(n + 3/2)$.

1.17.2 Practice Problems

Problem 1 A photon of wavelength $\lambda = 85$ nm is absorbed by a hydrogen atom thereby exciting it from the lowest energy ground state to the first excited state. The atom was initially at rest. The mass of hydrogen atom is 1.008 Da, where 1 Da (Dalton) = $1.6605 \cdot 10^{-27}$ kg.

(a) Use momentum conservation to find the final velocity of the hydrogen atom.
(b) Use energy conservation to find the final velocity of the hydrogen atom.
(c) Which answer, (a) or (b) above, is to be trusted? Why?

Problem 2 The wave function of an electron is given by $\psi(x, t) = A(ax - x^2)e^{-k|x|}e^{i\omega t}$ with $k = 0.5\,\text{Å}^{-1}$ and $a = 1\,\text{Å}$, and the value of the angular frequency ω is unimportant.

(a) Let $P_+ = \int_0^\infty dx |\psi(x, t)|^2$ be the probability to find the electron anywhere in the positive-x region and $P_- = \int_{-\infty}^0 dx |\psi(x, t)|^2$ be the probability to find the electron anywhere in the negative-x region. Find the ratio P_+/P_-.
(b) Determine the normalization constant A, including the units. Hint: You may want to use the integral formula $\int_0^\infty dx x^n e^{-x} = n!$.

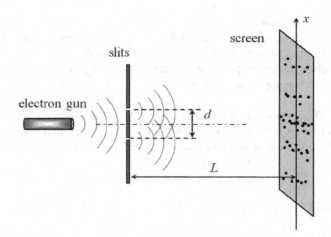

Fig. 1.4 Double slit experiment

Problem 3 Electrons of a given energy E are shot one by one onto a double slit, see Fig. 1.4. Assume the distance between the slits, d, and the distance from the slits to the screen, L, to be known. The screen is covered with a sensitive material that changes its brightness on the spots hit by electrons. Hence, after sufficiently many electrons are shot, a certain pattern is observed on the screen. Establish that pattern for $L \gg d$ and $L \gg x$, where x is the distance on the screen from the symmetry axis. Consider two cases:

(a) Assuming that electrons are classical particles, sketch the screen brightness $I(x)$.
(b) Assume electrons are plane waves, which interact with both slits simultaneously. Hence, each slit becomes a source of a partial wave, ψ_1 and ψ_2, where the subscripts 1 and 2 label the slits. The electron wave function at the screen is $\psi = \psi_1 + \psi_2$. Determine screen brightness $I(x) \sim |\psi(x, t)|^2$ up to a multiplicative constant.
Answer: $I(x) \sim 1 + \cos\left(k\frac{d}{L}x\right)$.

Problem 4 An electron is confined to the xy-plane, where it is placed inside a square potential well of side length L:

$$V(x, y) = \begin{cases} 0 \ \text{for } 0 < x < L \text{ and } 0 < y < L \\ \infty \ \text{otherwise.} \end{cases}$$

The time-independent Schrödinger equation reads

$$-\frac{\hbar^2}{2m_e} \left(\frac{\partial^2 \Psi(x, y)}{\partial x^2} + \frac{\partial^2 \Psi(x, y)}{\partial y^2} \right) = E\Psi(x, y)$$

inside the square. The boundary conditions are

$$\Psi(0, y) = \Psi(L, y) = \Psi(x, 0) = \Psi(x, L) = 0.$$

Note that the wave function must be normalized according to

$$\int_0^L dx \int_0^L dy \, |\Psi(x, y)|^2 = 1.$$

Find all possible wave functions $\Psi(x)$ and energies E that the electron can have.

Problem 5 The ground state wavefunction of a proton in a harmonic potential $V(x) = kx^2/2$ is given by $\Psi(x) = Ae^{-x^2/(2\sigma^2)}$. Assuming $k = 1\,\text{N/m}$, use Schrödinger's equation to find σ.

Problem 6 Two non-interacting spin-1/2 particles find themselves in a simple harmonic potential $V(x) = kx^2/2$. The energy of a single particle in the nth quantum state is $E_n = \hbar\omega(n + 1/2)$. Find the energies and degeneracies of the first 5 energy levels. Keep in mind that the two particles can have the same energy only if their spins are antiparallel.

Answer: $E_n = n\hbar\omega$ and $g_n = n + 2\lfloor n/2 \rfloor$, for $n = 1, 2, \ldots$, where $\lfloor \cdots \rfloor$ is the floor function, which equals the largest integer that does not exceed its argument. In particular, $g_1 = 1$, $g_2 = 4$; $g_3 = 5$; $g_4 = 8$; $g_5 = 9$, etc.

1.17.3 Solutions

Problem 1 The energy of the incident photon of wavelength 300 nm = 0.3 μm is

$$hf = \frac{1.24}{0.3} = 4.13\,\text{eV}.$$

Part of this energy, viz. $w = 3.5\,\text{eV}$, goes into helping the electron get out of the metal. The rest of the photon energy is converted into the electron's kinetic energy:

$$E_K = hf - w = 4.13 - 3.5 = 0.63\,\text{eV}.$$

The electron's wavelength is

$$\lambda_e = \sqrt{150/0.63} = 15.4\,\text{Å}.$$

Problem 2 In the presence of a magnetic field, an electron experiences a Lorentz force of magnitude

Fig. 1.5 Electron trajectory
in a magnetic field

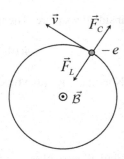

$$F_L = ev\mathcal{B}.$$

The direction of this force is perpendicular to both the velocity $\vec{v}$ and the magnetic field $\vec{\mathcal{B}}$. The Lorentz force is balanced by the centrifugal force,

$$F_C = \frac{m_e v^2}{r},$$

acting in the opposite direction, so the electron moves in a circle, see Fig. 1.5.

From the condition $F_L = F_C$, we find the first relation between the electron orbit radius r and its momentum $p = mv$:

$$ev\mathcal{B} = \frac{m_e v^2}{r}, \quad e\mathcal{B} = \frac{m_e v}{r} = \frac{p}{r}.$$

The second relation comes from the quantization condition that the circumference must accommodate an integer number of electron's de Broglie wavelengths $\lambda = h/p$:

$$2\pi r = n\lambda = n\frac{h}{p}, \quad n = 1, 2, 3, \ldots.$$

It is convenient to use the reduced Planck constant $\hbar = h/(2\pi)$, in terms of which this condition reads

$$rp = n\hbar.$$

From these equalities, we obtain

$$p^2 = n\hbar e\mathcal{B}.$$

Hence, the kinetic energy of the electron assumes discrete values

$$E_n = \frac{p^2}{2m_e} = n\frac{e\hbar}{2m_e}\mathcal{B} = n\mu_B\mathcal{B},$$

where the combination

$$\mu_B = \frac{e\hbar}{2m_e}$$

is called Bohr's magneton. A more precise quantum-mechanical treatment leads to the same result, but with n being an odd integer.

The electron moving in a circle can be viewed as a current loop with current flowing in the clockwise direction. Such a loop possesses a magnetic dipole moment μ pointing into the plane of the figure, i.e., in the direction opposite to $\vec{\mathcal{B}}$. Its magnitude is

$$\mu = IA,$$

where I is the electric current due to the circular motion of our electron and A is the loop area,

$$A = \pi r^2.$$

We find the radius of the electron's orbit from the equations obtained above. We get

$$r^2 = \frac{n\hbar}{e\mathcal{B}}.$$

Hence,

$$A = \pi\frac{n\hbar}{e\mathcal{B}}.$$

The electric current I is the amount of charge that is transferred through an arbitrarily chosen cross section of the loop per unit time. The electron's velocity is

$$v = \frac{p}{m_e} = \frac{\sqrt{ne\hbar\mathcal{B}}}{m_e},$$

and its rotation period is

$$T = \frac{2\pi r}{v}.$$

Every T time units, the charge transferred is $-e$. Hence, the magnitude of the current is

$$I = \frac{e}{T} = \frac{ev}{2\pi r} = \frac{e\sqrt{ne\hbar B/m_e}}{2\pi\sqrt{n\hbar/(eB)}} = \frac{e^2 B}{2\pi m_e},$$

and the magnetic moment

$$\mu = IA = \frac{e^2 B}{2\pi m_e} \cdot \frac{\pi n\hbar}{eB} = n\frac{\hbar e}{2m_e} = n\mu_B$$

is quantized in Bohr's magnetons.

Problem 3

(a) The normalization constant is found from the normalization condition

$$\int_{-\infty}^{\infty} dx\, |\psi(x,t)|^2 = 1.$$

Probability density to find the electron at the position x

$$|\psi(x,t)|^2 = \psi^*(x,t)\psi(x,t) = A^2 \sin^2(kx)e^{-2b|x|}.$$

We can find A from the equation

$$A^2 \int_{-\infty}^{\infty} dx\, \sin^2(kx)e^{-2b|x|} = 2A^2 \int_{0}^{\infty} dx\, \sin^2(kx)e^{-2bx},$$

where we used the fact that the integrand is an even function of x, allowing us to replace $\int_{-\infty}^{\infty}$ with $2\int_{0}^{\infty}$. We evaluate the integral using Euler's formulas:

$$e^{ikx} = \cos(kx) + i\sin(kx), \quad e^{-ikx} = \cos(kx) - i\sin(kx),$$

from which we obtain

$$\sin(kx) = \frac{e^{ikx} - e^{-ikx}}{2i}.$$

Plugging this result into the integral, we get

$$2A^2 \int_{0}^{\infty} dx\, \frac{\left(e^{ikx} - e^{-ikx}\right)^2}{(2i)^2} e^{-2bx} = -\frac{A^2}{2} \int_{0}^{\infty} dx\, \left(e^{2ikx} - 2 + e^{-2ikx}\right) e^{-2bx},$$

where we used the fact that $i^2 = -1$. What we now need to integrate are the exponential functions. Using the identity $\int_0^\infty dx e^{-px} = \frac{1}{p}$, where the real part of p is positive, we have

$$A^2 \left(\frac{1}{2b} - \frac{1}{4(b+ik)} - \frac{1}{4(b-ik)} \right) = \frac{A^2}{2} \left(\frac{1}{b} - \frac{b}{b^2 + k^2} \right) = \frac{A^2}{2b} \frac{k^2}{b^2 + k^2} = 1.$$

Finally, the normalization constant is

$$A = \frac{1}{k} \sqrt{2b(b^2 + k^2)}.$$

Numerically, $A = \sqrt{2 \cdot 2 \cdot (4+1)} = 2\sqrt{5}\,\text{Å}^{-1/2}$.

(b) The electron is most likely to be found at the position x that maximizes the probability density $|\psi(x,t)|^2$. We find these positions by differentiation:

$$\frac{d}{dx} \left(A^2 \sin^2(kx) e^{-2b|x|} \right) = 0,$$

as the expression for $|\psi(x,t)|^2$ is given in part (a). Differentiation gives

$$2 \sin(kx) \cos(kx) k + \sin^2(kx)(-2b)\text{sign}(x) = 0$$
$$\Rightarrow \quad k \cos(kx) = b \sin(kx)\text{sign}(x),$$

or simply

$$\text{sign}(x)\, \tan(kx) = \frac{k}{b}.$$

This equation determines the positions of maxima and minima of $|\psi(x,t)|^2$. The largest maxima, i.e., the positions where the particle is most likely to be found, are located at

$$x_{\max} = \pm \frac{1}{k} \tan^{-1} \left(\frac{k}{b} \right).$$

Problem 4 We look for the solution of Schrödinger's equation in the form

$$\Psi(x) = A \sin(kx) + B \cos(kx)$$

with unknown coefficients A and B and the wave number k. Substitution of the second derivative

$$\frac{d^2\Psi(x)}{dx^2} = -k^2\left(A\,\sin(kx) + B\,\cos(kx)\right) = -k^2\Psi(x)$$

into Schrödinger's equation gives

$$-\frac{\hbar^2}{2m_e}\frac{d^2\Psi}{dx^2} = \frac{\hbar^2 k^2}{2m_e}\Psi = E\Psi,$$

allowing us to relate the wave number k to the energy E:

$$\frac{\hbar^2 k^2}{2m_e} = E, \quad k = \frac{\sqrt{2m_e E}}{\hbar}.$$

To find A and B, we use the boundary conditions and normalization. First, we note that

$$\Psi(0) = A\,\sin(0) + B\,\cos(0) = 0 \;\Rightarrow\; B = 0,$$

and hence, there is no cosine term in $\Psi(x) = A\sin(kx)$. The second boundary condition,

$$\Psi(L) = A\,\sin(kL) = 0,$$

suggests that kL must be an integer multiple of π,

$$kL = n\pi, \quad n = 1, 2, 3, \ldots,$$

because $\sin(n\pi) = 0$. We do not include $n = 0$ because this would give $\Psi(x) = 0$ (no electron at all). We do not consider negative values of n because a wave function with negative n represents the same quantum state as the wave functions with positive n: the two such wave functions differ only by the sign.

Now that we have established that $k = \frac{n\pi}{L}$, we find the energy in the quantum state n:

$$E_n = \frac{\hbar^2 k^2}{2m_e} = n^2\frac{\hbar^2\pi^2}{2m_e L}.$$

It remains to determine the normalization constant A from the condition $\int_0^L dx\,|\Psi(x)|^2 = 1$, giving

$$A^2 \int_0^L dx\,\sin^2\left(\frac{n\pi x}{L}\right) = 1.$$

The integral can be found using either the identity $\sin^2(\theta) = \frac{1-\cos(2\theta)}{2}$ or the identity $\sin(\theta) = \frac{e^{i\theta}-e^{-i\theta}}{2i}$, where $\theta = \frac{n\pi x}{L}$. The value of the integral turns out to be $L/2$, giving

$$A^2 \frac{L}{2} = 1 \;\Rightarrow\; A = \sqrt{\frac{2}{L}}$$

for all quantum numbers n.

Problem 5

(a) The total wave function consists of two counterpropagating plane waves with the wave numbers

$$k_1 = \frac{1}{0.2} = 5\,\text{nm}^{-1}, \quad k_2 = \frac{1}{-1} = -1\,\text{nm}^{-1}$$

and the corresponding wave lengths found from $\lambda_{1,2} = \frac{2\pi}{k_{1,2}}$:

$$\lambda_1 = 0.4\,\pi\,\text{nm} = 4\,\pi\,\text{Å}, \quad \lambda_2 = 2\,\pi\,\text{nm} = 20\,\pi\,\text{Å}.$$

The energies corresponding to these wave lengths are found from $\lambda_{1,2} = \sqrt{\frac{150\,\text{eV}\cdot\text{Å}^2}{E_{1,2}}}$. On the other hand, energy is related to the angular frequency by $E_{1,2} = \hbar\omega_{1,2}$. Hence,

$$\lambda_{1,2} = \sqrt{\frac{150\,\text{eV}\cdot\text{Å}^2}{\hbar\omega_{1,2}}} \;\Rightarrow\; \omega_{1,2} = \frac{150\,\text{eV}\cdot\text{Å}^2}{\hbar\lambda_{1,2}^2}.$$

Numerically, $\hbar = 6.58 \cdot 10^{-16}\,\text{eV}\cdot\text{s}$. Thus,

$$\omega_{1,2} = \frac{2.28 \cdot 10^{17}\,\text{Å}^2 \cdot \text{s}^{-1}}{\lambda_{1,2}^2}.$$

After plugging in the values of $\lambda_{1,2}$ (in Å), we get $\omega_1 = 1.44 \cdot 10^{15}\,\text{s}^{-1}$, $\omega_2 = 5.77 \cdot 10^{13}\,\text{s}^{-1}$.

(b) The total wave function $\psi(x,t)$ consists of two plane-wave functions, corresponding to wave numbers k_1 and k_2:

$$\psi(x,t) = c_1\psi_1(x,t) + c_2\psi_2(x,t)$$

with

$$\psi_{1,2} = \frac{1}{L}\,e^{i(k_{1,2}x - \omega_{1,2}t)}.$$

The respective momentum values are found as $p_{1,2} = \hbar k_{1,2}$ with $\hbar = 1.055 \cdot 10^{-34}\,\text{J}\cdot\text{s}$:

$$p_1 = 5.275 \cdot 10^{-25} \frac{\text{kg} \cdot \text{m}}{\text{s}}, \quad p_2 = -1.055 \cdot 10^{-25} \frac{\text{kg} \cdot \text{m}}{\text{s}}.$$

The probabilities $P_{1,2}$ to measure momenta $p_{1,2}$ are related to the factors $c_{1,2}$ that multiply the plane waves $\psi_{1,2}$ as

$$P_{1,2} = |c_{1,2}|^2.$$

Numerically,

$$P_1 = \left(\frac{\sqrt{3}}{2}\right)^2 = \frac{3}{4} = 75\%, \quad P_2 = \frac{1-i}{2\sqrt{2}} \cdot \frac{1+i}{2\sqrt{2}} = \frac{1-i^2}{8} = \frac{2}{8} = \frac{1}{4} = 25\%.$$

As expected, $P_1 + P_2 = 1$.

(c) The average momentum is

$$\langle p \rangle = P_1 p_1 + P_2 p_2 = (5.275 \cdot 0.75 - 1.055 \cdot 0.25) \cdot 10^{-25} = 3.69 \cdot 10^{-25} \text{ kg} \cdot \text{m/s}.$$

Problem 6

(a) Keeping in mind the definition of degeneracy as the total number of combinations of quantum numbers that correspond to the same energy, we need to find the number of combinations of the non-negative integers $n_1, n_2 = 0, 1, 2, \ldots$ such that their sum equals n:

$$n_1 + n_2 = n.$$

To each value of one of those numbers, say, n_1 between 0 and n, there corresponds a single value of $n_2 = n - n_1$. Hence, there are $n + 1$ combinations of n_1 and n_2 such that the sum of these numbers is n. The degeneracy of the nth energy level is $n + 1$:

$$g_n = \sum_{n_1=0}^{n} 1 = n + 1.$$

(b) This time, we need to find the number of combinations of three non-negative integers n_1, n_2, n_3 such that their sum has a fixed value n:

$$n_1 + n_2 + n_3 = n.$$

Let one of these numbers, say, n_1, assume the values between 0 and n. The values of n_1 larger than n are not allowed because the sum $n_1 + n_2 + n_3$ will also be larger than n. For each value of n_1, another number, say, n_2, may assume the values between 0 and $n - n_1$. Again, higher values of n_2 are not allowed for the same reason. For each combination of n_1 and n_2, the third quantum number has a fixed value $n - n_1 - n_2$.

The degeneracy is the number of combination of n_1 between 0 and n and n_2 between 0 and $n - n_1$. For each value of n_1, the number n_2 assumes $n - n_1 + 1$ possible values. The degeneracy is then

$$g_n = \sum_{n_1=0}^{n} \sum_{n_2=0}^{n-n_1} 1 = \sum_{n_1=0}^{n} (n-n_1+1) = \sum_{n_1=0}^{n} (n+1) - \sum_{n_1=0}^{n} n_1 = (n+1)^2 - \sum_{n_1=0}^{n} n_1.$$

The last term is just the sum of an arithmetic series. It equals

$$\sum_{n_1=0}^{n} n_1 = \frac{n(n + 1)}{2}.$$

Finally,

$$g_n = (n + 1)^2 - \frac{n(n + 1)}{2} = (n + 1)\left(n + 1 - \frac{n}{2}\right) = \frac{(n + 1)(n + 2)}{2}.$$

Chapter 2
Crystal Structure of Solids

2.1 Periodic Table of Elements

We can now apply what we have learned about the energy levels of hydrogen, the concept of spin, and Pauli's exclusion principle to understand the periodic table of elements (Mendeleev, 1869). We may consider the electrons in an atom as approximately non-interacting because they tend to avoid each other due to their Coulomb repulsion. Then, the one-electron states in an atom are affected by the positive point-like nucleus to a larger degree than by the remaining electrons. The conclusion is that if the atomic number Z is not too high, the electron quantum states in an atom will be similar to Hydrogen quantum states, i.e., they will be characterized by the principal quantum number $n = 1, 2, \ldots$, orbital quantum number $l = 0, 1, \ldots, n$, magnetic quantum number $m = -l, \ldots, l$, and spin quantum number $s = \pm 1/2$, see Sect. 1.11. The single-electron states $\Psi_{nlms}(\vec{r})$ are then to be filled from bottom up taking into account Pauli's exclusion principle.

When filling the table, one should keep in mind that the nth energy shell can accommodate $2n^2$ electrons, each occupying a state with its own values of l, m, and s. The common notation is to denote the orbitals with $l = 0$ by the letter s (not to be confused with spin); the orbitals with $l = 1$ are denoted with the letter p (not to be confused with momentum); the $l = 2$ orbitals are denoted with the letter d; the $l = 3$ orbitals with letter f. The electron orbitals for the first 10 elements are shown in Table 2.1.

Consider silicon as an example, whose electron configuration is $1s^2\ 2s^2\ 2p^6\ 3s^2\ 3p^2$. Its deepest shell with $n = 1$ can accommodate two electrons with $l = m = 0$ and opposite spins; hence, we have $1s^2$. The second shell, $n = 2$, can accommodate $2 \cdot 2^2 = 8$ electrons. Those electrons can occupy an s-orbital with $l = 0$, which accommodates two electrons with opposite spins (hence $2s^2$) and three p-orbitals with $l = 1$ and $m = 1, 0, 1$, with two electrons on each, giving $2 \cdot 3 = 6$ electrons in the 2p-states (hence, $2p^6$). Thus, the $n = 2$ shell is completely filled. The electrons belonging to a partially filled shell are called the valence electrons. In Si, the valence

© The Author(s), under exclusive license to Springer Nature Switzerland AG 2022
M. Evstigneev, *Introduction to Semiconductor Physics and Devices*,
https://doi.org/10.1007/978-3-031-08458-4_2

Table 2.1 Electron orbitals for the first 10 elements of the periodic table

H (Hydrogen), $Z = 1$	$1s^1$
He (Helium), $Z = 2$	$1s^2$ (full shell)
Li (Lithium), $Z = 3$	$1s^2 \, 2s^1$
Be (Beryllium), $Z = 4$	$1s^2 \, 2s^2$
B (Boron), $Z = 5$	$1s^2 \, 2s^2 \, 2p^1$
C (Carbon), $Z = 6$	$1s^2 \, 2s^2 \, 2p^2$
N (Nitrogen), $Z = 7$	$1s^2 \, 2s^2 \, 2p^3$
O (Oxygen), $Z = 8$	$1s^2 \, 2s^2 \, 2p^4$
F (Fluorine), $Z = 9$	$1s^2 \, 2s^2 \, 2p^5$
Ne (Neon), $Z = 10$	$1s^2 \, 2s^2 \, 2p^6$ (full shell)

electrons belong to the $n = 3$ shell. It contains 2 electrons with opposite spins in the s-state with $l = m = 0$, and two electrons in the p-state with $l = 1$. These two electron states can have any value of $m = 1, 0, 1$ and spin $s = -1/2, 1/2$, with at least one of these quantum numbers in the two states being different.

If one considers atoms with a higher atomic number ($Z > 18$), i.e., with a higher number of electrons, then this simple scheme stops working because the approximation of non-interacting electrons does not apply anymore. Interaction between the electrons starts to play bigger role.

2.2 Chemical Bonding

Chemical bonding can be understood based on the general premise that the system of electrons generally occupies the lowest energy state, called the ground state. When two or more atoms are brought together, the ground state of the electrons belonging to all these atoms will be modified, depending on the distance between the atomic nuclei. The ground state energy of the electrons may either increase or decrease if the atoms are close to each other. If the ground state energy increases, the atoms of interest will not form molecules, and if it decreases, they will bind together.

Finding the ground state of a system consisting of several atoms is a difficult task, which can be tackled only numerically. But the general mechanisms that lead to qualitative understanding of atomic bonding can be described relatively easily. There are four basic binding mechanisms:

Van der Waals Bonding Van der Waals bonding is the weakest one of all four, but van der Waals forces responsible for this type of bonding act between all neutral atoms and molecules. While atoms are electrically neutral, the center of mass of the negative electron charge randomly fluctuates in time around the center of mass of the positive nuclear charge. These quantum fluctuations generate a randomly fluctuating electric field around an atom, and this field slightly displaces the electron cloud in another atom, thereby polarizing the other atom. Of course, the second atom polarizes the first one as well. The mutual polarization is such that the two atoms tend to attract each other. The resulting interaction potential is known to decrease with the distance, d, between the two atoms as the sixth power, $1/d^6$. For the noble gases, whose energy shells are completely filled (He, Ne, Ar, Kr, Xe, Rn), van der Waals forces are the only forces that can bind those atoms together at low temperature and high pressure.

Ionic Bonding Consider two atoms from the opposite ends of the period table. Atom A has all its inner energy shells filled, and the next partially filled shell contains only one electron (examples: A = Li, Na, or K). The other atom B lacks just one electron to have all its energy shells filled (examples: B = H, F, or Cl). In other words, the electron configuration of the atom A is the same as in some noble gas atom plus an extra electron; for the atom B, it is the same as some noble gas atom minus an electron from the outermost shell. When A and B are brought close to each other, it will be energetically favorable for the atom A to give its extra electron to the atom B, thus filling up all shells of both atoms. This turns A into a positive ion and B into a negative ion. The two ions attract each other electrostatically, so that they form a polar molecule AB. If there are macroscopically many atoms A and B, they will form an ionic crystal, such as the salts NaCl, KCl, NaF, LiBr, etc.

Covalent Bonding Covalent bonding acts between both similar and dissimilar atoms. Each electron state characterized by the same quantum numbers n, l, and m can admit two electrons with opposite spins. In a covalent bond, the valence electrons are shared between atoms. The sharing of electrons allows each atom to attain the equivalent of a full outer shell. For example, a Si atom has 4 valence electrons in the $n = 2$ shell, which can accommodate $2n^2 = 8$ electrons. In crystalline silicon, each atom has 4 neighbors to share the valence electrons with, thus giving $4 + 4 = 8$ electrons in the outer shell. A similar bonding mechanism acts in other important semiconductor materials.

Metallic Bonding Metallic bonding is a mechanism by which not molecules, but crystals are formed. Each atom in a metal contributes its valence electrons to the atoms of the whole crystal, thereby binding them together. The valence electrons detach from their host atoms and become free to move within the crystal. These free electrons are responsible for the high electric conductivity of metals.

2.3 Crystal Lattices

2.3.1 Atomic Order in Solids

If many atoms are brought together, then it can be energetically favorable for them to arrange themselves into a crystal, i.e., a pattern that periodically repeats itself in space. If the size of the crystal lattice exceeds a few hundred micrometers, one speaks about a single crystal. A polycrystalline material consists of a large number of crystalline grains, whose linear size may vary from tens of nanometers to a few hundred micrometers. Note that the typical interatomic distance in a solid is several Angstroms, so that even a nanometer-sized grain contains thousands of atoms arranged periodically. Finally, an amorphous solid does not have any degree of crystalline order even on the length scales of a few nanometers.

2.3.2 Bravais Lattices

A Bravais lattice is a periodic arrangement of structureless lattice points in space, such that every lattice point can be reached from any other lattice point by adding an integer number of the lattice translation vectors. The coordinate of a lattice point can be expressed in terms of the three translation vectors, denoted as $\vec{a}$, $\vec{b}$, and $\vec{c}$, as

$$\vec{r} = i\vec{a} + j\vec{b} + k\vec{c}$$

with integers i, j, and k. The translation vectors are not necessarily mutually orthogonal and do not necessarily have the same length.

At a first glance, it might seem that the number of such possible lattices is huge. But from the point of view of the symmetry, this intuition is not quite correct. It turns out that there are only 14 three-dimensional Bravais lattices, named after a French crystallographer Auguste Bravais, who summarized them in 1850.

Depending on the relation between the lengths of the translation vectors and the angles between them, 6 lattice families are possible, with the number of members in each family varying from one to four. The triclinic lattice is the least symmetric one, in which all translation vectors are of different lengths and all angles are arbitrary but not equal 90° (see Fig. 2.1a). The most symmetric is the cubic lattice, where all translation vectors are the same and all angles are 90° (see Fig. 2.1b).

Lattice constants are the parameters that are needed to completely characterize the Bravais lattice. A triclinic lattice requires six lattice constants: three lengths a,

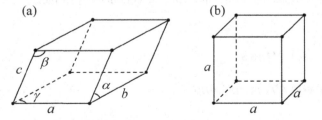

Fig. 2.1 Primitive cells of (**a**) a triclinic and (**b**) a cubic Bravais lattice

b, and c, and three angles α, β, and γ. For a cubic lattice, only one lattice constant, the side length a, is needed.

The difference between the members of the same family is that some of them may have additional lattice points not only in the corners, but also in the center, on the two opposite bases, or on all six bases. Then, the lattices are called body-centered, base-centered, and face-centered, respectively. If none of these possibilities are realized, the lattice is called simple. An exception is the hexagonal family, which can be of two types: simple and rhombohedral. All 14 Bravais lattices are described in standard textbooks on solid state physics and in numerous online resources; hence, we will not discuss this topic in much detail here.

2.3.3 Unit Cell, Primitive Cell, and Crystal Basis

A unit cell is a geometric figure whose periodic repetition in space generates the whole lattice without empty spaces in between. Its choice is not unique: it may have many shapes and sizes. A unit cell of the smallest volume is called primitive. The shape of a primitive cell is also not unique, but its volume is fixed: it contains exactly one lattice point.

Figure 2.2 illustrates these definitions on a two-dimensional oblique lattice, where the cells A and B are non-primitive, and the remaining cells C, D, and E are different choices of a primitive unit cell.

Depending on the position of a lattice point, it may be shared by several unit cells. For example in the two-dimensional lattice from Fig. 2.2, the lattice points in the corners of cell A are shared by four unit cells; with three points inside the cell, this gives $4 \times 1/4 + 3 = 4$ lattice point per cell A. Cell B has four lattice points in the corners, each shared by four similar cells, and two lattice points on the opposite sides, each shared by two cells; altogether, this gives $4 \times 1/4 + 2 \times 1/2 = 2$ lattice points per cell B.

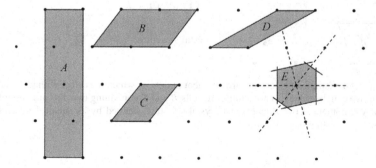

Fig. 2.2 An oblique lattice with two possible choices of the non-primitive cells A and B and three primitive unit cells C, D, and E, the last of which is the Wigner–Seitz cell

In contrast, a primitive cell contains exactly one lattice point. Cells C, D, and E in Fig. 2.2 are primitive.

A special type of a primitive cell is the so-called Wigner–Seitz cell, defined as a locus of all points in space that are closer to a given lattice point than to any other lattice point. A Wigner–Seitz cell is built by connecting a given lattice point to all its neighbor points with straight lines and then drawing planes perpendicular to each of those connecting lines. The Wigner–Seitz cell is enclosed inside the so-obtained planes around the selected lattice point. Cell E in Fig. 2.2 illustrates this procedure for a two-dimensional lattice.

Not every periodic arrangement of points can have the smallest unit cell containing just one lattice point. Consider, e.g., the two-dimensional lattice from Fig. 2.3a. Here, the cell A contains one dot; however, this cell cannot cover the whole lattice by periodic repetition in the plane. The smallest unit cell that reproduces the whole lattice must contain two dots, such as cell B or C.

In order to be able to describe this kind of lattices, we need to introduce an important concept of crystal basis. A lattice point itself is a structureless geometric entity; the basis is a structure associated with each lattice point. This structure may be, e.g., two points. For the lattice from Fig. 2.3a, this would be, e.g., one dot in the corner of cell C and the one inside this cell. Then, the lattice from Fig. 2.3a is a square lattice with the primitive cell C; its lattice points are endowed with a basis of two dots.

In order to describe a real crystal, the basis of its Bravais lattice must contain information about the type, number, and mutual arrangement of atoms that occupy each lattice point. A crystal structure is completely determined by its Bravais lattice and by the basis. This is illustrated by a two-dimensional crystal structure in Fig. 2.3b, showing a square Bravais lattice and a basis consisting of two different atoms.

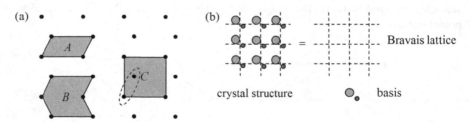

Fig. 2.3 (a) An example of a periodic arrangement of dots, in which the cell A with just one dot is not a primitive cell, because it is not a unit cell; cells B and C containing two dots are possible unit cells. (b) A hypothetical two-dimensional crystal structure, defined by its square Bravais lattice and the basis consisting of two atoms

2.3.4 Volume Density and Atomic Packing Fraction

Assume for simplicity's sake that the basis is formed by just one atom. Then, the volume of its primitive cell, built on the primitive vectors $\vec{a}$, $\vec{b}$, and $\vec{c}$, is $v = \vec{a} \cdot (\vec{b} \times \vec{c})$. The volume density of atoms is

$$n = \frac{1}{v} = \frac{1}{\vec{a} \cdot (\vec{b} \times \vec{c})}.$$

If, instead of a primitive cell, one works with a larger unit cell built on the translation vectors $\vec{a}$, $\vec{b}$, and $\vec{c}$, then

$$n = \frac{N_{cell}}{\vec{a} \cdot (\vec{b} \times \vec{c})},$$

where N_{cell} is the number of atoms per unit cell.

The physical boundary of an atom is not well defined. Nevertheless, one can approximately view atoms in a crystal lattice as solid impenetrable spheres of such radii that neighboring spheres are just touching each other. For a lattice made of just one type of atoms, the atomic radius r is half the distance d between the neighboring atoms in it, or $d = 2r$. The lattice constant a and the atomic radius r are thus proportional to each other, $a \propto d = 2r$, with the proportionality coefficient depending on the crystal structure.

The concept of atomic radius is useful because it allows one to approximately predict the lattice constant of a material if the atomic radii of the components forming it are known. For example, the lattice constant of the diamond modification of carbon $a_C = 3.567\,\text{Å}$, from which the atomic radius of carbon can be found as $r_C = 0.772\,\text{Å}$ with the interatomic distance $d_C = 1.544\,\text{Å}$. Silicon likewise forms diamond crystal lattice with the lattice constant $a_{Si} = 5.431\,\text{Å}$, which corresponds to the atomic radius $r_{Si} = 1.176\,\text{Å}$ and interatomic distance $d_{Si} = 2.352\,\text{Å}$. From this, we can expect that in a diamond-like silicon carbide lattice SiC, the distance between the neighbor silicon and carbon atoms is $d_{SiC} = r_C + r_{Si} = 1.948\,\text{Å}$. The expected lattice constant of this structure is $a_{SiC} \approx a_{Si} \times \frac{d_{SiC}}{d_{Si}} = \frac{1}{2}(a_{Si} + a_C) = 4.5\,\text{Å}$. This value is in reasonable agreement with the known measured value of $a_{SiC} = 4.36\,\text{Å}$, exceeding it by only 3%.

Atomic packing fraction (APF) is the fraction of volume in a crystal occupied by the atoms, which are viewed as solid spheres that touch each other. Specific examples of its calculation will be given below.

2.4   Basic Cubic Structures

Most semiconductor materials belong to the cubic crystal family, whose three members—the simple cubic (sc), body-centered cubic (bcc), and face-centered cubic (fcc)—and the related zincblende and diamond structures (see Fig. 2.4) are briefly reviewed here.

The Simple Cubic (sc) The simple cubic (sc) lattice is the simplest member of the cubic family. The side length of its unit cell, a, is called the lattice constant. Because each atom of the primitive cell is shared by eight cells with a common corner, it contributes 1/8 of its volume to a given cell. Because there are 8 corners, there are $8 \cdot \frac{1}{8} = 1$ atoms per unit cell. The volume density of atoms is

$$n_{sc} = \frac{1}{a^3}.$$

Each atom in an sc lattice has 6 nearest neighbors. If each atom is viewed as a solid sphere that touches the neighbor spheres, its radius would be $r = a/2$ and its volume $V_{sphere} = \frac{4}{3}\pi(a/2)^3$. Since there is just one atom per primitive cell, the atomic packing fraction

$$APF_{sc} = \frac{V_{sphere}}{a^3} = \frac{\pi}{6} = 0.52 = 52\%.$$

The only material with this structure is polonium. Simple cubic structure is rarely found in nature because of its small atomic packing fraction, which is the lowest out of all structures of cubic family. To minimize the total energy of the crystal, atoms tend to come close together, thereby maximizing the APF. The reason why polonium is an exception from this rule has to do with the specific configuration of its six valence electrons' orbitals that favor bonding to six nearest neighbors.

The Body-Centered Cubic (bcc) The body-centered cubic (bcc) lattice is obtained by placing an extra atom in the center of each cube of an sc lattice. The primitive cell of a bcc lattice can be built on the three vectors connecting a corner atom with

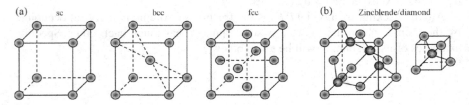

Fig. 2.4 (**a**) The basic cubic structures: simple cubic (sc), body-centered cubic (bcc), and face-centered cubic (fcc). (**b**) The zincblende/diamond structure (left) and a tetrahedron as its building block (right)

three neighboring central atoms. The primitive cell does not have cubic symmetry and is therefore inconvenient to work with. Hence, the conventional unit cell of a bcc lattice is a cube with one atom in the center.

One can think of it as an sc lattice with the basis consisting of two atoms: one in the cube corner and the other one in the center. Alternatively, one may view a bcc lattice as two sc lattices shifted with respect of each other along the main diagonal of the primitive cell by one-half of its length. Therefore, the volume density of a bcc lattice is

$$n_{\mathrm{bcc}} = \frac{2}{a^3}$$

because there are 2 atoms per unit cell. Each atom in a bcc structure has 8 nearest neighbors. The atomic packing fraction

$$APF_{\mathrm{bcc}} = 68\%$$

is notably higher than in the sc structure. Some metals form bcc lattices, such as chromium, molybdenum, tantalum, and tungsten.

The Face-Centered Cubic (fcc) The face-centered cubic (fcc) lattice is formed by placing six atoms in the centers of the six faces of an sc lattice. The primitive cell of an fcc structure is built on the vectors connecting a corner atom with its three neighboring atoms on the faces. The primitive cell does not have cubic symmetry and is inconvenient to work with. The conventional cell of an fcc lattice is a cube with six atoms on its six faces.

Each of the eight corner atoms is shared by 8 cells and therefore contributes 1/8 of its volume to a given unit cell. Each atom on the six faces contributes 1/2 of its volume to a given unit cell. The number of atoms per unit cell is therefore $\frac{1}{8} \cdot 8 + \frac{1}{2} \cdot 6 = 4$. The number of nearest neighbors of an atom in an fcc lattice is 12. The volume density of the atoms is then

$$n_{\mathrm{fcc}} = \frac{4}{a^3}.$$

The atomic packing fraction

$$APF_{\mathrm{fcc}} = 74\%.$$

This is the highest value of all cubic lattices. Another structure that has the same APF is hexagonal closed-packed (hcp). Due to their high degree of packing, fcc and hcp structures are the ones favored by most metals. Specifically, materials with the fcc structure include aluminum, copper, nickel, gold, silver, and platinum.

Zincblende and Diamond Structures   Zincblende and diamond structures can be described as two interpenetrating fcc lattices shifted with respect to each other along the main diagonal by 1/4 of the diagonal length. If the two lattices are made of the same type of atoms (e.g., C, Si, Ge, or α-Sn), the structure is a diamond. If the two fcc lattices are made of different atoms (e.g., GaAs, InSb, InAs, or ZnS), the structure is called zincblende, which is another name for zinc sulfide, ZnS. Note that the materials that prefer to form this structure are either non-metals (C, Si, As, S) or poor metals (Ga, In, Sb, Zn). The building block of this structure is formed by a tetragonal arrangement of atoms, see Fig. 2.4b.

The conventional unit cell of a diamond or a zincblende structure is a cube. The number of atoms in this cell is 8, i.e., $n = 8/a^3$. Each atom has 4 nearest neighbors. The distance between the neighboring atoms is $a\sqrt{3}/4$. The $APF = 34\%$.

Not all semiconductors form diamond or zincblende structure. Examples are numerous: ZnSb, ZnAs, ZnO, BeO, CdSb, CdAs, etc. Some materials may assume more than one different crystal structure, called polymorphs. For example, ZnS, ZnSe, CdS, and CdSe may form zincblende or wurtzite lattices, whereas silicon carbide (SiC) has 250 different crystalline forms, zincblende being one of them.

2.5   Formation of Diamond Structure

Quantum mechanics explains diamond structure as follows. Consider an element belonging to group IV of the periodic table, such as Si. All elements from group IV have 4 valence electrons, two in the s-state and two in the p-state. When many such atoms are brought together, a process called hybridization of orbitals occurs. In this process, one s-electron is promoted into a p-state, resulting in one s-orbital and three p-orbitals. Then, the s-orbital combines with the three p-orbitals to form four orbitals of the new type, the so-called sp^3 hybrid orbitals.

Each of the four sp^3-orbitals is elongated, with an angle of 109.5° formed between any two of them, see Fig. 2.5. Furthermore, each sp^3-orbital can be occupied by two electrons of opposite spins. Thus, when many group-IV atoms

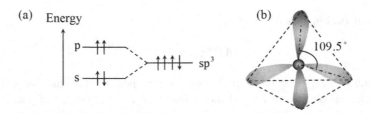

Fig. 2.5   (a) Formation of an sp^3 hybrid orbital. **(b)** A sketch of the electron density distribution in an sp^3 hybrid orbital accommodating four electrons

are brought together, they arrange themselves so that each atom has four nearest neighbors, thus forming tetragons, the building blocks of the diamond lattice.

To understand the zincblende structure formed by atoms belonging to groups III and V (the III–V semiconductors), we note that an atom from group III (e.g., Ga) has 3 valence electrons, and an atom from group V (e.g., As) has 5 valence electrons. An As atom can give one of its electrons to a neighboring Ga atom, so that both atoms now have 4 valence electrons. Then, the sp^3-hybridization proceeds as described above.

In a similar way, zincblende structure can be formed by some of the II-VI semiconductors, such as InSb, CdTe, etc. In these materials, a group-VI element gives two of its valence electrons to a group-II element, resulting in 4 valence electrons in the atoms of each type.

2.6 Miller Indices

2.6.1 Determination of Miller Indices

Miller indices (William H. Miller, 1839) are a standard convenient method of indicating various crystallographic directions and planes in a given crystal lattice.

To indicate a direction, a vector $\vec{d}$ in that direction has to be used. However, simply stating the components of $\vec{d}$ is not enough. One must also give a reference to a particular coordinate system and indicate how the crystal lattice is oriented relative to that system. In order to avoid the necessity of giving this additional information, it is more convenient to express $\vec{d}$ in terms of the lattice translation vectors, $\vec{a}$, $\vec{b}$, and $\vec{c}$.

Depending on the convenience considerations, those vectors may or may not be primitive. The direction vector $\vec{d}$ may have any length. It is decomposed into the translation vectors

$$\vec{d} = i\vec{a} + j\vec{b} + k\vec{c},$$

such that the three numbers i, j, and k are integers without a common denominator. The direction is then specified by placing these three integers into square brackets, $[i\,j\,k]$. If one of these numbers is negative, the corresponding Miller index is denoted with an overbar, e.g., $[1\,1\,\bar{1}]$. Figure 2.6a shows Miller indices for some directions of a cubic lattice.

Some crystallographic directions possess the same symmetry. To indicate the whole family of such symmetry-related directions, one representative member of this family is placed into triangular brackets. For instance, in a cubic lattice, one such family is $\langle 1\,0\,0 \rangle$, consisting of $[1\,0\,0]$, $[0\,1\,0]$, $[0\,0\,1]$, $[\bar{1}\,0\,0]$, $[0\,\bar{1}\,0]$, and $[0\,0\,\bar{1}]$ directions.

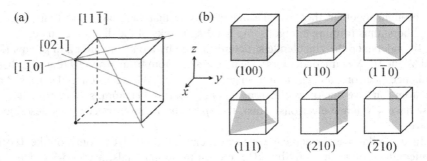

Fig. 2.6 Several crystallographic (**a**) directions and (**b**) crystal planes with the Miller indices indicated in the brackets. The translation vectors $\vec{a}$, $\vec{b}$, and $\vec{c}$ are aligned with the coordinate axes x, y, and z

To indicate a crystallographic plane, the following algorithm is used. First, the intercepts of the plane with the coordinate axes built on the translation vectors, $\vec{a}$, $\vec{b}$, and $\vec{c}$, are expressed in terms of the lengths of these vectors as pa, qb, and sc. The numbers p, q, and s cannot be zero, i.e., the plane cannot go through the origin. Then, the numbers are inverted to give $1/p$, $1/q$, and $1/s$. Each of these inverted numbers is multiplied by the smallest number, n, such that the triplet $i = n/p$, $j = n/q$, and $k = n/s$ is formed by integers i, j, and k. Those integers are the Miller indices for the plane of interest.

If one of the intercepts is infinite, then the corresponding Miller index is zero. Negative Miller indices are denoted with an overbar. Miller indices for planes are indicated in round brackets. Figure 2.6b shows Miller indices for several planes in a cubic lattice.

A family of symmetry-related planes is indicated by placing Miller indices for one representative member of this family into curly brackets. For instance, for cubic lattices, the $\{1\,0\,0\}$ family consists of $(1\,0\,0)$, $(0\,1\,0)$, $(0\,0\,1)$, $(\bar{1}\,0\,0)$, $(0\,\bar{1}\,0)$, and $(0\,0\,\bar{1})$ planes.

2.6.2 Miller Indices for Cubic Structures

As stated earlier, it is convenient to treat bcc, fcc, and diamond/zincblende lattices as cubic, with several atoms per unit cell (2 in bcc, 4 in fcc, and 8 in diamond). A nice feature of cubic lattices is that a direction with Miller indices $[i\,j\,k]$ is perpendicular to the plane with the same Miller indices $(i\,j\,k)$.

Let us find the distance between two successive $(i\,j\,k)$ planes. We note that an $(i\,j\,k)$ plane crosses the x-axis at the point $x = na/i$, the y-axis at the point $y = na/j$, and the z-axis at the point $z = na/k$, see Fig. 2.7. Here, n is an integer. Then, the two vectors

Fig. 2.7 An $(i\,j\,k)$ plane of a cubic lattice

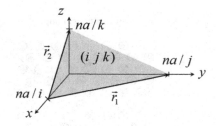

$$\vec{r}_1 = \frac{na}{j}\vec{e}_y - \frac{na}{i}\vec{e}_x, \quad \vec{r}_2 = \frac{na}{k}\vec{e}_z - \frac{na}{i}\vec{e}_x$$

belong to the $(i\,j\,k)$ plane. On the other hand, a unit vector in the $[i\,j\,k]$ direction is, by definition,

$$\vec{e} = \frac{i\vec{e}_x + j\vec{e}_y + k\vec{e}_z}{\sqrt{i^2 + j^2 + k^2}}.$$

Now, it is easy to verify that $\vec{r}_1 \cdot \vec{e} = \vec{r}_2 \cdot \vec{e} = 0$, and thus the unit vector in the $[i\,j\,k]$ direction is perpendicular to the two non-parallel vectors in the $(i\,j\,k)$ plane. This means that $[i\,j\,k]$ direction is perpendicular to that plane, as stated.

To find the distance between two $(i\,j\,k)$ planes, we note that the closest such plane intersects the coordinate axes at the following points: the x-axis at the point $x = (n+1)a/i$, the y-axis at the point $y = (n+1)a/j$, and the z-axis at the point $z = (n+1)a/k$. Then, a vector

$$\vec{d} = \frac{n+1}{i}a\vec{e}_x - \frac{n}{i}\vec{e}_x = \frac{1}{i}a\vec{e}_x$$

begins on the nth plane and ends on the $(n+1)$st plane. The distance between the two planes is the projection of this vector onto a unit vector perpendicular to the $(i\,j\,k)$ planes:

$$d = \vec{d} \cdot \vec{e} = \frac{a}{\sqrt{i^2 + j^2 + k^2}}.$$

2.7 Imperfections and Impurities in Solids

From the electrical properties' point of view, single crystals are preferred over poly-crystalline materials because the grain boundaries in the latter tend to deteriorate the electrical characteristics. However, even in a single crystal, deviations from perfect long-range order are possible for various reasons.

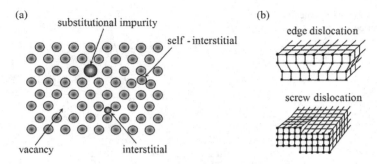

Fig. 2.8 (**a**) Point and (**b**) line defects in a crystal lattice

One obvious reason for such deviations is the unavoidable random thermal vibrations of atoms. However, the average positions around which the atoms vibrate do form a crystal structure.

Other causes of deviations from perfect crystallinity are the formation of defects of various types. Point defects, see Fig. 2.8a, result in a local distortion of the crystal lattice. They may form as vacancies, i.e., missing atoms on same lattice sites, or as interstitial defects, where an atom occupies the space of the crystal structure at which no atom should be present. Other types of point defects are possible, e.g., Frenkel defects formed by closely located vacancy and interstitial defects. Impurity atoms are considered point defects as well.

Line defects may extend over macroscopic distances. They appear as dislocations, which are of two main types: edge and screw dislocations, see Fig. 2.8b. In the former, one of the atomic planes terminates in the middle of the crystal, resulting in a "bending" of the adjacent planes around that half-plane. A screw dislocation is a bit harder to visualize, but basically it is a twisting of the crystal lattice.

Other types of defects are planar defects (grain boundaries, stacking faults, or twin boundaries) and bulk defects such as cracks, pores, voids, or impurity precipitates.

2.8 Problems

2.8.1 Solved Problems

Problem 1 Consider a two-dimensional honeycomb lattice formed by regular hexagons, see Fig. 2.9a. Can the hexagon corners serve as lattice points of a Bravais lattice? If no, how can they be grouped into the basis? What are the translation vectors of the Bravais lattice? What does the Wigner–Seitz cell look like?

Problem 2 Find the atomic packing fraction of a diamond lattice.

Problem 3 A certain plain of a cubic lattice intersects the coordinate axes at the points $x = 2a$, $y = -3a$, and $z = a$, where $a = 7\,\text{Å}$ is the lattice constant. Find the distance between two neighboring planes of this family.

Problem 4 Find the number of atoms per unit area in the (1 1 1) plane of a sc, bcc, and fcc lattices with the lattice constant $a = 4\,\text{Å}$.

2.8.2 Practice Problems

Problem 1 Find the atomic packing fraction of fcc and bcc lattices.

Problem 2 GaAs forms the zincblende lattice with the density of 5.318 g/cm^3. The atomic masses of the atoms forming this structure are $m_{Ga} = 69.7$ and $m_{As} = 74.9$ amu (1 amu $= 1.6605 \cdot 10^{-27}$ kg). Determine the lattice constant of this structure and find its concentration of valence electrons that form the Ga–As bonds.

Problem 3 Find the surface density of atoms (SDA), i.e., the number of atoms per cm^2, in the (1 1 0) plane of a diamond lattice with the lattice constant $a = 5.65\,\text{Å}$.

Problem 4 Consider a body-centered cubic lattice with the lattice constant $a = 6\,\text{Å}$. What are the Miller indices of the plane shown in Fig. 2.9b? What is the surface density of atoms in that plane?

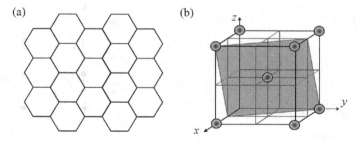

Fig. 2.9 (a) A honeycomb lattice from Problem 1. (b) A crystallographic plane in a bcc lattice from Problem 4

2.8.3   Solutions

Problem 1 If we align one translation vector $\vec{a}$ along the lower side of the hexagon and another vector $\vec{b}$ at an angle of 120° to $\vec{a}$, then a linear combination $i\vec{a} + j\vec{b}$ will cover not only the corners, but also the centers of each hexagon. The hexagon centers are not the lattice points of this structure. Hence, the corners of the hexagons cannot be used as lattice points of a Bravais lattice. But two corners can be grouped together to form the basis of the Bravais lattice, see Fig. 2.10. The resulting lattice points form a hexagonal lattice with the lattice vectors shown in the figure. The original hexagons are Wigner–Seitz cells of this lattice.

Problem 2 The diamond structure can be thought of as two fcc lattices shifted relative to each other along the longest diagonal of the unit cell by one-quarter of its length $d = a\sqrt{3}$, see Fig. 2.11. The distance between the nearest neighbor atoms is, then, $\frac{d}{4} = a\frac{\sqrt{3}}{4}$. If the atoms are viewed as solid spheres of radius r that just touch each other, then the distance between the centers of the neighboring spheres is $2r = d/4$, allowing us to relate the radius to the lattice constant as

$$r = a\frac{\sqrt{3}}{8}.$$

Now, let us count the number of atoms in a diamond structure unit cell. Each of its 8 corner atoms is shared by 8 cells; hence, each corner atom contributes 1/8 of its volume to a given cell. Each of its 6 atoms on the 6 faces is shared by 2 unit cells;

Fig. 2.10 Lattice points of a hexagonal Bravais lattice

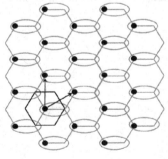

Fig. 2.11 Diamond (or zincblende) unit cell

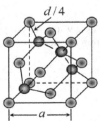

hence, the face atoms contribute 1/6 of their volume to a given cell. Finally, there are 4 atoms that are completely inside a given cell, see Fig. 2.9. Hence, the total number of atoms in a unit cell is

$$N = 8 \cdot \left(\frac{1}{8}\right) + 6 \cdot \left(\frac{1}{2}\right) + 4 = 1 + 3 + 4 = 8.$$

The volume occupied by a single atom is $V_0 = \frac{4}{3}\pi r^3$. Hence, the volume occupied by all 8 atoms in a cell is

$$V_{atoms} = N V_0 = \frac{8 \cdot 4}{3}\pi a^3 \frac{(\sqrt{3})^3}{8^3} = a^3 \pi \frac{\sqrt{3}}{16} \cong 0.34\, a^3.$$

Since the unit cell volume is a^3, the atomic packing fraction is

$$APF = \frac{V_{atoms}}{a^3} = 0.34 = 34\%.$$

Problem 3 We begin by finding the Miller indices of this plane. For this, we invert the intersection point coordinates divided by a:

$$\frac{a}{x} = \frac{1}{2}, \quad \frac{a}{y} = -\frac{1}{3}, \quad \frac{a}{z} = 1$$

and multiply these three numbers by the smallest positive integer ($= 6$) that will turn them into integers. The resulting triplet is the Miller indices:

$$(i\ j\ k) = (3\,\bar{2}\,6).$$

The distance between two neighbor planes with the same Miller indices is (for cubic family lattices) given by

$$d = \frac{a}{\sqrt{i^2 + j^2 + k^2}} = \frac{7}{\sqrt{9 + 4 + 36}} = 1\,\text{Å}.$$

Problem 4 Consider the sc lattice first with the (1 1 1) plane shown in Fig. 2.12. In this plane, the atoms form a triangular lattice consisting of equilateral triangles of side length $d = a\sqrt{2}$. Since each atom is shared by 6 triangles, there are $3/6 = 1/2$ of an atom per triangle. It is less than 1 because a triangle is not a unit cell of this two-dimensional lattice. The unit cell is a rhombus (shaded area) of side length d and height h found using Pythagoras' theorem:

$$h^2 + \left(\frac{d}{2}\right)^2 = d^2 \quad \Rightarrow \quad h = d\frac{\sqrt{3}}{2}.$$

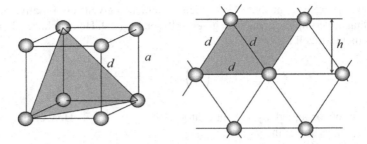

Fig. 2.12 Arrangement of atoms in the (111) plane of a sc lattice

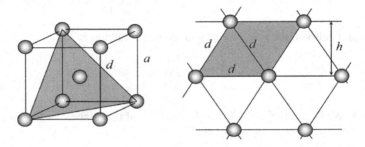

Fig. 2.13 Arrangement of atoms in the (111) plane of a bcc lattice

The area of such a rhombus is

$$A = dh = d^2 \frac{\sqrt{3}}{2} = a^2 \sqrt{3},$$

and, because there is just one atom per rhombus, the surface density of atoms is

$$SDA = \frac{1}{A} = \frac{1}{a^2\sqrt{3}} = \frac{1}{16\sqrt{3}} \text{Å}^{-2} = 3.61 \cdot 10^{-2} \text{Å}^{-2} = 3.61 \cdot 10^{14} \text{cm}^{-2}.$$

A unit cell of a bcc lattice contains an extra atom right in the center, see Fig. 2.13. However, and, perhaps, counterintuitively, this central atom does not happen to be in the (1 1 1) plane. The easiest way to see this is to imagine a (1 1 0) plane, which does contain that central atom. Since the (1 1 1) plane can be obtained by tilting the (1 1 0) around one of the cube's edges, the (1 1 1) plane does not contain the central atom of the bcc unit cell. Therefore, the arrangement of atoms in this plane is exactly the same as in the sc lattice, with the same surface density of atoms $3.61 \cdot 10^{14}$ cm^{-2}.

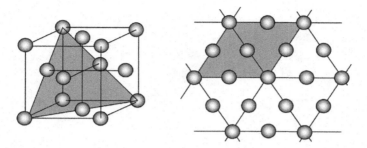

Fig. 2.14 Arrangement of atoms in the (111) plane of an fcc lattice

Finally, an fcc lattice has extra atoms located in the middle of each face. The arrangement of atoms in the (1 1 1) plane is the same as in the sc case, but with atoms in the middle of each side of the triangles, see Fig. 2.14. A unit cell (shaded area in the figure) contains $1 + 4 \cdot \left(\frac{1}{2}\right) + 1 = 4$ atoms. This means that the SDA is 4 times as big as in the sc case: $SDA = 4 \cdot 3.61 \cdot 10^{14} = 1.44 \cdot 10^{15} \, cm^{-2}$.

Chapter 3
Equilibrium Statistical Mechanics

3.1 Microstates and Macrostates

Physical systems consist of a huge number of particles. The most complete information about the system is contained in the system's microstate (microscopic state). In classical mechanics, the microstate is specified by a huge number of coordinates and velocities of all particles comprising the system. It evolves in time according to Newton's second law. In quantum mechanics, the microstate is given by the collective wave function, which depends on all the system's coordinates. It evolves in time according to the time-dependent Schrödinger equation.

Information about the microstate is complete in the following sense: If the system is perfectly isolated from the rest of the Universe, then its microstate will evolve in time according to the system's autonomous equations of motion. Then, the microstate at the initial moment of time determines the system's microstate at any future time according to these equations.

It is a hopeless task to solve the equations that govern the evolution of the system's microstate. There are several reasons for this:

1. The initial microstate of the system is unknown.
2. The equations that govern the system's microstate evolution are deterministic, but their solution is usually chaotic (random). This means that the tiniest deviation of the initial state will result in a completely different trajectory of the system.
3. The system is never completely isolated from the rest of the world. In particular, it is impossible to switch off gravity, or perfectly isolate the system from external electromagnetic waves. Those interactions cannot be completely accounted for. However, small, they affect the dynamics of the systems microstate, see the previous point.
4. Last but not least, the system may be simply too big to be treated theoretically. Solving 10^{30} or so equations of motion is practically impossible. Because of these reasons, the microstate changes in time in a random, unpredictable manner.

M. Evstigneev, *Introduction to Semiconductor Physics and Devices*,
https://doi.org/10.1007/978-3-031-08458-4_3

Luckily, information about the system's microstate is practically never required. What is of interest is a set of the coarse-grained macroscopic parameters that characterize the system as a whole, such as pressure, volume, energy, temperature, concentration of atoms, polarization, magnetization, etc.

Because the system's microstate changes randomly in time, we may introduce the probability p_α to find the system in a microstate α, in which the system's macroscopic parameter of interest, Y, has the value Y_α. As the microstate changes, so does its property Y. These random changes are called fluctuations.

A measurement of the parameter Y takes finite time t_Y. This time is of the same order of magnitude as the reaction time of the measuring device used. The time of measurement is many orders of magnitude longer than the characteristic time over which the system's microstate changes. For instance, oscillation period of an atom in a solid is about 10^{-13} s, much shorter than the reaction time of any device.

This means that during a measurement, the system visits an astronomically huge number of its microstates. The total amount of time it spends in a microstate α during the measurement can be estimated as $t_\alpha = p_\alpha t_Y$. When the system is in the microstate α, its property Y has the value Y_α. The result of a measurement will be the time-averaged value of Y:

$$\langle Y \rangle = \frac{1}{t_Y} \sum_\alpha Y_\alpha t_\alpha = \sum_\alpha Y_\alpha p_\alpha.$$

A macrostate of a system is specified by a relatively small number of the macroscopic parameters that describe some measurable property of the system. Those parameters can be either the control parameters that can be fixed by an experimentalist or they can be the average values of some fluctuating properties of the system. For example, if we consider a gas in a container, then the volume of the container V is part of the macrostate description. If the gas finds itself inside a cylinder with a movable piston, then the gas volume is not fixed; however, applying a constant pressure to the system fixes the average volume $\langle V \rangle$, which must be used as a macrostate variable. Other examples of the macrostate variables are the number of particles, temperature, pressure, etc.

3.2 Thermal Equilibrium

The microstate probabilities p_α, in general, change with time. As an example, consider a container whose right half is initially filled with gas and the left half is empty. Very quickly the gas will distribute itself uniformly within the container. The initial state and the final state are characterized by different sets of microstate probabilities. Corresponding to this transient process from the initial to the final states, the average value of a physical parameter Y will also be time-dependent: $\langle Y \rangle(t) = \sum_\alpha Y_\alpha p_\alpha(t)$.

But if we wait long enough, the system will eventually enter a special macroscopic state, in which the probabilities p_α do not change anymore. This state is called thermal equilibrium, or thermodynamic equilibrium.

Suppose two systems, initially isolated from each other, are in equilibrium states. They do not need to have the same values of such parameters as temperature or pressure. Now, we bring the two systems in contact with each other, allowing them to exchange energy and particles. Then, energy will flow from the hotter system to colder one and atoms will migrate from the system with the higher pressure to the one with the lower pressure. Eventually, the two systems will come to equilibrium with each other, as their temperature and pressure will become equal.

We may also consider a single system in some non-equilibrium initial state. As the system equilibrates, any non-uniformity of its temperature and pressure cause energy and particles to redistribute themselves inside the system until the system reaches equilibrium. In thermodynamic equilibrium, temperature and pressure are uniform. Also, in equilibrium, there is no transfer of energy and particles.

One may think (erroneously) that in thermal equilibrium, energy and particles do not flow between different parts of the system. In fact, thermal equilibrium should not be imagined as something static. Rather, it is a dynamic state, in which different parts of the system do exchange particles and energy, but the fluxes of those quantities are perfectly balanced by the respective counterfluxes. As a result, no overall transport occurs and the average amount of energy and the average number of particles in each part of the system remains constant.

3.3 Postulate of Equal A Priori Probabilities

Consider a system that is isolated to the best of our abilities. Its extensive properties—the volume V, the number of particles N, and the energy E—are held fixed and do not change in time. The microstate of the system, on the other hand, constantly changes. In each microstate α visited by the system, these extensive parameters have fixed values: $V_\alpha = V$, $N_\alpha = N$, $E_\alpha = E$.

The postulate of equal a priori[1] probabilities is a drastic statement about the system's dynamics in thermodynamic equilibrium. It states that in thermodynamic equilibrium all microstates which are consistent with the given fixed extensive parameter values are equally probable.

Mathematically, this can be expressed as follows: If, for two microstates α and α' of an isolated system we have $V_\alpha = V_{\alpha'} = V$, $N_\alpha = N_{\alpha'} = N$, and $E_\alpha = E_{\alpha'} = E$, then $p_\alpha = p_{\alpha'}$.

No rigorous proof of this statement exists. At the same time, all experimental evidence speaks in favor of this postulate. We accept simply because it successfully

---

[1] The Latin expression "a priori" means "from the earlier." It denotes those properties of an object that do not depend on the empirical evidence, but follow from common sense and logic.

explains all phenomena observed in the systems containing many particles. The success of statistical physics can be taken as an a posteriori[2] proof that the postulate of equal a priori probabilities is correct.

3.4 Grand Canonical Distribution

Consider a system, which can exchange energy and particles with a much bigger system called the heat bath. The energy E and the number of particles n in the system of interest is not fixed, because the system exchanges these parameters particles with the heat bath. The role of the heat bath is to maintain constant the average energy $\langle E \rangle$ and the average number of particles $\langle n \rangle$ in the system. If at some moment of time, the number of particles in the system is greater than $\langle n \rangle$, then the particles are more likely to leave the system and enter the heat bath. If $n < \langle n \rangle$, then the particles are more likely to enter the system from the heat bath; same applies to energy.

To each number of particles, n, in the system, there corresponds a certain set of states α. The energy of the system with n particles in a state α is denoted as $E_{n\alpha}$.

We assume that the properties of the system—namely, its quantum states and energies $E_{n\alpha}$—are known. The question that we attempt to answer here is: What is the probability $p_{n\alpha}$ for the system to have n particles and be in the microscopic state α?

At a first glance, we do not have enough information to answer this question. For instance, we do not know the nature of the heat bath and the mechanism of the system-bath coupling, which makes the energy and particle exchange between the two objects possible.

But this question can be answered with the help of the postulate of equal a priori probabilities. Namely, if, for a fixed number of particles, n, two different microstates α and α' have the same energy, $E_{n\alpha} = E_{n\alpha'}$, then they are equally probable, $p_{n\alpha} = p_{n\alpha'}$, independent of the nature of the system. In other words, for a fixed number of particles n, the microstate probability $p_{n\alpha}$ is simply the inverse of the total number of microstates with the energy $E_{n\alpha}$.

This implies that the probability $p_{n\alpha}$ can only depend on the number of particles n in the system and the energy $E_{n\alpha}$ of the state α:

$$p_{n\alpha} = \frac{1}{Z} G(n, E_{n\alpha}),$$

---

[2] The Latin phrase "a posteriori" means "from the later" and refers to the statements that follow from experience or observation of the physical world.

where the function $G(n, E_{n\alpha})$ is to be established, and the constant Z in the denominator is introduced in order to make sure that the total probability is normalized to one,

$$\sum_n \sum_\alpha p_{n\alpha} = 1,$$

meaning that

$$Z = \sum_n \sum_\alpha G(n, E_{n\alpha}).$$

The constant Z is called the grand partition function.

To find the function $G(n, E_{n\alpha})$, we consider two identical systems that are coupled to the same heat bath. The two systems are sufficiently far away from each other to be statistically independent. The probability for one system to have n particles and to be in the state α is $p_{n\alpha}$. The probability for the other system to have n' particles and to be in the state α' is $p_{n'\alpha'}$. The combined probability for the first system to be in the state (n, α) and the other system to be in the state (n', α') is given by the product of the individual probabilities:

$$p_{n\alpha} p_{n'\alpha'} = \frac{1}{Z^2} G(n, E_{n\alpha}) G(n', E_{n'\alpha'}).$$

On the other hand, this composite probability must depend on the total energy and the total number of particles in the two systems via the function G:

$$p_{n\alpha} p_{n'\alpha'} = \frac{1}{\tilde{Z}} G(n + n', E_{n\alpha} + E_{n'\alpha'})$$

with the normalization constant

$$\tilde{Z} = \sum_n \sum_{n'} \sum_\alpha \sum_{\alpha'} G(n + n', E_{n\alpha} + E_{n'\alpha'}).$$

Equating the natural logarithms of the last two expressions for $p_{n\alpha} p_{n'\alpha'}$, we conclude that the function $G(n, E_{n\alpha})$ must be such that

$$\ln G(n + n', E_{n\alpha} + E_{n'\alpha'}) = \ln G(n, E_{n\alpha}) + \ln G(n', E_{n'\alpha'}) + \ln(\tilde{Z}/Z^2).$$

But this is possible if and only if $\ln G(n, E_{n\alpha})$ depends linearly on both arguments, n and $E_{n\alpha}$. Hence, we introduce two parameters, A and B, to express the linear relation between the function $G(n, E_{n\alpha})$ and its arguments:

$$\ln G(n, E_{n\alpha}) = An + BE_{n\alpha} \quad \Rightarrow \quad G(n, E_{n\alpha}) = e^{An + BE_{n\alpha}}.$$

Note that the condition $\tilde{Z} = Z^2$ is then fulfilled automatically, because

$$Z = \sum_{n\alpha} e^{An + BE_{n\alpha}},$$

$$\tilde{Z} = \sum_{n} \sum_{n'} \sum_{\alpha} \sum_{\alpha'} e^{A(n+n') + B(E_{n\alpha} + E_{n'\alpha'})}$$

$$= \sum_{n\alpha} e^{An + BE_{n\alpha}} \sum_{n'\alpha'} e^{An' + BE_{n'\alpha'}} = Z^2.$$

Actually, it is customary to express the parameters A and B in terms of two other parameters, denoted as T and μ: $B = -\frac{1}{kT}$ and $A = \frac{\mu}{kT}$. Here, T is the temperature (measured in Kelvins), the parameter μ is called the chemical potential and,

$$k = 1.3806\ldots \cdot 10^{-23} \text{ J/K} = 8.6173\ldots \cdot 10^{-5} \text{ eV/K},$$

is Boltzmann's constant.

We thus arrive at the standard expression for the grand canonical distribution for the system to have n particles and find itself in the state α:

$$p_{n\alpha} = \frac{1}{Z} \exp\left(\frac{\mu n - E_{n\alpha}}{kT}\right), \quad Z = \sum_{n} \sum_{\alpha} \exp\left(\frac{\mu n - E_{n\alpha}}{kT}\right).$$

The constants μ and T are determined from the requirements that the system has a given average number of particles and a given average energy:

$$\langle n \rangle = \sum_{n} \sum_{\alpha} n p_{n\alpha}, \quad \langle E \rangle = \sum_{n} \sum_{\alpha} E_{n\alpha} p_{n\alpha}.$$

As stated earlier, thermodynamic equilibrium state is the state of no microscopic currents, by means of which different parts of the system would exchange energy or particles. Consider now a spatially extended large system containing a very large number of particles. In this system, we focus on two small volumes ΔV, one around some point $\vec{r}$ and the other around a point $\vec{r}\,'$. The average number of particles and the average energy in both small volumes must be the same, independent of the choice of their positions $\vec{r}$ and $\vec{r}\,'$. But $\langle n \rangle$ and $\langle E \rangle$ are uniquely determined by the chemical potential μ and temperature T. Hence, we conclude that in thermodynamics equilibrium, the chemical potential μ and temperature T are constant everywhere inside a spatially extended system.

3.5 Fermi-Dirac Distribution

In the previous discussion, we did not specify the nature of the system. Therefore, the main result obtained there—namely, the grand canonical distribution—is very general and applies to systems of any nature.

In particular, consider electrons in a crystal. Assume that the interaction between electrons can be neglected. Hence, each electron may occupy a single-electron state α. Nothing can stop us from treating this quantum state as the system and to apply the grand canonical distribution to get its occupation probability.

The energy of this system can be either the quantum state energy E_α if it is occupied by an electron, or 0 if it is empty:

$$E_{n\alpha} = nE_\alpha, \quad n = 0, 1.$$

The number of electrons n in the state α can assume only two values, 0 and 1, according to Pauli's exclusion principle.

We can determine the average number of electrons in the state α by using the grand canonical distribution. If the state is empty, then $n = 0$, $E_{0\alpha} = 0$, and $p_{0\alpha} = \frac{1}{Z}$. If the state is filled with an electron, then $n = 1$, $E_{1\alpha} = E_\alpha$, and $p_{1\alpha} = \frac{e^{(\mu-E_\alpha)/kT}}{Z}$. The average number of electrons in the state α state is, thus,

$$\langle n_\alpha \rangle = 0 \cdot p_{0\alpha} + 1 \cdot p_{1\alpha} = p_{1\alpha} = \frac{1}{Z} e^{(\mu-E_\alpha)/kT}.$$

The partition function is

$$Z = \sum_{n=0,1} e^{n(\mu-E_{n\alpha})/kT} = 1 + e^{(\mu-E_\alpha)/kT}.$$

Then

$$\langle n_\alpha \rangle = \frac{e^{(\mu-E_\alpha)/kT}}{1 + e^{(\mu-E_\alpha)/kT}} = \frac{1}{1 + e^{(E_\alpha-\mu)/kT}}.$$

This number, which is between 0 and 1, is the probability $p_{1\alpha}$ that the state α is occupied. This formula is called the Fermi-Dirac distribution.

The letter μ is a standard notation for the chemical potential adopted in thermodynamics. In solid-state physics, a different name and notation is used for the chemical potential. Namely, in solid-state physics, the chemical potential is called the Fermi energy and is denoted as E_F:

$$\mu \equiv E_F.$$

This is the convention that we also will be using. Hence, the probability of occupation of a particular quantum state α is given by the Fermi-Dirac distribution:

$$f(E_\alpha) = \frac{1}{1 + e^{(E_\alpha - E_F)/kT}}.$$

The probability for this state to be empty is given by a similar expression:

$$1 - f(E_\alpha) = 1 - \frac{1}{1 + e^{(E_\alpha - E_F)/kT}} = \frac{e^{(E_\alpha - E_F)/kT}}{1 + e^{(E_\alpha - E_F)/kT}} = \frac{1}{1 + e^{(E_F - E_\alpha)/kT}},$$

note the change of sign in the exponential function in the denominator.

A plot of the Fermi-Dirac distribution vs. energy (with the state subscript α omitted) is shown in Fig. 3.1a. At $E < E_F$, Fermi-Dirac distribution approaches the value 1; at $E > E_F$, it goes to zero very quickly. The transition between these two extremes occurs near Fermi energy, at which the distribution assumes the value

$$f(E_F) = \frac{1}{2}.$$

The width of the transition region depends on the temperature. At $T = 0$, the distribution $f(E)$ is a sharp step going from the value 1 at $E < E_F$ to 0 at $E > E_F$ (dotted line). Increasing the temperature makes the transition of $f(E)$ near $E = E_F$ from 1 to 0 more smooth.

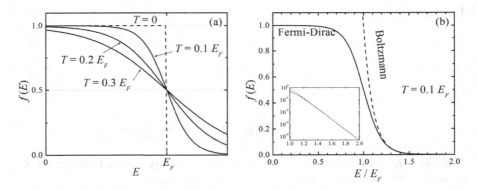

Fig. 3.1 (a) Fermi-Dirac distribution at $T = 0$ (dashed line) and for three non-zero temperatures (solid curves). (b) Fermi-Dirac distribution at $T = 0.1\,E_F$ (solid curve) and its Boltzmann approximation $f(E) = e^{-(E - E_F)/kT}$ (dashed curve). The inset shows the same plot at $E > E_F$ with the logarithmic scale on the y-axis chosen for the ease of comparison of the curves

3.6 Boltzmann Approximation

For energies exceeding E_F by a few thermal energies kT, Fermi-Dirac distribution can be approximated by the so-called Boltzmann's distribution, obtained by neglecting the 1 in the denominator:

$$f(E) \approx e^{-(E-E_F)/kT}.$$

This approximation is valid for energies, at which $e^{(E-E_F)/kT}$ is much bigger than 1. Typically, the term "much bigger" means "bigger by at least one order of magnitude," i.e.,

$$e^{(E-E_F)/kT} > 10 \gg 1 \quad \Rightarrow \quad E - E_F > kT \ln(10) = 2.3\,kT.$$

The accuracy of Boltzmann approximation at large energies is seen in Fig. 3.1b showing Fermi-Dirac distribution and Boltzmann approximation at $T/E_F = 0.1$. Boltzmann approximation is quite accurate in this case for $E/E_F > 1.2$, see inset.

The probability for a given state below Fermi energy to be empty (see Sect. 3.5) can be approximated in a similar way:

$$1 - f(E) = \frac{1}{1 + e^{(E-E_F)/kT}} \approx e^{-(E-E_F)/kT}.$$

This approximation is valid provided that

$$E_F - E > kT \ln(10) = 2.3\,kT.$$

At the room temperature $T = 300\,\mathrm{K}$, the thermal energy $kT = 0.02585\,\mathrm{eV}$, and we see that Boltzmann's approximation applies if the Fermi energy differs from a particular energy E by at least $0.06\,\mathrm{eV}$.

3.7 Fermi Energy at Low Temperatures

Fermi energy depends on temperature. Its temperature dependence should be found from the condition that the sum of the occupation probabilities for all quantum states be the same as the total number of electrons N. This gives an equation for E_F:

$$\sum_\alpha f(E_\alpha) \equiv \sum_\alpha \frac{1}{1 + e^{(E_\alpha - E_F)/kT}} = N.$$

Fermi energy E_F at any temperature can be determined from this equation, provided that the energies of the quantum states E_α are known. It is often convenient to take the sum not over the quantum states, but over the energy levels E_n, each of which is degenerate and contains g_n quantum states. In this formulation, Fermi energy is to be found from

$$\sum_n \frac{g_n}{1 + e^{(E_n - E_F)/kT}} = N.$$

At $T > 0$, the solution of this equation needs to be performed numerically or using analytical approximations. The resulting E_F does not need to coincide with any energy level E_n of the system.

At absolute zero temperature, all states with energy below Fermi level are occupied and all states above Fermi level are empty. Let us denote the energy of the highest occupied energy level at $T = 0$ as E_1 and its degeneracy as g_1. If the number of particles N is known together with the energies and degeneracies of all energy levels of the system, then the highest occupied energy level is relatively easy to figure out. One just needs to fill all energy levels with particles, starting from the ground state and up, until all particles are exhausted.

Let us also denote the energy and the degeneracy of the lowest energy level which is empty at $T = 0$ as E_2 and g_2. In this section, we wish to find the Fermi energy at the thermal energy slightly higher than zero, but well below the energy difference between the lowest empty and the highest occupied levels, $kT \ll E_2 - E_1$.

We need to consider the cases of complete and partial occupancy of the highest occupied level at $T = 0$ separately:

$$\text{Case 1: } n_1 = g_1 ;$$

$$\text{Case 2: } n_1 < g_1,$$

where n_1 is the number of electrons in the level E_1 at $T = 0$ (Fig. 3.2).

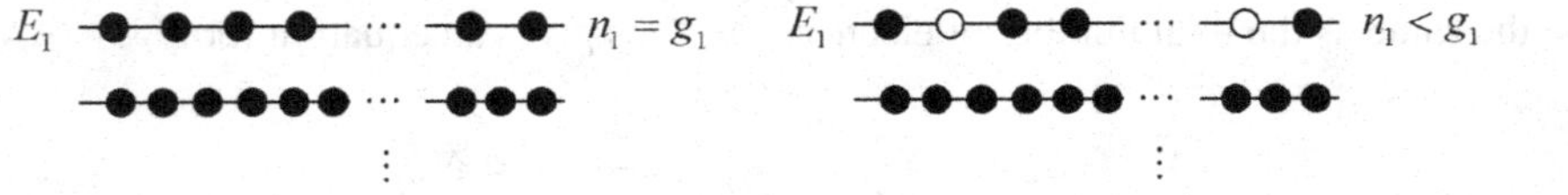

Fig. 3.2 Occupancies of the levels E_1 and E_2 at (**a**) $T = 0$ and (**b**) $T > 0$. Filled circles represent occupied quantum states, empty circles are empty states. It is assumed that at $T = 0$, all states belonging to the level E_1 and all levels below E_1 are occupied, and all states belonging to the levels E_2 and higher are empty

Case 1 At $T \ll (E_2 - E_1)/k$, only the excitation of particles from the level E_1 to the level E_2 plays a role. In other words, the g_1 particles that belonged to the level E_1 at $T = 0$ get redistributed between the levels E_1 and E_2 at $T > 0$. In this case, the chemical potential at low temperatures is found from

$$\frac{g_1}{1 + e^{(E_1 - E_F)/kT}} + \frac{g_2}{1 + e^{(E_2 - E_F)/kT}} = g_1,$$

because there are $n_1 + n_2 = g_1$ particles that belong to the levels E_1 and E_2. Defining

$$x = e^{(E_1 - E_F)/kT}, \quad \Delta E = E_2 - E_1,$$

we have:

$$\frac{g_1}{1 + x} + \frac{g_2}{1 + xe^{\Delta E/kT}} = g_1.$$

Multiplying both sides by the product of the denominators and rearranging the terms, we obtain a quadratic equation

$$x^2 g_1 e^{\Delta E/kT} + x(g_1 - g_2) - g_2 = 0,$$

whose solution reads

$$x = \frac{g_2 - g_1}{2g_1} e^{-\Delta E/kT} + \sqrt{\frac{(g_2 - g_1)^2}{4g_1^2} e^{-2\Delta E/kT} + \frac{g_2}{g_1} e^{-\Delta E/kT}}.$$

In view of the condition $kT \ll \Delta E$, we can arrange the exponentials as

$$e^{-2\Delta E/kT} \ll e^{-\Delta E/kT} \ll e^{-\Delta E/(2kT)}.$$

This allows us to neglect the first term in the square root relative to the second one, and then to neglect the first term in the right-hand side relative to the square root. Then,

$$x = e^{(E_1 - E_F)/kT} = \sqrt{\frac{g_2}{g_1}} e^{-(E_2 - E_1)/(2kT)}.$$

Solving this for the Fermi energy, we obtain:

$$E_F(T) = \frac{E_1 + E_2}{2} + \frac{kT}{2} \ln \frac{g_1}{g_2}.$$

On heating, the Fermi energy shifts toward that energy level which has the lower degeneracy.

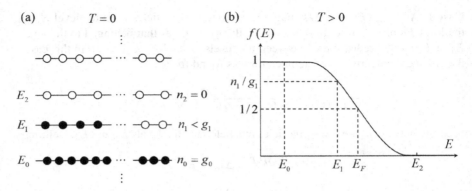

Fig. 3.3 (a) Same as in Fig. 3.2a, but this time with $n_1 < g_1$. (b) Fermi-Dirac distribution at $T > 0$

Case 2 This time, we have $n_1 < g_1$ particles in the level E_1 at $T = 0$. At $T > 0$, particles can get transferred from the level E_1 to the higher level E_2 but also from the lower level $E_0 < E_1$ to the level E_1, see Fig. 3.3a. That is, now we have $g_0 + n_1$ particles that get redistributed among three levels E_0, E_1, and E_2, where g_0 is the degeneracy of the level E_0, which was completely filled at $T = 0$. Accordingly, Fermi level must be established from the equation

$$\frac{g_0}{1 + e^{(E_0 - E_F)/kT}} + \frac{g_1}{1 + e^{(E_1 - E_F)/kT}} + \frac{g_2}{1 + e^{(E_2 - E_F)/kT}} = g_0 + n_1.$$

A rigorous solution of this equation for E_F is difficult, but still possible. But since we are interested in the temperatures $T \ll E_2 - E_1$ and $E_1 - E_0$, we can simplify the calculations considerably with the following reasoning:

The Fermi-Dirac occupation probability of a quantum state with the energy E,

$$f(E) = \frac{1}{1 + e^{(E - \mu)/T}},$$

makes a sharp transition from 1 to 0 at $E = \mu$. At $T = 0$, the occupation probability of a quantum state belonging to the energy level 0 is $f(E_0) = 1$; for the quantum states belonging to the level 2 it is $f(E_2) = 0$; and for a quantum state belonging to the level E_1 it is n_1/g_1, as n_1 states out of g_1 are occupied. Hence, at $0 < T \ll E_1 - E_0$ and $E_2 - E_1$, the occupation probabilities should not be very different from these values, see Fig. 3.3b. Therefore, we can approximate the equation that determines the chemical potential by setting the first term in the left-hand side to g_0 and the third term to 0, giving

$$\frac{g_1}{1 + e^{(E_1 - \mu)/T}} = n_1.$$

Solving this for the chemical potential, we obtain:

$$E_F(T) = E_1 - T \ln \frac{g_1 - n_1}{n_1}.$$

As the temperature increases, the Fermi-Dirac distribution broadens around the energy E_F. The Fermi energy adjusts itself so as to maintain the occupation probability of any quantum state belonging to the level E_1 equal to n_1/g_1, see Fig. 3.3b. If more than half of the states in the level E_1 are filled, then $(g_1 - n_1)/n_1 < 1$ and the Fermi energy increases; for $n_1 < g_1/2$, it decreases. If $n_1 = g_1/2$, it stays equal to E_1 same to the first order in T.

3.8 Problems

3.8.1 Solved Problems

Problem 1 A certain disease is known to affect 1 in 10,000 people. The medical test for this disease is 99.9% accurate, i.e., the probability of a false results (no matter if positive or negative) is 0.1%. What is the probability that a person who got a positive result of that test actually has this disease?

Problem 2 What is the smallest number of students in a class such that at least two students have birthdays on the same day with at least 50% probability? Assume that a year has 365 days, i.e., neglect leap years.

Problem 3 Consider a gas of bosons at temperature T and chemical potential μ. What is the average number of bosons in some quantum state with energy E? Note that, like fermions, bosons are indistinguishable particles, but there is no restriction on the number of bosons in the same quantum state.

Problem 4 The degeneracy of a particular energy level $g = 6$. The probability for a particular quantum state belonging to this level to be occupied by an electron is $f = 0.1$. What is the probability P_n for this level to have n electrons with $n = 0, \ldots, 6$? What is the average number of electrons occupying this level?

Problem 5 A particular energy level E has the degeneracy 6. What should be the energy difference $\Delta E = E - E_F$ for this energy level to be empty with probability $1/3$ at $T = 300$ K?

Problem 6 A system consists of $N = 15$ non-interacting spin-1/2 particles with the single-particle energies $E_{ij} = \Delta E \cdot (i^2 + j^2)$, where $\Delta E = 1$ eV and the quantum numbers i and j can assume the non-negative integer values $1, 2, 3, \ldots$. Determine the Fermi energy E_F at zero temperature. Does the Fermi energy increase, decrease, or stay the same as the system's temperature is increased from zero? Answer these questions for $N = 16$ and $N = 20$ particles as well.

3.8.2 Practice Problems

Problem 1 Three fair dices are rolled. Calculate the probabilities p_n to obtain n dots on their faces, with $n = 3, 4, \ldots, 18$.

Problem 2 The energy of a harmonic oscillator with the natural frequency ω is quantized as $E_n = \hbar\omega(n + 1/2)$ with $n = 0, 1, 2, \ldots$. At temperature T, what is the probability for the oscillator to be in the quantum state n? What is its average energy $\langle E \rangle$? You may need the formula for the sum of a geometric series, $\sum_{n=0}^{\infty} x^n = (1 - x)^{-1}$, valid for $|x| < 1$.

Problem 3 A particular energy level has the energy 0.04 eV above Fermi energy. It can accommodate at most 2 electrons of opposite spins. The temperature is 300 K with $kT = 0.02585$ eV. What is the probability for this level to be empty? What is the probability for it to be occupied by just one electron? By two electrons?

Problem 4 The degeneracy of a particular energy level $g = 100$. Its energy is 0.1 eV above Fermi energy, E_F, whose temperature dependence can be neglected. At what temperature will this energy level be occupied by at least one electron with 1% probability?

Problem 5 A gas of $N = 1000$ electrons may occupy the energy levels $E_n = \hbar\omega(n + 1/2)$ of a harmonic oscillator with the natural frequency such that $\hbar\omega = 1$ eV and $n = 0, 1, 2, \ldots$. Not more than two electrons can occupy the same energy level. Find the Fermi energy of this system at $T = 0$.

3.8.3 Solutions

Problem 1 The wrong way of solving this problem is the obvious one: The probability to get a false positive result is 0.1%. The probability to be sick is then $100\% - 0.1\% = 99.9\%$.

The right way is a bit more intricate. Let $p_S = 1/10\,000 = 10^{-4}$ be the probability of having the disease and $p_F = 0.1\% = 10^{-3}$ be the probability of getting a false test result. Suppose a large number N of people took the test, and N_P people tested positive. This number consists of two groups: (1) the people who are actually sick and were tested correctly, and (2) the healthy people who got false positive results. The number of people in group 1 is given by the product of the number of sick people, $p_S N$, and the probability of test being correct, $1 - p_F$. Thus,

$$N_1 = (1 - p_F) \, p_S \, N.$$

The number of people in group 2 is the product of the total number of healthy people, $N - N_S = N(1 - p_S)$, and the probability that the testing went wrong, p_F. Thus,

$$N_2 = (1 - p_S) \, p_F \, N.$$

The probability for a person with a positive test result to belong to group 1 (i.e., to be sick) is

$$p = \frac{N_1}{N_1 + N_2} = \frac{(1 - p_F) \, p_S}{(1 - p_F) \, p_S + (1 - p_S) \, p_F} = \frac{1}{1 + \frac{(1 - p_S) \, p_F}{(1 - p_F) \, p_S}} = \frac{1}{1 + \frac{0.9999 \cdot 10^{-3}}{0.999 \cdot 10^{-4}}},$$

which is 9.08%. The probability that a person with a positive test result is actually not sick is almost 91%.

Problem 2 Let N be the number of students in the class. The probability P_0 that all N students have birthdays on different days is

$$P_0 = \frac{364}{365} \cdot \ldots \cdot \frac{366 - N}{365} = \prod_{i=1}^{N-1} \frac{365 - i}{365}.$$

We need to find the smallest N, at which $1 - P_0 > 1/2$, or $P_0 < 1/2$, or

$$\prod_{i=1}^{N-1} \frac{365 - i}{365} < \frac{1}{2}.$$

One can attack this problem numerically, e.g., by building an Excel table and calculating the value of the product in the left-hand side of this inequality for different values of N. A more elegant way to find the answer is to perform an analytical calculation. Taking the natural logarithm of both sides,

$$\sum_{i=1}^{N-1} \ln \frac{365 - i}{365} < -\ln(2) ; \quad \sum_{i=1}^{N-1} \ln \left(1 - \frac{i}{365} \right) < -\ln(2).$$

The sum of the logarithms can be approximately found using the approximation

$$\ln(1 + x) \approx x \quad \text{for } |x| \ll 1.$$

We assume that $N \ll 365$, meaning that $i/365 \ll 1$ for $i = 1, \ldots, N - 1$. We will have to check the validity of this assumption after obtaining the value of N. We have

$$\sum_{i=1}^{N-1} \ln\left(1 - \frac{i}{365}\right) \approx -\sum_{i=1}^{N-1} \frac{i}{365} = -\frac{1}{365} \sum_{i=1}^{N-1} i = -\frac{1}{365} \frac{(N-1)N}{2},$$

where we recognized an arithmetic series to obtain the last equality.

We now need to find the smallest N, for which

$$\frac{1}{365} \frac{N(N-1)}{2} > \ln(2) \implies N(N-1) > 2 \cdot 365 \cdot \ln(2) \approx 506.$$

We obtain

$$N = 23.$$

Problem 3 The probability to have n particles in a quantum state with the energy E is given by the grand canonical distribution:

$$p_n = \frac{e^{-n(E-\mu)/kT}}{Z}.$$

The value of the constant Z is such that the probabilities to have any number of particles add up to one:

$$Z = \sum_{n=0}^{\infty} e^{-n(E-\mu)/kT} = \sum_{n=0}^{\infty} \left(e^{-(E-\mu)/kT}\right)^n = \frac{1}{1 - e^{-(E-\mu)/kT}},$$

where we used the sum of the geometric series formula, $1 + x + x^2 + \ldots = 1/(1-x)$ for $|x| < 1$, in the last equality.

The average number of particles in this quantum state is

$$\langle n \rangle = \sum_{n=0}^{\infty} n p_n = \frac{\sum_{n=0}^{\infty} n e^{-n\epsilon}}{Z}, \quad \epsilon = \frac{E - \mu}{kT}.$$

The numerator can be evaluating by noting that

$$\sum_{n=0}^{\infty} n e^{-n\epsilon} = -\sum_{n=0}^{\infty} \frac{d e^{-n\epsilon}}{d\epsilon} = -\frac{d}{d\epsilon} \sum_{n=0}^{\infty} e^{-n\epsilon} = -\frac{d}{d\epsilon} \frac{1}{1 - e^{-\epsilon}} = \frac{e^{-\epsilon}}{\left(1 - e^{-\epsilon}\right)^2}.$$

Dividing this result by $Z = 1/(1 - e^{-\epsilon})$, we obtain:

$$\langle n \rangle = \frac{e^{-\epsilon}}{1 - e^{-\epsilon}} = \frac{1}{e^{\epsilon} - 1} = \frac{1}{e^{(E-\mu)/kT} - 1}.$$

This result for the average number of bosons in a particular quantum state is known as the Bose–Einstein distribution. It differs from the fermionic counterpart, the Fermi-Dirac distribution, in that the unity is subtracted from the exponential rather than added to it. When $(E - \mu)/kT$ is bigger than 1 by at least 2 or 3 times, the unity in the denominator can be neglected, and the Bose–Einstein distribution can be approximated with Boltzmann's formula, $\langle n \rangle \approx e^{-(E-\mu)/kT}$.

Problem 4 Each of the $g = 6$ states belonging to this energy level can be occupied with probability $f = 0.1$ and empty with probability $1 - f = 0.9$. The probability for all states to be empty is

$$P_0 = (1 - f)^g = 0.9^6 = 0.5314.$$

The probability for the energy level to have exactly one electron is

$$P_1 = gf(1 - f)^{g-1} = 6 \cdot 0.1 \cdot 0.9^5 = 0.3543.$$

Here, the factor f gives the probability for one state to be occupied, and the factor $(1 - f)^{g-1}$ gives the probability for the remaining $g - 1$ states to be empty. Since there are g states that can be occupied, the product $f(1 - f)^{g-1}$ is multiplied by g.

The probability for the energy level to have exactly n electrons should be proportional to $f^n(1 - f)^{g-n}$. The first term in the product is the probability for some specific n states to be occupied and for the remaining $g - n$ states to be empty. The proportionality coefficient, denoted as C_n^g, is the number of ways to select n occupied states out of g available states:

$$P_n = C_n^g f^n (1 - f)^{g-n}.$$

The number of selections of n filled states out of g available states is called the binomial coefficient and is given by

$$C_n^g = \frac{g!}{n!(g - n)!}.$$

Here is the explanation of this formula. Imagine two bins, one labeled "occupied" and the other labeled "unoccupied," into which we put objects called "states." These objects are labeled: $1, 2, 3, \ldots, g$. We put n states into the "occupied" bin, and the remaining $g - n$ states into the "unoccupied" bin. We are interested in the number of such arrangements which are distinct. For example, an arrangement that puts states $1, 2, 3, \ldots, n$ into the first bin is the same as the arrangement $2, 1, 3, \ldots, n$, as the two arrangements just differ by a permutation of the states 1 and 2 within the first "occupied" bin. Now, the total number of permutations of all states is $g!$.

Some of those permutations swap the states between bins, which results in a new arrangement, while some permutations just swap the states within the same bin, which does not result in a new arrangement. Hence, the total number of permutations needs to be divided by the number of permutations of n states in the "occupied" bin and by the number of permutations of $g - n$ states in the "unoccupied" bin, i.e., by $n!$ and $(g - n)!$. The result is the expression for C_n^g above.

In particular,

$$C_0^6 = C_6^6 = 1, \quad C_1^6 = C_5^6 = 6, \quad C_2^6 = C_4^6 = \frac{6!}{2!4!} = 15, \quad C_3^6 = \frac{6!}{(3!)^2} = 20.$$

We obtain:

$$P_2 = 15 \cdot 0.1^2 \cdot 0.9^4 = 0.09842,$$

$$P_3 = 20 \cdot 0.1^3 \cdot 0.9^3 = 0.01458,$$

$$P_4 = 15 \cdot 0.1^4 \cdot 0.9^2 = 0.001215,$$

$$P_5 = 6 \cdot 0.1^5 \cdot 0.9 = 0.000054,$$

$$P_6 = 0.1^6 = 0.000001.$$

The average number of electrons in this level is

$$\langle n \rangle = \sum_{n=0}^{g} n P_n = 0 \cdot P_0 + 1 \cdot P_1 + \ldots + 6 \cdot P_6 \cong 0.6.$$

Problem 5 The probabilities for a particular quantum state to be occupied and to be empty are, respectively,

$$f = \frac{1}{1 + e^{\Delta E/kT}}, \quad 1 - f = 1 - \frac{1}{1 + e^{\Delta E/kT}} = \frac{1}{1 + e^{-\Delta E/kT}}.$$

The probability for all $g = 6$ quantum states belonging to this level to be empty is

$$P_0 = (1 - f)^g = \frac{1}{(1 + e^{-\Delta E/kT})^g}.$$

With $P_0 = 1/3$, we can solve this equation to find ΔE:

$$\frac{1}{1 + e^{-\Delta E/kT}} = P_0^{1/g} \Rightarrow 1 + e^{-\Delta E/kT} = P_0^{-1/g} \Rightarrow e^{-\Delta E/kT} = P_0^{-1/g} - 1.$$

Taking the natural logarithm in both sides of the last equation, we finally obtain:

$$\Delta E = -kT \ln(P_0^{-1/g} - 1).$$

At $T = 300\,K$, the thermal energy is $kT = 0.02585\,eV$. Plugging in all other numerical values,

$$\Delta E = -0.02585 \ln((1/3)^{-1/6} - 1) = -0.02585 \ln(3^{1/6} - 1) = 0.0415\,eV.$$

Problem 6 Quantum states are labeled by three quantum numbers: i and $j = 1, 2, \ldots$, and spin $s = \pm 1/2$. Each energy level characterized by a particular combination of the quantum numbers i and j can thus accommodate two particles of opposite spins. We form a table showing the energies and degeneracies of the lowest few energy levels:

Level	(i, j)	Degeneracy	Energy [eV]
1	(1, 1)	2	2
2	(1, 2) and (2, 1)	4	5
3	(2, 2)	2	8
4	(1, 3) and (3, 1)	4	10
5	(2, 3) and (3, 2)	4	13
6	(1, 4) and (4, 1)	4	17
7	(3, 3)	2	18

The degeneracy of each level is obtained by multiplying by two the number of the (i, j) combinations that result in the same energy, as there are two possible spin orientations.

For $N = 15$, all particles end up in the first five levels at zero temperature: $2 + 4 + 2 + 4 + 3 = 15$. The highest occupied energy level is filled with $n_5 = 3$ particles and has the degeneracy $g_5 = 4$. Because $n_5 < g_5$, we have for the Fermi energy $E_F = E_5 - kT \ln \frac{g_5 - n_5}{n_5} = 13\,eV + kT \ln(3)$. It has the value of $13\,eV$ at $T = 0$ and increases with temperature.

For $N = 16$, all particles occupy the first five energy levels, as the sum of their degeneracies, $2 + 4 + 2 + 4 + 4 = 16$, equals the number of particles. The highest occupied energy level is now completely filled: $n_5 = g_5 = 4$. The lowest empty energy level E_6 has the degeneracy $g_6 = 4$ and energy $E_6 = 17\,eV$. The Fermi energy at low temperatures is given by $E_F = \frac{E_5 + E_6}{2} + \frac{kT}{2} \ln \frac{g_5}{g_6} = \frac{13 + 17}{2} = 15\,eV$ independent of temperature at low T.

Finally, for $N = 20$, the particles occupy the first six energy levels ($2 + 4 + 2 + 4 + 4 + 4 = 20$) with the last one being completely filled: $n = g_6 = 4$. The lowest empty energy level has $g_7 = 2$ and $E_7 = 18\,eV$. The Fermi energy is $E_F = \frac{E_6 + E_7}{2} + \frac{kT}{2} \ln \frac{g_6}{g_7} = 17.5\,eV + kT \ln(2)/2$, an increasing function of temperature with $E_F(T = 0) = 17.5\,eV$.

Chapter 4
Band Theory of Solids

4.1 Bloch's Theorem

In the absence of an external electric field, an electron in a solid interacts with the positively charged point-like nuclei located in the nodes of the crystal lattice, as well as with a huge number of other electrons. The positive nuclei and the negative electrons together produce an electrostatic potential, $U(\vec{r})$, which is periodic in space, having the periodicity of the crystal lattice. This means that a displacement by any translation vector $\vec{d} = i\vec{a} + j\vec{b} + k\vec{c}$ of the lattice will not result in a change of the potential energy:

$$U(\vec{r} + \vec{d}) = U(\vec{r}).$$

Of course, perfect periodicity of this potential is broken by the presence of defects, impurities, and the thermal vibrations of the crystal lattice. We assume that all these effects are weak, so that we can treat the potential $U(\vec{r})$ as periodic to sufficient accuracy.

Now, we must solve the time-independent Schrödinger equation for one electron in a periodic potential:

$$E\Psi(\vec{r}) = -\frac{\hbar^2}{2m}\vec{\nabla}^2\Psi(\vec{r}) + U(\vec{r})\,\Psi(\vec{r}).$$

We assume for simplicity that the crystal lattice has cubic symmetry. An infinite crystal is modeled as a large cube of side length L with the periodic boundary conditions imposed on the wave function:

$$\Psi(\vec{r} + \vec{e}L) = \Psi(\vec{r}),$$

where $\vec{e}$ is a unit vector in the x-, y-, or z-direction. That is, we do the same trick as in Sect. 1.7, when we discussed the wave function of a free particle.

In the special case of no potential, $U(\vec{r}) = 0$, the solution of the time-independent Schrödinger equation is a plane wave explained in Sect. 1.7:

$$\text{for } U(\vec{r}) = 0 \ : \ \Psi(\vec{r}) = \frac{1}{L^{3/2}} e^{i\vec{k}\cdot\vec{r}},$$

where the wave vector $\vec{k}$ can have only discrete values,

$$\vec{k} = \frac{2\pi}{L}(k_1 \vec{e}_x + k_2 \vec{e}_y + k_3 \vec{e}_z)$$

with all three components k_1, k_2, and k_3 being integer numbers.

When the periodic potential is present, the wave function should reflect its translational symmetry. But the wave function itself is not physically measurable, only its modulus squared is. Therefore, it should have the property:

$$\text{for } U(\vec{r} + \vec{d}) = U(\vec{r}) \neq 0 \ : \ |\Psi(\vec{r} + \vec{d})|^2 = |\Psi(\vec{r})|^2,$$

where $\vec{d}$ is a translation vector of the lattice.

There is an important result about the wave function of an electron in a periodic potential, derived by Felix Bloch in 1928. Bloch's theorem states that if the periodic potential $U(\vec{r})$ is present, then the solution of the time-independent Schrödinger equation takes the form of a modulated plane wave

$$\Psi(\vec{r}) = \Psi_{n\vec{k}}(\vec{r}) = \frac{1}{L^{3/2}} e^{i\vec{k}\cdot\vec{r}} u_{n\vec{k}}(\vec{r}),$$

where the functions $u_{n\vec{k}}(\vec{r})$ have the lattice periodicity,

$$u_{n\vec{k}}(\vec{r}) = u_{n\vec{k}}(\vec{r} + \vec{d}).$$

In the absence of the periodic potential, they assume the value one:

$$\text{for } U(\vec{r}) = 0 \ : \ u_{n\vec{k}}(\vec{r}) = 1.$$

The integer n is an additional quantum number that labels the wave functions $\Psi(\vec{r}) = \Psi_{n\vec{k}}(\vec{r})$. It is called the band index. The band index n and the wave vector $\vec{k}$ define Bloch states with the wave function $\Psi_{n\vec{k}}(\vec{r})$.

The derivation of Bloch's theorem is given in Appendix A.2. Here, we would like to point out that the Bloch wave function $\Psi_{n\vec{k}}(\vec{r})$ has the desired properties above.

Namely, it becomes a plane wave in the absence of the periodic potential, and its modulus squared is periodic with the periodicity of the potential:

$$\text{for } U(\vec{r}) \neq 0 \; : \; |\Psi_{n\vec{k}}(\vec{r} + \vec{d})|^2 = |u_{n\vec{k}}(\vec{r} + \vec{d})|^2 = |u_{n\vec{k}}(\vec{r})|^2 = |\Psi_{n\vec{k}}(\vec{r})|^2.$$

4.2 Energy Bands

4.2.1 Physical Origin of the Energy Bands

Bands are formed by those Bloch states $\Psi_{n\vec{k}}(\vec{r})$ that are characterized by the same value of the band index n, but different wave vectors $\vec{k}$. To each wave vector $\vec{k}$ there corresponds the energy $E_n(\vec{k})$, which is a continuous function of $\vec{k}$ for a given value of the band index n. Following the wave–particle duality argument, we can identify the product $\hbar\vec{k}$ with the electron momentum $\vec{p}$ in the crystal; it is called the crystal momentum, or the quasimomentum.[1] In this way, we find that a single energy-momentum relation of a free electron gets replaced with many energy-momentum relations, $E_n(\hbar\vec{k})$, that depend on the band index n and are more complicated than the simple quadratic formula for a free electron, $E_{free}(\vec{p}) = \frac{\vec{p}^2}{2m_e}$.

So far, the emergence of bands looks like a purely formal result that follows mathematically from Schrödinger's equation with a periodic potential. Let us now try and understand the physical origin of the bands.

The building block of a crystal is its primitive cell, see Sect. 2.3.3. A primitive cell is equipped with one or several atoms that form the basis of the crystal lattice. Now, imagine that we could somehow control the lattice constant a of the crystal without affecting the mutual arrangement of atoms inside each primitive cells. At large lattice constant, the primitive cells resemble isolated molecules, whereas at small a, they form the crystal, see Fig. 4.1a and b for a two-dimensional illustration.

At large intercell separation a, the electron states in each cell are obtained by solving Schrödinger's equation with the potential generated by those several nuclei that belong to a particular cell. Those electron states are discrete, i.e., they can be labeled with a quantum number, which we call n. Each such state can be filled with at most 2 electrons of opposite spins. The energy in the state n is denoted as E_n.

If we now reduce the distance between the cells so that the electron wave functions of neighboring cells start to overlap, then the energies E_n will be affected by the neighboring atoms. Because electrons are indistinguishable, one cannot assign electrons to their host cells anymore. In other words, each electron starts to belong not to an individual cell, but to the whole crystal consisting of N cells. This implies that a single-cell state with a given set of quantum number n can

---

[1] From Latin quasi = apparent.

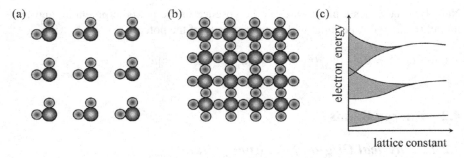

Fig. 4.1 (**a**) Periodic arrangement of primitive cells in space with the symmetry of the crystal lattice, but with the lattice so large that the intercell interaction is negligible. (**b**) An arrangement of the same primitive cells in a crystal. (**c**) Schematic illustration of the splitting of electron states into bands as the lattice constant is decreased

now be occupied by $2N$ electrons. But this is clearly impossible by Pauli exclusion principle, which allows only two electrons of opposite spins per state.

In order to comply with Pauli's exclusion principle and to accommodate all $2N$ electrons in a quantum state at the same time, each formerly single-cell state must broaden into N closely spaced substates with two electrons per substate, see Fig. 4.1c. Those substates form energy bands. Then, a new quantum number emerges, which allows one to distinguish between different substates within the same band. This quantum number is the wave vector $\vec{k}$ in the Bloch's wave function $\Psi_{n\vec{k}}(\vec{r})$. The energies now depends on two quantum numbers, n and $\vec{k}$, i.e., $E = E_n(\vec{k})$. The number of possible $\vec{k}$-states in the same energy band must be equal to the number N of the primitive cells in the crystal.

4.2.2 The First Brillouin Zone

Brillouin Zones

In the beginning of Sect. 4.1 we said that each component of the wave vector $\vec{k}$ is given by an integer multiple of $2\pi/L$, where L is the linear size of the crystal with periodic boundary conditions. Then, in the end of the last section, we stated that the number of different $\vec{k}$-values in each band equals the number of primitive cells in the crystal.

The two statements apparently contradict each other. Indeed, the number of primitive cells, N, no matter how large, is still finite, whereas the number of integers is infinite.

The contradiction is resolved by the fact that in each band, there is finite a collection of N different wave vectors $\vec{k}$ that result in physically distinct Bloch functions $\Psi_{n\vec{k}}(\vec{r})$. This collection of N physically distinct wave vectors is called the first Brillouin zone. All wave vectors outside of this collection yield the wave functions that are physically equivalent to those obtained from the wave vectors in the first Brillouin zone. They belong to other Brillouin zones: the second, the third, etc. Each Brillouin zone contains the same number N of wave vectors as the first one.

One-Dimensional Crystal

To see how this works, consider a single electron in a one-dimensional model crystal obtained by a periodic arrangement of atoms along the x-axis. The electron finds itself in a periodic potential $U(x+a) = U(x)$. The lattice constant of this crystal is a, the number of primitive cells (i.e., the number of periods of $U(x)$) is N, and the length is $L = Na$. The periodic boundary conditions mean that the wave function of an electron satisfies $\Psi(x+L) = \Psi(x)$. According to Bloch's theorem, it is given by the product

$$\Psi_{nk}(x) = \frac{1}{\sqrt{L}} e^{ikx} u_{nk}(x),$$

where n is the band index, $\sqrt{L}$ serves the purpose of the plane-wave part normalization to one: $\int_0^L dx \left| \frac{e^{ikx}}{\sqrt{L}} \right|^2 = \int_0^L \frac{dx}{L} = 1$, and the wave number is an integer multiple of $2\pi/L$, see Sect. 1.7:

$$k = j\frac{2\pi}{L}, \quad j = 0, \pm 1, \pm 2, \dots.$$

For most values of j, the plane-wave part of the Bloch function, viz., $\frac{1}{\sqrt{L}} e^{ikx}$, does not have the lattice periodicity. But notice what happens if k is increased by $\frac{2\pi}{a}$. Then, the plane-wave part becomes

$$\frac{1}{\sqrt{L}} e^{i\left(k+\frac{2\pi}{a}\right)x} = \frac{1}{\sqrt{L}} e^{ikx} e^{i\frac{2\pi}{a}x}.$$

The function $e^{i\frac{2\pi}{a}x}$ does have the periodicity of the lattice, because increasing x by a does not change its value: $e^{i\frac{2\pi}{a}(x+a)} = e^{i\frac{2\pi}{a}x}$ in view of the fact that $e^{i2\pi} = 1$, see Appendix A.1.

Now, the function

$$e^{i\frac{2\pi}{a}x} \Psi_{nk}(x) = \frac{1}{\sqrt{L}} e^{i\left(k+\frac{2\pi}{a}\right)x} u_{nk}(x)$$

has the same modulus squared as $\Psi_{nk}(x)$; therefore, it is physically equivalent to $\Psi_{nk}(x)$, see Sect. 1.10. However, its plane-wave part has the wave number

$$k' = k + \frac{2\pi}{a}.$$

This means that two Bloch states whose band index n is the same, but the wave numbers differ by $2\pi/a$, are identical. This also implies that the energy $E_n(k)$ is a periodic function of the wave number with the periodicity

$$E_n(k + 2\pi/a) = E_n(k).$$

Alternatively, we can express energy in terms of momentum $p = \hbar k$ and state its periodicity as

$$E_n(p + h/a) = E_n(p),$$

as $h = 2\pi\hbar$.

The first Brillouin zone of a one-dimensional crystal consists of N wave numbers that have the smallest magnitude:

$$\text{1st B.z.} : \quad k = j\frac{2\pi}{Na} \text{ with } -\frac{N}{2} < j < \frac{N}{2}; \text{ hence } -\frac{\pi}{a} < k < \frac{\pi}{a}.$$

In other words, the first Brillouin zone is a line of length $2\pi/a$ centered around $k = 0$. Here, we assumed N to be an odd number without loss of generality. The second Brillouin zone consists of the wave numbers in the range

$$\text{2nd B.z.} : \quad -2\frac{\pi}{a} < k < -\frac{\pi}{a} \text{ and } \frac{\pi}{a} < k < 2\frac{\pi}{a}, \text{ or simply } \frac{\pi}{a} < |k| < 2\frac{\pi}{a}.$$

In general, the nth Brillouin zone is defined as

$$n\text{th B.z.} : \quad (n-1)\frac{\pi}{a} < |k| < n\frac{\pi}{a}.$$

To get an idea of the energy-momentum curves in the first Brillouin zone, we first take the effect of the lattice periodicity into account. For this, we take the energy-momentum curve of a free electron, $E(p) = p^2/(2m_e)$, see Fig. 4.2a, and periodically replicate it along the momentum axis with the periodicity $2\pi\hbar/a$ corresponding to the wave number periodicity of $2\pi/a$, see Fig. 4.2b. By combining different parts of parabolas, one can build periodic $E(p)$ curves with the periodicity $\hbar 2\pi/a$. Those curves will have cusps at the odd or even integer multiples of $\hbar\pi/a$.

Next, we take into account the periodic potential. Without going into detail of solving the corresponding Schrödinger's equation, the main result of the periodic potential is rounding off of those cusps to produce smooth periodic energy-momentum curves $E(p)$ for all momenta in the so-called extended zone scheme, see Fig. 4.2c. Finally, keeping in mind that only the first Brillouin zone matters, we

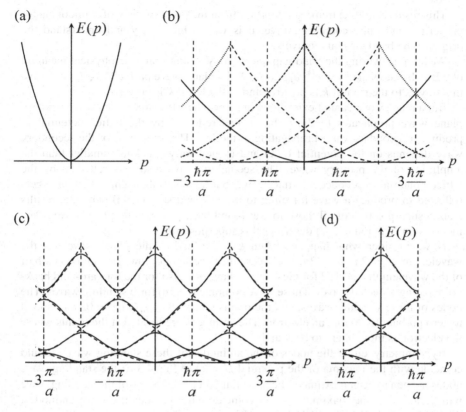

Fig. 4.2 (**a**) Energy-momentum relation of a free electron. (**b**) The graph from (**a**) is periodically repeated horizontally with the periodicity $\hbar 2\pi/a$. (**c**) The effect of the lattice potential $U(x)$ is to round off the cusps of the energy-momentum curves belonging to different bands. (**d**) Energy-momentum curves for the first 4 bands in the reduced zone scheme

restrict the momentum values to the range $-\hbar\pi/a < p < \hbar\pi/a$, corresponding to the reduced zone scheme, see Fig. 4.2d.

Band Gap

The main result of the interaction between the electron and the periodic potential $U(x)$ is rounding off of the cusps at the edges of the Brillouin zones and the development of the bandgaps. A band gap is the smallest energy interval that separates two bands. An electron in a crystal cannot have energy within the band gap, no matter what its momentum is.

This result is both extremely important from the point of view of semiconductor properties and quite counterintuitive. It is worthwhile to try and understand the origin of the bandgaps qualitatively.

We start answering the band gap question by looking into the physical meaning of a Bloch state $\Psi_{nk}(x) = e^{ikx}u_{nk}(x)/\sqrt{L}$. Its plane-wave part, $e^{ikx}/\sqrt{L}$, is a wave that travels to the right if k is positive and to the left if k is negative.

But a plane wave cannot exist alone in a periodic structure. Let us say, a primary plane wave propagates to the right. It gets reflected by the lattice potential to produce a secondary wave propagating to the left. The amplitude of the secondary wave depends on the potential $U(x)$ and is, generally speaking, smaller than the amplitude of the primary wave. The secondary wave also gets reflected by the lattice potential to produce a tertiary wave that travels to the right, which also gets reflected to produce a wave traveling to the left, and so on. A Bloch state results from a superposition of all those forward- and back-propagating plane waves. The modulation function $u_{nk}(x)$ describes this superposition.

Now, consider what happens when $k = \pi/a$, i.e., the plane wave has the wavelength $\lambda = 2\pi/k = 2a$. Each period of the lattice now accommodates half of the wavelength $a = \lambda/2$ for the original wave and also for the forward- and back-propagating reflected waves. Those wave combine so as to form standing waves. The nodes of these standing waves, i.e., the points where $|\Psi_{n,\pi/a}(x)|^2 = 0$, correspond to zero probability to find an electron. The antinodes are located at the points where the electron is most likely to be found.

By symmetry, either the nodes or the antinodes of these standing waves should coincide with the maxima of the potential $U(x)$, see Fig. 4.3. If the standing wave nodes happen to be at maxima of the potential $U(x)$, the energy of the standing wave will be lower; if the maxima of $U(x)$ coincide with the standing wave antinodes, the energy of the wave will be higher. The energy difference between these two situations is the band gap.

Fig. 4.3 A sketch of the periodic lattice potential $U(x)$ (upper graph), and the electron probability density in the two possible standing waves that correspond to the wave number $k = \pi/a$

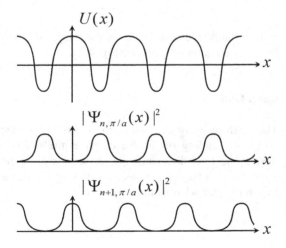

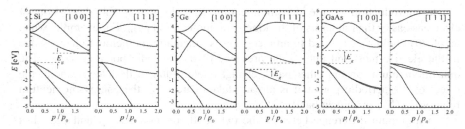

Fig. 4.4 Band structure of silicon, germanium, and gallium arsenide, showing the energy vs. momentum curves for a few bands and for two crystallographic directions. The momentum is normalized to the momentum p_0 of a free electron with the kinetic energy $p_0^2/(2m_e) = 1 \, \text{eV}$

The Energy-Momentum Diagrams in Three Dimensions

In three-dimensional crystals, the first Brillouin zone consists of the $\vec{k}$-vectors inside a certain polyhedron centered around $\vec{k} = 0$. The shape of that polyhedron is chosen based on the symmetry of the crystal structure. Finding the electron energies vs. momentum for different bands in a crystalline solid is a hard mathematical and computational task. Nevertheless, the band structure calculations are performed routinely for many materials of practical significance. The structure of a few bands in Si, Ge, and GaAs is shown schematically in Fig. 4.4 as an example.

It is seen the function $E_n(\vec{p})$ depends not only on the magnitude but also on the direction of $\vec{p}$. This is so, because the atoms of the crystal are arranged differently along the crystallographic directions that do not belong to the same family. For the $\vec{p}$-vectors that are oriented along the crystallographic directions of the same family, the $E_n(p)$ curves are identical.

The highest maximum and the lowest minimum of $E_n(\vec{p})$ at a fixed n are called the top and the bottom of the band, or the band edges. The energy difference between the top of a lower band and the bottom of a higher band is called the band gap, E_g. It has a typical value of the order of 1 eV. The band gap values of three representative semiconductors, Si, Ge, and GaAs, at $T = 300 \, \text{K}$ are given in Appendix A.3. Due to the thermal expansion of the crystal lattice, the band gap depends weakly on temperature. The presence of a band gap plays an important role in the formation of the electronic properties of semiconductors.

Quasimomentum

The momentum-like quantity $\vec{p} = \hbar \vec{k}$ of an electron in a crystal is called quasimomentum, or the crystal momentum. Quasimomentum is in many respects similar to the kinematic momentum $\vec{p}_{kin} = m_e \vec{v}$ of a free electron. For example, the quasimomentum of an electron in a crystal is conserved, just like the kinematic momentum of a free electron does not change in time.

The introduction of the quasimomentum actually makes the discussion of electrons easier than in terms of the kinematic momentum $\vec{p}_{kin}$. For example, when an electric field $\mathcal{E}$ is applied to an electron, its kinematic momentum changes due to the electric field and due to the presence of the periodic potential $U(\vec{r})$. The quasimomentum, on the other hand, changes solely due to the electric field; the effect of the periodic potential is already taken care of in the energy-momentum relation $E_n(\hbar\vec{k})$ in a given band.

An important difference between the two kinds of momenta, $\vec{p}$ and $\vec{p}_{kin}$, lies in the fact that the energy of an electron in a crystal is a periodic function of the quasimomentum, so that only $\vec{p}$ inside the first Brillouin zone are physically relevant. In a one-dimensional crystal, adding an integer multiple of $2\pi/a$ to the quasimomentum results brings the electron to a quantum state physically identical to the state with the quasimomentum p.

In this book, we will omit the prefix "quasi" and will refer to $\vec{p} = \hbar\vec{k}$ simply as momentum, keeping these differences from the kinematic momentum in mind.

4.2.3 Phase Velocity vs. Group Velocity

Phase Velocity

So far, we described an electron in a periodic potential as a Bloch wave. The full time-dependent Bloch wave function is (see Sect. 1.8.2)

$$\psi_{n\vec{k}}(\vec{r}, t) = \Psi_{n\vec{k}}(\vec{r}) \, e^{-i\omega_n(\vec{k})t} = u_{n\vec{k}}(\vec{r}) \frac{1}{L^{3/2}} \, e^{i(\vec{k}\cdot\vec{r}-\omega_n(\vec{k})t)}, \quad \omega_n(\vec{k}) = \frac{E_n(\vec{k})}{\hbar},$$

where $\omega_n(\vec{k})$ is the angular frequency corresponding to the energy $E_n(\vec{k})$. The phase of the plane-wave part of this wave function,

$$\theta(\vec{r}, t) = \vec{k} \cdot \vec{r} - \omega_n(\vec{k})t$$

has a constant value at the point $\vec{r}$ given by

$$\vec{r}(t) = \vec{r}_0 + \frac{\omega_n(\vec{k})}{|\vec{k}|^2}\vec{k}\, t.$$

Indeed,

$$\theta(\vec{r}(t), t) = \vec{k} \cdot \vec{r}(t) - \omega_n(\vec{k})t = \vec{k} \cdot \vec{r}_0 = \text{const.}$$

The time-dependent position $\vec{r}(t)$ above describes motion from the starting point $\vec{r}(0) = \vec{r}_0$ in the direction of a unit vector $\vec{k}/|\vec{k}|$ with the velocity

$$v_{ph}(\vec{k}) = \frac{\omega_n(\vec{k})}{|\vec{k}|} = \frac{E_n(\vec{k})}{\hbar|\vec{k}|} = \frac{E_n(\vec{p})}{|\vec{p}|},$$

called the phase velocity. In the second equality, we switched from the description in terms of the wave vector $\vec{k}$ to the one in terms of the momentum $\vec{p} = \hbar\vec{k}$.

Even though the phase of the Bloch wave function moves with the velocity v_{ph}, this motion does not result in any observable effect. The reason is that the wave function is not a measurable quantity, only it modulus squared is. The probability density corresponding to the time-dependent state $\psi_{n\vec{k}}(\vec{r}, t)$ is stationary:

$$|\psi_{n\vec{k}}(\vec{r}, t)|^2 = \frac{1}{L^3}|u_{n\vec{k}}(\vec{r})|^2.$$

The probability to find an electron is periodically modulated, but the electron remains delocalized in the following sense. The probability to find an electron inside a unit cell located at the point $\vec{R}$ is obtained by integrating the probability density over the volume of the unit cell $V_{cell}(\vec{R})$) located at $\vec{R}$:

$$P(\vec{R}) = \int_{V_{cell}(\vec{R})} d^3r \, |\psi_{n\vec{k}}(\vec{r}, t)|^2 = \frac{\int_{V_{cell}(\vec{R})} d^3r \, |u_{n\vec{k}}(\vec{r}, t)|^2}{L^3} = \text{const.}$$

Because the function $u_{n\vec{k}}(\vec{r})$ has the periodicity of the crystal lattice, the integral does not depend on the location $\vec{R}$ of the unit cell. Hence, an electron can be found in any unit cell with the same probability.

Group Velocity

The wave–particle duality principle says the same object may behave as a particle or as a wave, depending on the situation. Here, we will try to understand how electrons in a crystal can behave as particles.

An object manifests its properties when it interacts with other objects. Electrons in a crystal interact with the crystal defects, with the thermal vibrations of the crystal lattice, and with other electrons. These interactions usually are more efficient for some values of the wave vector $\vec{k}$ and energy $E_n(\vec{k})$ than for others. This is similar to the interaction of a simple harmonic oscillator with an external force: it is most efficient if the force changes in time periodically with the natural frequency of the oscillator.

Consider a group of Bloch states that belong to the same band n and whose wave vectors have the values close to $\vec{k}_0$. These states are described by a collective wave function given by a linear combination

$$\phi(\vec{r}, t) = \sum_{\vec{k}} A(\vec{k}) \, \psi_{n\vec{k}}(\vec{r}, t)$$

with some function $A(\vec{k})$, called the envelope function. This linear combination of Bloch states is called the wave packet.

The collective wave function $\phi(\vec{r}, t)$ does not describe a state of definite energy. Rather, it is a superposition of quantum states of many energies, $E_n(\vec{k})$, which are selected by the envelope function $A(\vec{k})$. Because of that, $\phi(\vec{r}, t)$ does not satisfy the time-independent Schrödinger equation. However, it satisfies the time-dependent Schrödinger equation

$$i\hbar \frac{\partial \phi(\vec{r}, t)}{\partial t} = -\frac{\hbar^2}{2m_e} \vec{\nabla}^2 \phi(\vec{r}, t) + U(\vec{r})\phi(\vec{r}, t).$$

Therefore, $\phi(\vec{r}, t)$ is a possible time-dependent state of an electron.

Because the discrete wave vectors $\vec{k}$ are closely spaced, we can replace the sum by an integral and write

$$\phi(\vec{r}, t) = \int d^3k \, A(\vec{k}) \, u_{n\vec{k}}(\vec{r}) \, e^{i(\vec{k} \cdot \vec{r} - \omega_n(\vec{k})t)}.$$

We have absorbed the constant $1/L^{3/2}$ and whatever other constants arise in this replacement into the envelope function $A(\vec{k} - \vec{k}_0)$.

Suppose a particular interaction is most efficient for the wave vectors around some specific value $\vec{k}_0$. Correspondingly, let us choose the envelope function $A(\vec{k})$ such that it has a sharp peak at $\vec{k} = \vec{k}_0$ and quickly decays to zero as the modulus of the difference $|\vec{k} - \vec{k}_0|$ gets larger. To understand the behavior of the wave packet, let us consider the simpler case of a one-dimensional crystal; the insight that we will obtain in its analysis will be applicable to the three-dimensional crystals as well.

Let us choose the envelope function to be a box of side $2\Delta k$ around the wave number k_0:

$$A(k) = \begin{cases} A_0 & \text{if } k_0 - \Delta k < k < k_0 + \Delta k, \\ 0 & \text{otherwise.} \end{cases}$$

The one-dimensional wave packet then becomes

$$\phi(x, t) = A_0 \int_{k_0 - \Delta k}^{k_0 + \Delta k} dk \, u_{nk}(x) \, e^{i(kx - \omega_n(k)t)}.$$

For small Δk, the term $u_{nk}(x)$ can be replaced with $u_{nk_0}(x)$ and taken out of the integral, because k_0 is not an integration variable. The angular frequency as a function of k can be approximated as a linear function:

$$\omega_n(k) = \omega_n(k_0) + \omega_n'(k_0)(k - k_0) + \dots,$$

where $\omega_n'(k_0) = \frac{d\omega_n(k)}{dk}\big|_{k=k_0}$. With these replacements, we have

$$\phi(x, t) = A_0 \, u_{nk_0}(x) \int_{k_0-\Delta k}^{k_0+\Delta k} dk \, e^{i \, [kx - (\omega_n(k_0) + \omega_n'(k_0)(k-k_0))t]}$$

$$= A_0 \, u_{nk_0}(x) \, e^{-i(\omega_n(k_0) - \omega_n'(k_0) \, k_0)t} \int_{k_0-\Delta k}^{k_0+\Delta k} dk \, e^{ik(x - \omega_n'(k_0)t)}.$$

In the second step, we moved the part of the exponential that does not depend on k outside the integral. The integration is easy to perform:

$$\int_{k_0-\Delta k}^{k_0+\Delta k} dk \, e^{ik(x - \omega_n'(k_0)t)} = \left. \frac{e^{ik(x - \omega_n'(k_0)t)}}{i(x - \omega_n'(k_0)t)} \right|_{k=k_0-\Delta k}^{k=k_0+\Delta k}$$

$$= e^{ik_0(x - \omega_n'(k_0)t)} \frac{e^{i\Delta k(x - \omega_n'(k_0)t)} - e^{-i\Delta k(x - \omega_n'(k_0)t)}}{i(x - \omega_n'(k_0)t)}$$

$$= 2e^{ik_0(x - \omega_n'(k_0)t)} \frac{\sin\left(\Delta k(x - \omega_n'(k_0)t)\right)}{x - \omega_n'(k_0)t},$$

where we used Euler's formula $\sin(\theta) = (e^{i\theta} - e^{-i\theta})/(2i)$ in the last step, see Appendix A.1. Substitution of this result into the last expression for $\phi(x, t)$ gives

$$\phi(x, t) = 2 \, A_0 u_{nk_0}(x) \, e^{i(k_0 x - \omega_n(k_0)t)} \frac{\sin\left(\Delta k(x - \omega_n'(k_0)t)\right)}{x - \omega_n'(k_0)t}.$$

It gives the probability density to find an electron

$$|\phi(x, t)|^2 = 4A_0^2 |u_{nk_0}(x)|^2 \, \Delta k^2 \, f(x, t),$$

where

$$f(x, t) = \frac{\sin^2\left(\Delta k(x - \omega_n'(k_0)t)\right)}{\Delta k^2 (x - \omega_n'(k_0)t)^2}.$$

If not for the function $f(x, t)$, the probability density $|\phi(x, t)|^2$ would have been just a periodic function of space. The function $f(x, t)$ makes all the difference. It has the maximal value[2]

$$f(x_{peak}(t), t) = 1 \quad \text{at} \quad x_{peak}(t) = \omega_n'(k_0)t$$

and decays to the left and to the right of that peak, simultaneously performing a few oscillations. But then the probability density $|\phi(\vec{x}, t)|^2$ to find an electron is

---

[2] Note that $\lim_{\epsilon \to 0} \frac{\sin \epsilon}{\epsilon} = 1$.

Fig. 4.5 The function
$f(x, t)$ at times $t_0 = 0$,
$t_1 > 0$, and $t_2 > t_1$

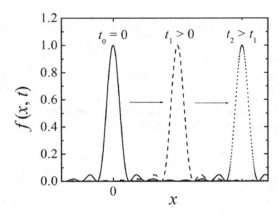

maximal at x_{peak} and decays to zero to the left and to the right of the peak. In other words, the electron is now localized at $x = x_{peak}$. It behaves as a particle.

The behavior of the function $f(x, t)$ at three different times is shown in Fig. 4.5. Its maximum moves with the velocity

$$v = \frac{dx_{peak}}{dt} = \omega'_n(k_0),$$

called the group velocity. Remarkably, the group velocity is independent of the width Δk of the envelope function $A(k)$. As a matter of fact, any other reasonable choice of $A(k)$ yields the same group velocity value.

The argument can be generalized to a three-dimensional crystal. In this case, the group velocity of a wave packet is a vector, whose components are given by

$$v_i = \frac{\partial \omega_n(\vec{k})}{\partial k_i} = \frac{\partial E_n(\vec{p})}{\partial p_i}, \quad i = x, y, z,$$

where we used the relations $E_n = \hbar\omega_n$ and $\vec{p} = \hbar\vec{k}$ and dropped the subscript 0 in the wave vector $\vec{k}$, around which the wave packet is centered. Note that for free classical particles with the standard energy-momentum relation $E(\vec{p}) = \vec{p}^2/(2m)$, the velocity components are likewise given by $v_i = \partial E/\partial p_i = p_i/m$.

4.2.4 Bloch Oscillations

Let us consider again a hypothetical one-dimensional crystal model with just one electron in the energy band n. Suppose that an external electric field $\mathcal{E}$ is applied to the crystal. The electron momentum p changes in time according to

Fig. 4.6 Bloch oscillations of the quasimomentum in the presence of a constant electric field

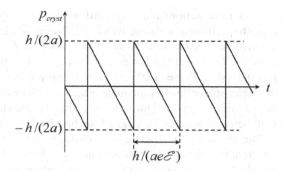

$$\frac{dp(t)}{dt} = -e\mathcal{E} \;\; \Rightarrow \;\; p(t) = -e\mathcal{E}t,$$

assuming $p(0) = 0$. Eventually, p will reach the left edge of the first Brillouin zone at $-\hbar\pi/a = -h/(2a)$. This state is physically identical to the one with the momentum $+\hbar\pi/a = h/(2a)$. Hence, the momentum suddenly changes its value to $h/(2a)$ and continues its linear decrease until it reaches the end of the Brillouin zone again. The electron momentum in the presence of a constant electric field performs saw-tooth-like oscillations with the amplitude $h/(2a)$ and period $h/(ae\mathcal{E})$, see Fig. 4.6. This periodic variation of the momentum is physically equivalent to its steady decrease $p(t) = -e\mathcal{E}t$ at all time in view of the equivalence of the Bloch states with momenta differing by h/a.

Consider now what happens to the electron group velocity, given by

$$v(t) = E'_n(-e\mathcal{E}t),$$

in a one-dimensional crystal. Since the energy is a periodic function of p with the period h/a, the electron group velocity will perform oscillations around the value $v = 0$ with the period

$$T_B = \frac{h}{ae\mathcal{E}}.$$

Also the position of the electron wave packet,

$$x(t) = \int_0^t dt' \, v(t') + x(0),$$

will be oscillating with the same period.

This is a remarkable result that demonstrates the completely different behavior of an electron in a crystal from a free electron. A constant electric field accelerates a free electron, but an electron in a crystal performs oscillator motion, whose period depends on the crystal's lattice constant. This effect is called Bloch oscillations.

Bloch oscillations are very hard to observe experimentally in real crystals. In order to perform at least one Bloch oscillation, an electron wave packet must remain unperturbed on the time scale of the order of T_B. For the electric field strength $\mathcal{E} = 1\,\text{kV/cm}$ and the lattice constant $a = 4\,\text{Å}$, the period of Bloch oscillations has the value $T_B = 1.03\,\mu\text{s}$. This is many orders of magnitude longer than the characteristic collision time of an electron with quanta of lattice vibrations and with impurities in a crystal. Due to these collisions, the electron wave packet is destroyed well before it reaches the edge of the first Brillouin zone.

But even in the absence of collisions, the wave packet gets destroyed for a different reason. Our derivation of the expression for the group velocity relied on the simplest linear approximation of the energy-momentum relation around the value p_0, the momentum of the wave packet: $E_n(p) = E_n(p_0) + E'_n(p_0)(p - p_0)$. A more accurate analysis should include higher-order terms in $p - p_0$. It shows that the wave packet width increases in time, which means that the wave packet eventually becomes delocalized.

Nevertheless, Bloch oscillations have been observed in the so-called superlattices—artificial periodic structures with large lattice constant—at high electric fields.

4.3 Conduction Types of Solids

4.3.1 Band Filling and Electrical Conductivity

After the band structure of a material is established, the electron states must be filled with electrons, with two electrons per state and $2N$ electrons per band, where N is the number of primitive cells. In doing so, the lower-energy bands will be completely filled. Naïve classical considerations would lead us to think (incorrectly) that the states with the lowest energy are characterized by zero electron velocity, and application of an electric field would lead to an overall velocity increase for those electrons, i.e., to an electric current.

A completely filled band contains electrons with a non-zero wave number, $\vec{k}$, and the momentum $\vec{p} = \hbar\vec{k}$, which may be very large in magnitude. In the absence of an electric field, there is no net current produced by the electrons in a completely filled band, because for each electron moving in one direction, the band contains another electron with the same magnitude of momentum, but moving in the opposite direction.

Even in the presence of an electric field, a completely filled band produces zero contribution to the net electric current. A single electron is accelerated by an electric field, i.e., it changes its momentum. In a crystal containing many electrons, an electric field attempts to put each electron into a different momentum state in the same band. But if all those states are occupied, then Pauli exclusion principle disallows an electron to change its momentum in response to the electric field. Then,

Fig. 4.7 Schematic illustrations of the energy-momentum diagrams of metals and semimetals. The shaded areas indicate the occupied states

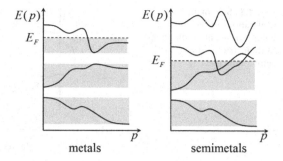

the band will continue to contain the same number of electrons moving against and along the electric field, producing zero contribution to the electric current.

It is only the bands that are partly filled that contribute to the electric current in response to an external electric field.

The band theory allows one to understand the conductivity of conductors (metals and semimetals) and the lack thereof in insulators. The type of a given material is determined by its band structure and the location of Fermi energy at zero temperature. Note that the band structure diagrams that appear in this section do not refer to any particular materials and are given for illustration purposes only.

4.3.2 Metals and Semimetals

In metals, such as Li, Au, Cu, etc., Fermi energy at zero temperature happens to be in the middle of one of the energy bands (Fig. 4.7). This means that the highest occupied band is filled only partly and contains both filled and empty states. If an electric field is applied to a metal, electrons in that partly filled band will change their momentum states, because there are empty states in the same band for its electrons to occupy while remaining in the same band. As a result, the field will create a disbalance of the electrons moving against and along the field, leading to the onset of an electric current.

In semimetals, such as As, Sb, Bi, α-Sn, etc., the two highest bands that contain electrons at zero temperature overlap, and both are partly filled. As a result, both bands contribute to the electric current when an external electric field is applied.

4.3.3 Dielectrics and Semiconductors

In dielectrics and semiconductors, Fermi energy at $T = 0$ happens to be inside the energy gap between two bands. As a result, the highest occupied band is completely

Fig. 4.8 Schematic illustrations of the energy-momentum diagrams of dielectrics and semiconductors

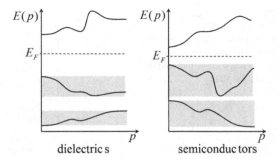

filled, while the next band is empty. Application of an electric field cannot change the momentum states of the electrons in the highest filled band, because there are no empty states in that band for the electrons to go to. Then, the electrons will not react to an applied electric field, and the material will not conduct electricity.

The difference between the dielectrics and semiconductors is in the size of the energy gap, E_g, separating the highest filled band from the lowest empty band, see Fig. 4.8. In semiconductors, the band gap is of the order of 1–4 eV, whereas in dielectrics it exceeds 5 eV. Because of this, the lowest empty band in dielectrics remains empty even at the temperature of a few hundred Kelvin, whereas in semiconductors, some electrons can go into it from the highest occupied band at the room temperature.

4.4 Conduction and Valence Bands

In semiconductors, the tail of the Fermi-Dirac distribution at $T > 0$ extends into the upper band that is empty at zero temperature, but becomes partly filled at $T > 0$. This band gets a non-negligible amount of free electrons which can change their momenta in response to an external electric field. Such electrons contribute to the electric conductivity and are called conduction electrons. The band where they "live" is called the conduction band. Because Fermi-Dirac distribution decreases exponentially with energy at its tail, conduction electrons occupy the bottom part of the conduction band.

The band from which the electrons were promoted at non-zero temperature is called the valence band. This band was completely filled at zero temperature, but at a non-zero temperature, some electrons from the upper part, or the top, of the valence band move into the conduction band. As a result, the top of the valence band lacks a non-negligible number of electrons. Because empty states exist in the valence band at $T > 0$, this band also contributes to the electric current in response to an external electric field.

Semiconductors in which the top of the valence band is at the same wave vector as the bottom of the conduction band are called direct-band gap semiconductors.

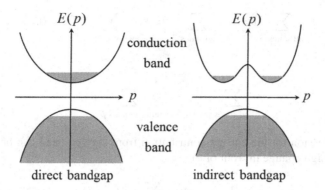

Fig. 4.9 Schematic illustrations of the energy-momentum diagrams in direct-band gap vs. indirect-band gap semiconductors. The shaded areas indicate the states occupied with electrons at $T > 0$

The typical example of direct-band gap semiconductors is GaAs. If the top of the valence band happens at a different momentum than the bottom of the valence band, the material is called an indirect-band gap semiconductor, the typical example being Si. Figure 4.9 illustrates this distinction.

The band gap of a semiconductor is measured experimentally by studying the optical absorption of light. Namely, the light with the photon energy hf exceeding the band gap energy E_g can promote electrons from the valence band to the conduction band and is, therefore, absorbed by the material. On the other hand, if the photon energy is below E_g, the light will pass through the material. A measurement of the light frequency f_{edge} at the absorption edge gives the band gap energy $E_g = hf_{edge}$.

4.5 Holes

The contribution to the net current caused by an electric field from a particular band is proportional to the sum of the velocities that correspond to those states of this band which are filled with an electron:

$$I_{band} \propto -e \sum_{\text{filled states } \vec{p}} \vec{v}_{\vec{p}}.$$

Note that the sum of velocities over all states, both filled and empty, from the same band is exactly zero. This is so, because to each state in the band with a particular velocity value there corresponds a state with the opposite velocity:

$$\sum_{\text{all states } \vec{p}} \vec{v}_{\vec{p}} = \sum_{\text{filled states } \vec{p}} \vec{v}_{\vec{p}} + \sum_{\text{empty states } \vec{p}} \vec{v}_{\vec{p}} = 0.$$

Thus,

$$\sum_{\text{filled states } \vec{p}} \vec{v}_{\vec{p}} = - \sum_{\text{empty states } \vec{p}} \vec{v}_{\vec{p}},$$

and the expression for the current contribution from a given band can be rewritten as (note the sign change in front of e)

$$I_{band} = e \sum_{\text{empty states } \vec{p}} \vec{v}_{\vec{p}}.$$

The number of empty states in the valence band of a semiconductor is usually much smaller than the number of the filled states. It is for this reason that the second expression for I_{band} as the sum over the empty states is more preferable to use than the first one.

Each empty state $\vec{p}$ in the valence band can be interpreted as a state filled with an imaginary particle, or a quasiparticle, of positive charge e and velocity $\vec{v}_{\vec{p}}$. These quasiparticles are called holes. While a hole means the absence of an electron in a particular momentum and spin state in a band, holes do behave like real particles. Their introduction makes the discussion of semiconductors much easier, whereas the discussion involving only electrons would be excessively cumbersome.

Therefore, the picture that we will adopt is this: There are two charge carriers in a semiconductor, the negatively charged electrons in the conduction band and the positively charged holes in the valence band.

The probability f_h that a particular state $\vec{p}$ with the energy $E(\vec{p})$ in the valence band is occupied by a hole is just the probability that it does not contain an electron. Because state occupation probability by electrons is given by the Fermi-Dirac distribution, $f_e(E)$,

$$f_h(E) = 1 - f_e(E) = 1 - \frac{1}{1 + e^{(E-E_F)/kT}} = \frac{1}{1 + e^{-(E-E_F)/kT}}.$$

The Fermi-Dirac distribution for holes, $f_h(E)$, looks almost the same as that for electrons, $f_e(E)$, except for the sign in front of the energy difference $E - E_F$ in the exponential. We can interpret this observation by stating that hole energies, $E^{(h)}(\vec{p})$, have the sign opposite to electron energy in the same quantum state $\vec{p}$, up to an additive constant ΔE:

$$E^{(h)}(\vec{p}) = -E^{(e)}(\vec{p}) + \Delta E,$$

see Fig. 4.10. The additive constant should be chosen such that holes and electrons have the same Fermi level:

Fig. 4.10 Electron and hole energy-momentum curves

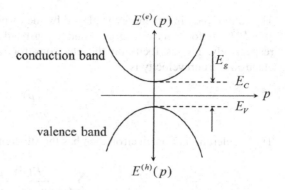

$$E_F^{(h)} = -E_F^{(e)} + \Delta E = E_F^{(e)},$$

that is,

$$\Delta E = 2E_F^{(e)}.$$

In this way, the Fermi-Dirac distribution for holes looks formally the same as for electrons:

$$f_h(E^{(h)}) = \frac{1}{1 + e^{(E^{(h)} - E_F^{(h)})/kT}}.$$

We will always give energy from the electrons' perspective and omit the superscript (e) by default. If we wish to indicate energy from the holes' viewpoint, we will attach the superscript (h) to the respective energy.

4.6 Effective Mass Tensor

In classical physics, the mass of an object is the proportionality coefficient between its acceleration and the force acting on it,

$$\vec{a} = \vec{F}/m.$$

Alternatively, it is a coefficient that relates the object's kinetic energy and momentum, $E_K = \vec{p}^2/(2m)$.

For electrons and holes in a crystal, the role of the momentum is played by the quasimomentum $\vec{p} = \hbar\vec{k}$, and an applied force results in its change at a rate

$$\frac{dp_i}{dt} = F_i, \quad i = x, y, z.$$

The role of the kinetic energy is played by the dispersion relation $E(\vec{p}) = E^{(e)}(\vec{p})$ and $E^{(h)}(\vec{p})$ for electrons in the conduction band and holes in the valence band, respectively. Let us focus on the electrons for now. The ith component of an electron's group velocity is

$$v_i = \frac{\partial E(\vec{p})}{\partial p_i}.$$

The acceleration of an electron then has the ith component

$$a_i = \frac{dv_i}{dt} = \frac{d}{dt}\frac{\partial E(\vec{p})}{\partial p_i} = \sum_{j=x,y,z} \frac{\partial^2 E(\vec{p})}{\partial p_i \, \partial p_j}\frac{dp_j}{dt} = \sum_{j=x,y,z} \frac{\partial^2 E(\vec{p})}{\partial p_i \partial p_j} F_j.$$

The coefficients that multiply the force components in the sum play the role of the inverse mass of an electron in a semiconductor. It is called the effective mass tensor. In contrast to the usual mass of a free electron m_e, the effective mass depends on momentum and is a 3×3 matrix. It is marked with a star to distinguish from the free electron mass. The components of its inverse are related to the second derivative of the energy-momentum relation:

$$a_i = \sum_{j=x,y,z} \left(\frac{1}{m_e^*(\vec{p})}\right)_{ij} F_j, \quad \left(\frac{1}{m_e^*(\vec{p})}\right)_{ij} = \frac{\partial^2 E(\vec{p})}{\partial p_i \partial p_j}.$$

We will be interested in the situations when the electrons' momenta do not differ much from the momentum $\vec{p}_C$ at the minimum of the conduction band. Then, the effective mass tensor is evaluated at a particular momentum $\vec{p} = \vec{p}_C$; as such, it has constant components independent of $\vec{p}$.

The next simplification has to do with a convenient choice of the coordinate axes. At not too big deviations $|\vec{p} - \vec{p}_C|$ from the band minimum, the surface of constant energy $E(\vec{p}) = $ const in the momentum space has the form of an ellipsoid, see Fig. 4.11a. Aligning the coordinate axes with its principal axes makes the matrix $(1/m_e^*)_{ij}$ diagonal, i.e., its elements with $i \neq j$ become zero. Hence, the effective mass has only three non-zero components:

$$\frac{1}{m_{e,i}^*} \equiv \left(\frac{1}{m_e^*}\right)_{ii} = \left.\frac{\partial^2 E(\vec{p})}{\partial p_i^2}\right|_{\vec{p}=\vec{p}_C}, \quad i = x, y, z.$$

This definition is valid only for a particular choice of the coordinate axes; any other choice will give non-zero values of $m_{e,ij}^*$ with $i \neq j$.

In many semiconductors with ellipsoidal constant-energy surfaces, including Si and Ge, two components of the effective mass tensor are the same (Fig. 4.12a). They are called the transverse (subscript t) effective masses. The remaining third component, which is different, is called the longitudinal (subscript l) effective mass:

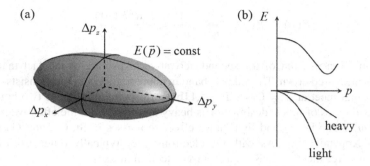

Fig. 4.11 (a) Constant-energy surface in the momentum space near the bottom of the conduction band; note that $\Delta \vec{p} = \vec{p} - \vec{p}_C$. (b) A sketch of the band structure of a semiconductor with two hole branches: heavy and light

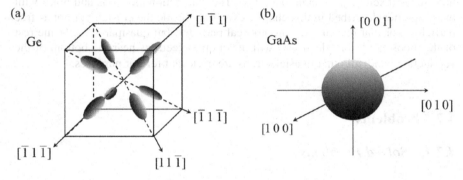

Fig. 4.12 Constant-energy surfaces in the momentum space near the bottom of the conduction band of (a) Ge and (b) of GaAs

$$m^*_{e,x} = m^*_{e,y} = m^*_{e,t} \ , \quad m^*_{e,z} = m^*_{e,l}.$$

The numerical values of the transverse and longitudinal effective masses are usually given as ratios to the free electron mass m_e. Depending on the material, the ratios $m^*_{e,t}$ and $m^*_{e,l}$ range from a few percent to a few. The values of the effective mass components in three representative semiconductors are given in Appendix A.3. In GaAs, which is a direct-band gap semiconductor, the constant-energy surface is a sphere centered at $\vec{p} = 0$; hence, all three effective mass components are the same in this semiconductor (Fig. 4.12b).

The effective mass of a hole m^*_h in many semiconductors is not a matrix, but an ordinary number, because the constant-energy surface in the valence band happens to be a sphere centered at $\vec{p} = 0$ at small p. The hole effective mass is found as the curvature of the $E(\vec{p})$ in the valence band near the valence band maximum:

$$E(p) = E_V - \frac{p^2}{2m_h^*}, \quad \frac{1}{m_h^*} = -\frac{d^2 E(p)}{dp^2}\bigg|_{p=0}.$$

The minus sign in front of the second derivative takes care of the fact that hole energy is measured down. The valence band in many semiconductors consists of two branches that meet at $\vec{p} = 0$, see Fig. 4.11b. Corresponding to these two branches are the two types of holes, designated as heavy and light, with the effective masses $m_{hh}^* > m_{lh}^*$. The heavy and light holes effective masses in Si, Ge, and GaAs are given in Appendix A.3. As with the electrons, they typically range from a few percent of a free electron mass to a few free electron masses.

A quasiparticle is a useful concept to describe particle-like elementary excitations in solids. Examples of quasiparticles are numerous and include the quanta of lattice vibrations (phonons), magnetization spin waves (magnons), bound electron–hole states (excitons), to name but a few. The conduction electrons and holes with the properties described in this chapter exist only inside the crystal, but not as free particles. For that reason, they are not real particles, but quasiparticles. In the rest of this book the term "electrons" will mean quasielectrons near the bottom of the conduction band; all other quasielectrons are irrelevant for our purposes.

4.7 Problems

4.7.1 Solved Problems

Problem 1 Show explicitly that a linear combination of Bloch states $\psi_{n\vec{k}}(\vec{r}, t)$, viz.,

$$\phi(\vec{r}, t) = \sum_{\vec{k}} A(\vec{k}) \, \psi_{n\vec{k}}(\vec{r}, t),$$

satisfies the time-dependent Schrödinger equation,

$$i\hbar \frac{\partial \phi(\vec{r}, t)}{\partial t} = -\frac{\hbar^2}{2m_e} \vec{\nabla}^2 \phi(\vec{r}, t) + U(\vec{r}) \, \phi(\vec{r}, t).$$

Problem 2 Consider a wave packet formed by the plane waves in one dimension:

$$\phi(x, t) = \int_{-\infty}^{\infty} dk \, A(k) \, \frac{e^{i(kx - \omega(k)t)}}{\sqrt{L}}.$$

The angular frequency depends on the wave number according to the free-particle energy-momentum relation:

$$\omega(k) = \frac{E(\hbar k)}{\hbar} = \frac{\hbar k^2}{2m},$$

where m is the particle's mass. The envelope function is a Gaussian peak centered at $k = k_0$ and having the width σ_k, viz.,

$$A(k) = A_0 e^{-(k-k_0)^2/(2\sigma_k^2)}.$$

(a) Find the particle's probability density $|\phi(x, t)|^2$ by performing a integration over k.
(b) Find the constant A_0 from the normalization condition $\int_0^L dx\, |\phi(x, t)|^2 = 1$.

Hint: use the identity $\int_{-\infty}^{\infty} dk\, e^{-\alpha(k-\kappa)^2} = \sqrt{\pi/\alpha}$, where α is an arbitrary complex number with $\operatorname{Re}\alpha > 0$ and κ may be complex-valued as well.

Problem 3 The energy-momentum dispersion relations near the edges of the two bands of a hypothetical one-dimensional semiconductor read

$$\text{conduction band}: \qquad E(p) = A\, e^{\gamma\, \cos(2\pi p/p_0)}$$

$$\text{valence band}: \qquad E(p) = B\, \cos^2(\pi p/p_0)$$

with positive constants $A = 3\,\text{eV}$, $B = 1.5\,\text{eV}$, $\gamma = 0.2$, and p_0 such that $\frac{p_0^2}{2m_e} = 1\,\text{eV}$.

(a) Is it a direct-band gap or an indirect-band gap semiconductor?
(b) Determine the band gap and the Fermi energy at $T = 0$ relative to the top of the valence band.
(c) Express the electron and hole effective masses in terms of the free electron mass, i.e., as ratios m_e^*/m_e, m_h^*/m_e.
(d) Assuming the Fermi energy to be in the middle of the band gap, what is the probability for the lowest energy state in the conduction band to be filled with an electron at $T = 300\,\text{K}$?

4.7.2 Practice Problems

Problem 1 An electron in a one-dimensional lattice with periodicity a finds itself in a Bloch state $\Psi_{nk}(x) = \frac{1}{\sqrt{L}} e^{ikx} u_{nk}(x)$ with $u_{nk}(x + a) = u_{nk}(x)$. Its average momentum is

$$\langle p \rangle = -i\hbar \int dx\, \Psi_{nk}^*(x) \frac{d\Psi_{nk}(x)}{dx},$$

see Sect. 1.9. Show that $\langle p \rangle = \hbar k$.

Problem 2 Imagine a one-dimensional semiconductor with just one electron in the conduction band. A constant electric field $\mathcal{E}$ is applied to the sample. The assumed energy-momentum relation in the conduction band is $E(p) = -E_0 \cos(ap/\hbar)$. The electron's wave function is not a pure Bloch state, but a wave packet of finite width. How does the group velocity of the electron evolves as it undergoes Bloch oscillations? What happens to its position?

4.7.3 Solutions

Problem 1 The time derivative of the wave packet wave function multiplied by $i\hbar$ is

$$i\hbar \frac{\partial \phi(\vec{r}, t)}{\partial t} = \sum_{\vec{k}} A(\vec{k}) \, i\hbar \frac{\partial \psi_{n\vec{k}}(\vec{r}, t)}{\partial t}.$$

Each time-dependent Bloch function satisfies the time-dependent Schrödinger equation,

$$i\hbar \frac{\partial \psi_{n\vec{k}}(\vec{r}, t)}{\partial t} = -\frac{\hbar^2}{2m_e} \vec{\nabla}^2 \psi_{n\vec{k}}(\vec{r}, t) + U(\vec{r}) \, \psi_{n\vec{k}}(\vec{r}, t).$$

Hence,

$$
\begin{aligned}
i\hbar \frac{\partial \phi(\vec{r}, t)}{\partial t} &= \sum_{\vec{k}} A(\vec{k}) \left(-\frac{\hbar^2}{2m_e} \vec{\nabla}^2 \psi_{n\vec{k}}(\vec{r}, t) + U(\vec{r}) \, \psi_{n\vec{k}}(\vec{r}, t) \right) \\
&= -\frac{\hbar^2}{2m_e} \vec{\nabla}^2 \sum_{\vec{k}} A(\vec{k}) \, \psi_{n\vec{k}}(\vec{r}, t) + U(\vec{r}) \sum_{\vec{k}} A(\vec{k}) \, \psi_{n\vec{k}}(\vec{r}, t) \\
&= -\frac{\hbar^2}{2m_e} \vec{\nabla}^2 \phi(\vec{r}, t) + U(\vec{r}) \phi(\vec{r}, t).
\end{aligned}
$$

Problem 2

(a) The wave function is given by the integral

$$\phi(x, t) = \frac{A_0}{\sqrt{L}} \int_{-\infty}^{\infty} dk \, \exp\left(-\frac{(k - k_0)^2}{2\sigma_k^2} + i\left(kx - \frac{\hbar}{2m} k^2 t \right) \right).$$

The argument of the exponential function is transformed so as to complete the square:

$$\frac{(k-k_0)^2}{2\sigma_k^2} - i\left(kx - \frac{\hbar}{2m}k^2 t\right)$$

$$= \frac{k^2 - 2k_0 k + k_0^2}{2\sigma_k^2} - ikx + i\frac{\hbar}{2m}tk^2$$

$$= \frac{1}{2}\left(\frac{1}{\sigma_k^2} + i\frac{\hbar}{m}t\right)k^2 - k\left(\frac{k_0}{\sigma_k^2} + ix\right) + \frac{k_0^2}{2\sigma_k^2}$$

$$= \frac{1}{2}\left(\frac{1}{\sigma_k^2} + i\frac{\hbar}{m}t\right)\left(k^2 - 2k\frac{k_0 + ix\sigma_k^2}{1 + i\frac{\hbar}{m}\sigma_k^2 t} + \frac{k_0^2}{1 + i\frac{\hbar}{m}\sigma_k^2 t}\right)$$

$$= \frac{1}{2}\left(\frac{1}{\sigma_k^2} + i\frac{\hbar}{m}t\right)\left(\left(k - \frac{k_0 + ix\sigma_k^2}{1 + i\frac{\hbar}{m}\sigma_k^2 t}\right)^2 - \left(\frac{k_0 + ix\sigma_k^2}{1 + i\frac{\hbar}{m}\sigma_k^2 t}\right)^2 + \frac{k_0^2}{1 + i\frac{\hbar}{m}\sigma_k^2 t}\right).$$

Performing integration over k, we obtain

$$\phi(x,t) = \frac{A_0}{\sqrt{L}}\sqrt{\frac{2\pi}{\frac{1}{\sigma_k^2} + i\frac{\hbar}{m}t}}$$

$$\times \exp\left[\frac{1}{2}\left(\frac{1}{\sigma_k^2} + i\frac{\hbar}{m}t\right)\left(\left(\frac{k_0 + ix\sigma_k^2}{1 + i\frac{\hbar}{m}\sigma_k^2 t}\right)^2 - \frac{k_0^2}{1 + i\frac{\hbar}{m}\sigma_k^2 t}\right)\right]$$

$$= \frac{A_0\sigma_k}{\sqrt{L}}\sqrt{\frac{2\pi}{1 + i\frac{\hbar}{m}\sigma_k^2 t}}\exp\left[\frac{1}{2\sigma_k^2}\left(\frac{(k_0 + ix\sigma_k^2)^2}{1 + i\frac{\hbar}{m}\sigma_k^2 t} - k_0^2\right)\right]$$

$$= \frac{A_0\sigma_k}{\sqrt{L}}\sqrt{\frac{2\pi}{1 + i\frac{\hbar}{m}\sigma_k^2 t}}\exp\left(-\frac{\sigma_k^2 x^2 - 2ik_0 x + i\frac{\hbar}{m}k_0^2 t}{2\left(1 + i\frac{\hbar}{m}\sigma_k^2 t\right)}\right).$$

Particle probability density is

$$|\phi(x,t)|^2 = \phi^*(x,t)\phi(x,t) = \frac{A_0^2}{L}\frac{2\pi}{\sqrt{\frac{1}{\sigma_k^2} + \frac{\hbar^2}{m^2}\sigma_k^2 t^2}}$$

$$\exp\left(-\frac{1}{2}\left[\frac{\sigma_k^2 x^2 - 2ik_0 x + i\frac{\hbar}{m}k_0^2 t}{1 + i\frac{\hbar}{m}\sigma_k^2 t} + \frac{\sigma_k^2 x^2 + 2ik_0 x - i\frac{\hbar}{m}k_0^2 t}{1 - i\frac{\hbar}{m}\sigma_k^2 t}\right]\right).$$

After some algebra, the expression in the square brackets simplifies to

$$[\ldots] = 2\sigma_k^2 \frac{\left(x - \frac{\hbar k_0}{m}t\right)^2}{1 + \frac{\hbar^2}{m^2}\sigma_k^4 t^2},$$

so that the particle probability density is

$$|\phi(x,t)|^2 = \frac{A_0^2}{L} \frac{2\pi}{\sqrt{\frac{1}{\sigma_k^2} + \frac{\hbar^2}{m^2}\sigma_k^2 t^2}} \exp\left(-\sigma_k^2 \frac{\left(x - \frac{\hbar k_0}{m}t\right)^2}{1 + \frac{\hbar^2}{m^2}\sigma_k^4 t^2}\right).$$

This is a Gaussian distribution,

$$|\phi(x,t)|^2 = \frac{A_0^2}{L} \frac{\sqrt{2\pi}}{\sigma_x(t)} \exp\left(-\frac{\left(x - \frac{\hbar k_0}{m}t\right)^2}{2\sigma_x^2(t)}\right).$$

The peak of this distribution is located at $x = \frac{\hbar k_0}{m}t$; it moves with the velocity $\frac{\hbar k_0}{m}$. Its width

$$\sigma_x(t) = \sqrt{\frac{1}{2}\left(\frac{1}{\sigma_k^2} + \frac{\hbar^2}{m^2}\sigma_k^2 t^2\right)}.$$

Initially (at $t = 0$), the width of the probability density is

$$\sigma_x(0) = \frac{1}{\sqrt{2}\sigma_k}.$$

The broader the distribution of the wave numbers that comprise the wave packet, the better it is initially localized in the x-direction. However, the probability distribution $|\phi(x,t)|^2$ becomes broader and broader in time. At large times it behaves as

$$\sigma_x(t) = \frac{\hbar \sigma_k}{\sqrt{2}m}t.$$

Its rate of spreading is proportional to the width of the wave packet σ_k in the k-space.

(b) Integrating the probability density and using the integral formula, we obtain

$$\int_{-\infty}^{\infty} dx \, |\phi(x,t)|^2 = \frac{A_0^2}{L} 2\pi^{3/2} = 1 \implies A_0 = \frac{1}{\pi^{3/4}} \sqrt{\frac{L}{2}}.$$

Without the constant A_0, the probability density is written as

$$|\phi(x,t)|^2 = \frac{1}{\sqrt{2\pi}\sigma_x(t)} \exp\left(-\frac{\left(x - \frac{\hbar k_0}{m}t\right)^2}{2\sigma_x^2(t)}\right).$$

Problem 3

(a) To establish the type of this hypothetical semiconductor, we need to find the energy minimum in the conduction band and energy maximum in the valence band. Let us begin with the conduction band. Since exponential is an increasing function of its argument, the minimum of $E_C(p)$ is found at that momentum p, at which $\cos(2\pi p/p_0)$ has the minimal value of -1. This happens at

$$p = p_{C,min} = p_0/2.$$

Let us look at the valence band. The maximum of $E_V(p)$ is found at

$$p = p_{V,max} = 0.$$

Since $p_{C,min} \neq p_{V,max}$, this is an indirect-band gap semiconductor.

(b) The band gap energy is the difference between the minimal energy in the conduction band and the maximal energy in the valence band:

$$E_g = E_C(p_{C,min}) - E_V(p_{V,max}) = A e^{-\gamma} - B = 3 \cdot e^{-0.2} - 1.5 = 0.956 \, \text{eV}.$$

At $T = 0$, the Fermi energy is right in between the band gap:

$$E_F(T = 0) - E_V = E_g/2 = 0.478 \, \text{eV}.$$

(c) To find the carriers' effective masses, we need to differentiate $E_C(p)$ near its minimum and $E_V(p)$ near its maximum twice and use the formulas

$$m_e^* = \frac{1}{\left.\frac{d^2 E_C}{dp}\right|_{p=p_{C,min}}}, \quad m_h^* = -\frac{1}{\left.\frac{d^2 E_V}{dp}\right|_{p=p_{V,max}}}.$$

Let us do it:

$$\frac{dE_C}{dp} = A\,e^{\gamma\,\cos(2\pi p/p_0)} \cdot \gamma \cdot \left(-\sin\left(\frac{2\pi p}{p_0}\right)\right) \cdot \frac{2\pi}{p_0}$$

$$= -\frac{2\pi\gamma A}{p_0}\,e^{\gamma\,\cos(2\pi p/p_0)}\,\sin\left(\frac{2\pi p}{p_0}\right),$$

$$\left.\frac{d^2E_C}{dp^2}\right|_{p=p_0/2} = -\frac{2\pi\gamma A}{p_0} \cdot \left(e^{\gamma\,\cos(2\pi p/p_0)} \cdot \frac{2\pi}{p_0}\,\cos\left(\frac{2\pi p}{p_0}\right)\right)\Bigg|_{p=p_0/2}.$$

We did not write the term proportional to $\sin(2\pi p/p_0)$, which appears after differentiating the exponential function, because this term goes to zero after substitution $p = p_0/2$. With $\cos(2\pi p/p_0)|_{p=p_0/2} = \cos(\pi) = -1$ we have:

$$\left.\frac{d^2E_C}{dp^2}\right|_{p=p_0/2} = \frac{1}{m_e^*} = \frac{(2\pi)^2\gamma A}{p_0^2}e^{-\gamma}.$$

This gives the electron effective mass

$$m_e^* = \frac{p_0^2 e^{\gamma}}{4\pi^2\gamma A}.$$

With $p_0^2/(2m_e) = C = 1\,\text{eV}$, we find $p_0^2 = 2m_eC$. Hence, the ratio of electron effective mass to the free electron mass can be finally found:

$$m_e^* = \frac{2m_e C e^{\gamma}}{4\pi^2\gamma A} \quad \Rightarrow \quad \frac{m_e^*}{m_e} = \frac{C e^{\gamma}}{2\pi^2\gamma A}.$$

Numerically,

$$\frac{m_e^*}{m_e} = \frac{(1\,\text{eV})\cdot e^{0.2}}{2\cdot\pi^2\cdot 0.2\cdot(3\,\text{eV})} = 0.103.$$

Now, we perform a similar calculation for the valence band near its maximum at $p = 0$. We have:

$$\frac{dE_V}{dp} = -B\cdot 2\frac{\pi}{p_0}\,\cos\left(\frac{\pi p}{p_0}\right)\sin\left(\frac{\pi p}{p_0}\right) = -\frac{\pi}{p_0}B\,\sin\left(\frac{2\pi p}{p_0}\right),$$

where we used the identity $2\sin(\theta)\cos(\theta) = \sin(2\theta)$. Taking the second derivative at $p = 0$,

$$\left.\frac{d^2E_V}{dp^2}\right|_{p=0} = -\frac{1}{m_h^*} = -\frac{\pi B}{p_0}\frac{2\pi}{p_0}\,\cos\left(\frac{2\pi p}{p_0}\right)\Bigg|_{p=0}$$

$$= -\frac{2\pi^2 B}{p_0^2} \quad \Rightarrow \quad m_h^* = \frac{p_0^2}{2\pi^2 B}.$$

As in the electron case, we plug in $p_0^2 = 2m_e C$, where $C = 1\,\mathrm{eV}$, and obtain

$$m_h^* = \frac{m_e C}{\pi^2 B} \quad \Rightarrow \quad \frac{m_h^*}{m_e} = \frac{C}{\pi^2 B}.$$

Numerically,

$$\frac{m_h^*}{m_e} = \frac{1}{\pi^2 \cdot 1.5} = 0.0675.$$

(d) The probability for the lowest energy state in the conduction band to be occupied by an electron is given by the value of the Fermi-Dirac distribution

$$f = \frac{1}{1 + e^{(E_C - E_F)/kT}}.$$

Since the Fermi energy is in the middle of the band gap, we have

$$E_C - E_F = E_g/2 = 0.478\,\mathrm{eV}.$$

At $T = 300\,\mathrm{K}$, the thermal energy is $kT = 0.02585\,\mathrm{eV}$. The probability f has the numerical value

$$f = \frac{1}{1 + e^{0.478/0.02585}} = 9.32 \cdot 10^{-9}.$$

Part II
Semiconductors in and out of Equilibrium

Chapter 5
Semiconductors in Equilibrium

5.1 Density of States

We are now ready to calculate the equilibrium electron concentration in the conduction band, n_0, and equilibrium concentration of holes in the valence band, p_0. The notation comes from the sign of the respective charges: negative for electrons and positive for holes. The subscript 0 indicates thermal equilibrium.

We will see later on in this chapter that equilibrium charge carrier concentration can be affected by adding various impurities that may also alter the band structure of the material, provided that impurity concentration is high. Here and in the following, we assume this not to be the case, i.e., impurity concentration is insufficient to change the band structure.

Let us start with the electrons. The most general expression for the concentration of the conduction electrons is

$$n_0 = \frac{2}{L^3} \sum_{\vec{p} \in C.B.} f(E(\vec{p})) \,,$$

where the sum is over all momentum states $\vec{p}$ in the conduction band $(C.B.)$, $E(\vec{p})$ is the electron energy in the conduction band as a function of electron momentum, and the prefactor 2 comes from the two possible spin orientations, $\pm 1/2$. The result is divided by L^3, the volume of the semiconductor, assumed to be a cube of side length L with periodic boundary conditions.

Evaluation of this sum is difficult because the dependence $E(\vec{p})$ is quite complicated and because Fermi–Dirac distribution $f(E)$ is not easy to deal with. We can overcome both difficulties by making use of the fact that because $f(E)$ decreases exponentially strongly with energy above Fermi energy, only the low-energy states in the conduction band are populated with appreciable probability. Hence, we can approximate the energy–momentum relation with a quadratic function

© The Author(s), under exclusive license to Springer Nature Switzerland AG 2022
M. Evstigneev, *Introduction to Semiconductor Physics and Devices*,
https://doi.org/10.1007/978-3-031-08458-4_5

$$E(\vec{p}) = E_C + \sum_{i=x,y,z} \frac{1}{2m^*_{e,i}} (p_i - p_{C,i})^2 ,$$

where E_C is the energy at the conduction band minimum and $\vec{p}_C$ is the momentum at which this minimum occurs. The coordinate axes are chosen such that only the diagonal components of the effective mass matrix are non-zero (see Sect. 4.6).

The value of the Fermi–Dirac distribution depends on the momentum $\vec{p}$ via the energy $E(\vec{p})$. Therefore, it is convenient in the expression for n_0 to convert the sum over momenta into the sum over energies in the conduction band. Furthermore, the typical spacing between two neighboring momentum states is of the order of h/L, which is a very small number if L is macroscopically large. Then, instead of a sum, we can use an integral over energies:

$$n_0 = \int_{E_C}^{\infty} dE \, g_C(E) \, f(E) .$$

The density of states, $g_C(E)$, is a function that tells us how many momentum and spin states are there in the conduction band per energy interval from E to $E + dE$ in a unit volume. This number is $g_C(E) \, dE$.

The functional form of $g_C(E)$ can be established arithmetically by performing dimensional analysis. The unit of any physical parameter is some combination of mass (measured in kg), time (measured in s), and length (measured in m). For instance, the dimension of energy is

$$[E] = J = \mathrm{kg\,m^2\,s^{-2}}$$

in SI units. The dimension of Planck's constant is

$$[h] = \mathrm{kg\,m^2\,s^{-1}} ,$$

and, obviously, the dimension of the effective mass is

$$[m^*_e] = \mathrm{kg} .$$

The density of states in the conduction band depends on these three parameters. The energy should be counted from the conduction band edge, i.e., we should be writing $E - E_C$ instead of just E. The density of states should be proportional to the combination

$$g_C(E) \propto (E - E_C)^\alpha \, m^{*\beta}_e \, h^\gamma$$

with the exponents α, β, and γ that are yet to be found. In terms of these exponents, the dimension of the density of states is

$$[g_C(E)] = (\text{kg m}^2 \text{ s}^{-2})^\alpha \text{ kg}^\beta (\text{kg m}^2 \text{ s}^{-1})^\gamma = \text{kg}^{\alpha+\beta+\gamma} \text{ m}^{2(\alpha+\gamma)} \text{ s}^{-(2\alpha+\gamma)} .$$

By its very definition, the dimension of the density of states—that is, the number of states per unit energy per unit volume—is the inverse of the product of energy and volume:

$$[g_C(E)] = [E^{-1}]\text{m}^{-3} = \text{kg}^{-1}\text{m}^{-5}\text{s}^2 .$$

Hence, the exponents α, β, and γ must be such that

$$\alpha + \beta + \gamma = -1 , \quad 2(\alpha + \gamma) = -5 , \quad -(2\alpha + \gamma) = 2 .$$

Solving these three equations for the three unknowns gives their values

$$\alpha = \frac{1}{2} , \quad \beta = \frac{3}{2} , \quad \gamma = -3 .$$

Then, the density of states in the conduction band is, up to an unknown constant C,

$$g_C(E) = C \frac{\sqrt{E - E_C} \, m_e^{*3/2}}{h^3} .$$

The non-dimensional number C cannot be established within the dimensional analysis. Neither does the dimensional analysis answer the question about the relevant effective mass m_e^*, which should be some combination of the components of the effective mass tensor $m_{e,i}^*$ with $i = x, y, z$. To figure out the constants C and m_e^*, a more rigorous treatment is needed.

We first find the total number $N(E)$ of the conduction band states whose energy is below a given value E. The possible momentum values of an electron are given by $\vec{p} = \frac{h}{L}\vec{n}$, where $\vec{n}$ is a vector whose all three components are integers. In the momentum space, each momentum state occupies a cube of volume

$$\Delta\Omega = (h/L)^3 .$$

To find $N(E)$, we need to find the number of such cubes inside an ellipsoid defined by

$$\frac{(p_x - p_{Cx})^2}{2m_{e,x}^*(E - E_C)} + \frac{(p_y - p_{Cy})^2}{2m_{e,y}^*(E - E_C)} + \frac{(p_z - p_{Cz})^2}{2m_{e,z}^*(E - E_C)} = 1 ,$$

see Fig. 4.11a. The volume of this ellipsoid is[1]

---

[1] An ellipsoid with the major axes oriented in the x-, y-, and z-directions is defined by an equation $\frac{x^2}{a^2} + \frac{y^2}{b^2} + \frac{z^2}{c^2} = 1$ with the parameters a, b, and c being half the distance between the opposite poles of the ellipsoid in the x-, y-, and z-directions. The volume of an ellipsoid is $V = \frac{4}{3}\pi abc$.

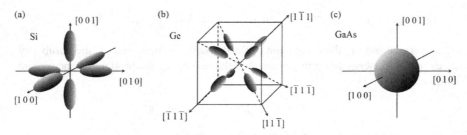

Fig. 5.1 Constant-energy surfaces in the momentum space near the bottom of the conduction band of (**a**) Si, (**b**) Ge, and (**c**) GaAs. Note that in the case of Ge, each of the eight valleys is shared by two Brillouin zones; as a result, the number of valleys in the first Brillouin zone of Ge is $\nu_C = 4$

$$\Omega(E) = \frac{4}{3}\pi \, (2(E - E_C))^{3/2} \sqrt{m_{e,x}^* m_{e,y}^* m_{e,z}^*}.$$

The number of the conduction band states of energy below E is obtained by dividing $\Omega(E)$ by $\Delta\Omega$. The result needs to be multiplied by 2 to account for two possible spin orientations and by the number ν_C of the equivalent energy minima in the conduction band. For example, in silicon, $\nu_C = 6$; in Ge, $\nu_C = 4$; in GaAs, $\nu_C = 1$, see Fig. 5.1. Hence,

$$N(E) = 2\nu_C \frac{\Omega(E)}{\Delta\Omega} = \frac{L^3}{h^3} \frac{8\nu_C}{3}\pi \, (2(E - E_C))^{3/2} \sqrt{m_{e,x}^* m_{e,y}^* m_{e,z}^*} \, .$$

The density of states per unit energy interval per unit volume is obtained by differentiating this quantity with respect to energy:

$$g_C(E) = \frac{1}{L^3}\frac{dN(E)}{dE} = \frac{8\sqrt{2}\pi\,\nu_C\sqrt{m_{e,x}^* m_{e,y}^* m_{e,z}^*}}{h^3}\sqrt{E - E_C} \, .$$

The concentration of holes is obtained by integrating over those states in the valence band (V.B.) that are empty:

$$p_0 = \frac{2}{L^3} \sum_{\vec{p} \in V.B.} (1 - f(E(\vec{p}))) = \int_{-\infty}^{E_V} dE \, g_V(E)\,(1 - f(E)) \, .$$

The hole density of states in the valence band is found by a similar analysis as for electrons, but with two modifications. First, we focus on the materials in which there is only one energy maximum in the valence band at $\vec{p} = 0$, and where the hole effective mass is an ordinary number (see Sect. 4.6), i.e., $m_{h,x}^* = m_{h,y}^* = m_{h,z}^*$. Second, we take into account the existence of two kinds of holes, the heavy and the light, with the effective masses m_{hh}^* and m_{lh}^*. Adding the density of states for the two kinds of holes together, we have the total density of states in the valence band

$$g_V(E) = \frac{8\pi \left((m_{hh}^*)^{3/2} + (m_{lh}^*)^{3/2}\right)}{h^3} \sqrt{E_V - E} \,.$$

5.2 Equilibrium Carrier Concentration

Boltzmann approximation for the Fermi–Dirac distribution is expected to hold quite well because, typically, the energy difference between E_F and the band edges E_C, E_V, is of the order of a few tenths of eV. At the same time, at room temperature $T = 300\,\text{K}$, we have $kT = 0.02585\,\text{eV}$, which is much smaller than $E_C - E_F$ and $E_F - E_V$.

With Boltzmann's approximation

$$f(E) \approx e^{-(E-E_F)/kT} \,, \quad 1 - f(E) \approx e^{(E-E_F)/kT}$$

and our expressions for the densities of states in the conduction and valence bands, the expression for the electron concentration becomes

$$n_0 = \int_{E_C}^{\infty} dE \, g_C(E) \, f(E)$$

$$\approx \frac{8\sqrt{2}\pi \, v_C \sqrt{m_{e,x}^* m_{e,y}^* m_{e,z}^*}}{h^3} \int_{E_C}^{\infty} dE \, \sqrt{E - E_C} \, e^{-(E-E_F)/kT} \,.$$

The integral is evaluated by changing the integration variables from E to $x = (E - E_C)/kT$:

$$\int_{E_C}^{\infty} dE \, \sqrt{E - E_C} \, e^{-(E-E_F)/kT} = (kT)^{3/2} \, e^{-(E_C-E_F)/kT} \int_0^{\infty} dx \, \sqrt{x} \, e^{-x}$$

$$= (kT)^{3/2} \frac{\sqrt{\pi}}{2} \, e^{-(E_C-E_F)/kT} \,,$$

where the x-integration is performed in Appendix A.4. This gives, finally,

$$n_0 = N_C \, e^{-(E_C-E_F)/kT} \,, \quad N_C = \frac{4\sqrt{2} v_C (\pi kT)^{3/2} \sqrt{m_{e,x}^* m_{e,y}^* m_{e,z}^*}}{h^3} \,.$$

The parameter N_C is called the electron effective density of states. This formula can be interpreted physically as the product of the concentration N_C of the conduction band states available for electrons and the probability of $f(E) = \frac{1}{1+e^{(E_C-E_F)/kT}} \approx e^{-(E_C-E_F)/kT}$ for a particular state near the bottom of the conduction band to be occupied. The concentration N_C of the available states is approximately the concentration of states whose energy does not exceed the energy E_C at the bottom of the conduction band by more than the thermal energy kT. Hence, N_C must increase with kT. Also, N_C must depend on the curvature of the energy–momentum curve near the conduction band minimum: the lower the curvature (i.e., the higher the effective mass m_e^*), the flatter the band near the bottom, and the more states it offers for the electrons to occupy. Hence, N_C also increases with the effective mass m_e^*.

A similar treatment for holes gives the concentration

$$p_0 = N_V \, e^{(E_V-E_F)/kT} \, , \quad N_V = \frac{4\sqrt{2}(\pi kT)^{3/2}\left((m_{hh}^*)^{3/2} + (m_{lh}^*)^{3/2}\right)}{h^3} \, .$$

The parameter N_V is called the hole effective density of states. The physical meaning of this formula is similar to that of the electron counterpart, except now we need to count the unoccupied states near the top of the valence band. The probability of a state at $E = E_V$ to be empty (i.e., to be occupied by a hole) is

$$1 - f(E_V) = 1 - \frac{1}{1 + e^{(E_V-E_F)/kT}} = \frac{e^{(E_V-E_F)/kT}}{1 + e^{(E_V-E_F)/kT}} = \frac{1}{1 + e^{-(E_V-E_F)/kT}}$$
$$\approx e^{-(E_V-E_F)/kT} \, .$$

The total concentration of holes is a sum of the heavy and light holes' concentrations. The contribution of each hole species is proportional to the respective effective mass raised to the power 3/2.

The typical values of the effective densities of states at room temperature $T = 300\,\text{K}$ are of the order of $10^{17} \ldots 10^{20}\,\text{cm}^{-3}$, depending on the material. Because the effective densities of states depend on temperature in proportion to $T^{3/2}$, its knowledge at some particular reference temperature T_0, say, $T_0 = 300\,\text{K}$, is sufficient to determine them at any other temperature as

$$N_{C,V}(T) = N_{C,V}(T_0) \left(\frac{T}{T_0}\right)^{3/2} \, .$$

The values of the effective densities of states at $T_0 = 300\,\text{K}$ for Si, Ge, and GaAs can be found in Appendix A.3.

5.3 Energy Probability Distribution

The function

$$P_e(E) = C\sqrt{E - E_C}\, e^{-(E-E_F)/kT} \quad \text{with} \ E > E_C$$

used above has the physical meaning of the energy probability distribution of electrons in the conduction band. The parameter C is the normalization constant, whose value is determined by the normalization condition

$$\int_{E_C}^{\infty} dE\, P(E) = C \int_{E_C}^{\infty} dE\, \sqrt{E - E_C}\, e^{-(E-E_F)/kT} = 1\,.$$

The integral has been evaluated previously; with this result,

$$C = \frac{2}{\sqrt{\pi}(kT)^{3/2}} e^{(E_C - E_F)/kT}\,,$$

and thus

$$P_e(E) = \frac{2}{\sqrt{\pi}(kT)^{3/2}} \sqrt{E - E_C}\, e^{-(E-E_C)/kT}\,.$$

This function is shown in the right panel of Fig. 5.2. By setting the derivative to zero, we find that the most probable electron energy is

$$\frac{dP_e(E)}{dE}\bigg|_{E=E_m} = 0 \ \Rightarrow \ E_m = E_C + \frac{kT}{2}\,.$$

It is also useful to find the average electron energy relative to the conduction band edge:

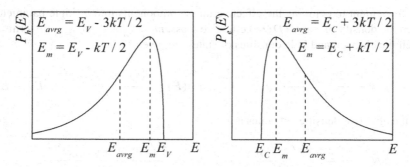

Fig. 5.2 Hole (left) and electron (right) energy probability distribution

$$E_{avrg} = E_C + \int_{E_C}^{\infty} dE\,(E - E_C)\,P_e(E)$$

$$= E_C + \frac{2}{\sqrt{\pi}(kT)^{3/2}} \int_{E_C}^{\infty} dE\,(E - E_C)^{3/2} e^{-(E-E_C)/kT} \ .$$

We change the integration variable to $x = (E - E_C)/kT$, giving

$$E_{avrg} = E_C + \frac{2}{\sqrt{\pi}}kT \int_0^{\infty} dx\,x^{3/2}\,e^{-x} \ .$$

The integral with respect to x is evaluated in Appendix A.4; its value is $3\sqrt{\pi}/4$. Hence, the average electron energy is

$$E_{avrg} = E_C + \frac{3kT}{2} \ .$$

The result for holes is similar. Hole energy probability distribution is

$$P_h(E) = \frac{2}{\sqrt{\pi}(kT)^{3/2}} \sqrt{E_V - E}\,e^{-(E_V - E)/kT} \quad \text{for} \ \ E < E_V \ .$$

Note that hole energy here is measured from the electron's perspective. The most probable and average energies of a hole are below the valence band edge by the amounts $kT/2$ and $3kT/2$, respectively, see Fig. 5.2, left panel.

5.4 Density of States Effective Mass vs. Conductivity Effective Mass

5.4.1 Density of States Effective Mass

It is convenient to combine the effective mass components into a single parameter, called the density of states (DOS) effective mass, $m^*_{e,DOS}$ and $m^*_{h,DOS}$, in terms of which the electron and hole densities of states are given by

$$g_C(E) = \frac{4\pi(2m^*_{e,DOS})^{3/2}}{h^3} \sqrt{E - E_C} \ , \quad g_V(E) = \frac{4\pi(2m^*_{h,DOS})^{3/2}}{h^3} \sqrt{E_V - E}$$

and the effective densities of states are

$$N_C = \frac{4\sqrt{2}(\pi kT\,m^*_{e,DOS})^{3/2}}{h^3} \ , \quad N_V = \frac{4\sqrt{2}(\pi kT\,m^*_{h,DOS})^{3/2}}{h^3} \ .$$

The electron and hole DOS effective masses are given by

$$m_{e,DOS}^* = v_C^{2/3}(m_{e,x}^* m_{e,y}^* m_{e,z}^*)^{1/3} = v_C^{2/3}(m_{e,t}^*)^{2/3}(m_{e,l}^*)^{1/3} \ ,$$

$$m_{h,DOS}^* = \left((m_{hh}^*)^{3/2} + (m_{hl}^*)^{3/2}\right)^{2/3} .$$

In the electron expression, we used the fact that two effective mass components are equal to the transverse effective mass and the third one equals the longitudinal effective mass, as discussed in Sect. 4.6.

5.4.2 Conductivity Effective Mass

Electrons

In silicon, which is an indirect-band gap semiconductors, the conduction band has six equivalent valleys, see Fig. 5.1a. To each constant-energy ellipsoid, we can attach its own three coordinate axes aligned along its principal axes. A particular crystallographic direction, say, [1 0 0], is now aligned with the x-axes of two valleys, the y-axes of two other valleys, and the z-axes of the remaining two valleys.

From the point of view of electric properties of the material, we are interested in the reaction of the electrons in all six valleys to the electric field. Therefore, if an electric field $\vec{\mathcal{E}}$ is applied, say, in the [1 0 0] direction, the acceleration of an electron should be averaged over all six valleys. This gives the average acceleration in the [1 0 0] direction

$$a_{[100]} = -e\mathcal{E}_{[100]} \frac{1}{3}\left(\frac{1}{m_{e,x}^*} + \frac{1}{m_{e,y}^*} + \frac{1}{m_{e,z}^*}\right) .$$

Same is true about the two other crystallographic directions, [0 1 0] and [0 0 1]. Therefore, on average, an electron responds to an electric field according to

$$\vec{a} = -e\vec{\mathcal{E}}/m_{e,cond}^*$$

where the conductivity effective mass is defined by

$$\frac{1}{m_{e,cond}^*} = \frac{1}{3}\left(\frac{1}{m_{e,x}^*} + \frac{1}{m_{e,y}^*} + \frac{1}{m_{e,z}^*}\right) .$$

In other indirect-band gap semiconductors, such as germanium, there are eight constant-energy surfaces oriented along the 8 directions of the $\langle 1\,1\,1 \rangle$ family, see Fig. 5.1b. Nevertheless, the rule to obtain the average conductivity effective mass turns out to be the same as in silicon. The conductivity effective mass of Si and Ge is, then,

$$m_e^* = \frac{3}{(m_{e,l}^*)^{-1} + 2(m_{e,t}^*)^{-1}}$$

because $m_{e,x}^* = m_{e,y}^* = m_{e,t}^*$ and $m_{e,z}^* = m_{e,l}^*$.

In GaAs, a direct-band gap semiconductor, Fig. 5.1c, conductivity effective mass is the same as the density of states effective mass, $m_{e,cond}^* = m_{e,DOS}^* = m_e^*$.

Holes

It is clear from the discussion of Sect. 5.2 that the fractions of heavy and light holes among all holes are

$$\frac{p_{0,hh}}{p_0} = \frac{(m_{hh}^*)^{3/2}}{(m_{hh}^*)^{3/2} + (m_{lh}^*)^{3/2}} , \quad \frac{p_{0,lh}}{p_0} = \frac{(m_{lh}^*)^{3/2}}{(m_{hh}^*)^{3/2} + (m_{lh}^*)^{3/2}} ,$$

where $p_{0,hh}$ and $p_{0,lh}$ are the equilibrium heavy and light hole concentrations, respectively.

In the electric current measurements, one does not distinguish the currents produced by each hole type. Rather, one measures the total current produced by both heavy and light holes. Therefore, the accelerations of both holes due to an external electric field $\vec{\mathcal{E}}$, viz.,

$$\vec{a}_{hh} = \frac{e\vec{\mathcal{E}}}{m_{hh}^*} , \quad \vec{a}_{lh} = \frac{e\vec{\mathcal{E}}}{m_{lh}^*} ,$$

should be averaged according to their contributions to the net hole concentrations. We thus obtain the average hole acceleration

$$\vec{a}_h = \vec{a}_{hh}\frac{p_{0,hh}}{p_0} + \vec{a}_{lh}\frac{p_{0,lh}}{p_0} = \frac{e\vec{\mathcal{E}}}{m_{h,cond}^*} ,$$

where the hole conductivity effective mass is

$$m_{h,cond}^* = \frac{(m_{hh}^*)^{3/2} + (m_{lh}^*)^{3/2}}{(m_{hh}^*)^{1/2} + (m_{lh}^*)^{1/2}} .$$

5.4.3 Thermal Velocity

Before we proceed, it is instructive to address the question about the typical group velocity of an electron and a hole at a given temperature T. By "typical," we mean that the kinetic energy of a particle equals the average thermal energy, E_{avrg}, which is above the conduction band edge or below the valence band edge by the amount $3kT/2$, see Sect. 5.3.

Let us consider the electrons first. For the quadratic energy–momentum relation (see Sect. 5.1), the group velocity components are

$$v_i = \frac{\partial E}{\partial p_i} = \frac{p_i - p_{C,i}}{m^*_{e,i}} , \quad i = x, y, z .$$

The electron energy near the bottom of the conduction band can then be expressed in terms of the group velocity components as

$$E(\vec{v}) = E_C + \sum_i \frac{m^*_{e,i} v_i^2}{2} .$$

The thermal velocity is defined as the velocity of an electron whose energy exceeds the conduction band edge energy by $3kT/2$. More precisely, its three components are given by the identities

$$\frac{m^*_{e,i} v^2_{e|th,i}}{2} = \frac{kT}{2} ,$$

and the thermal velocity of an electron is then

$$v_{e|th} = \sqrt{v^2_{e|th,x} + v^2_{e|th,y} + v^2_{e|th,z}} .$$

Combining the last two equations together and recalling the definition of the electron conductivity effective mass, we have

$$v_{e|th} = \sqrt{\frac{3kT}{m^*_{e,cond}}} .$$

Similarly, we define the thermal velocities of the heavy and light holes as

$$v_{hh|th} = \sqrt{\frac{3kT}{m^*_{hh}}} , \quad v_{lh|th} = \sqrt{\frac{3kT}{m^*_{lh}}} .$$

Their combination according to the relative contribution of the two kinds of holes to the total hole concentration gives

$$v_{h|th} = v_{hh|th}\frac{p_{0,hh}}{p_0} + v_{lh|th}\frac{p_{0,lh}}{p_0} = \sqrt{\frac{3kT}{m^*_{h,cond}}} \ .$$

In the rest of this book, we will omit the subscripts DOS and $cond$ in the effective masses m^*_e and m^*_h. Which effective mass type is meant will be clear from the context of the discussion.

5.5   Intrinsic Semiconductors

Intrinsic semiconductors are materials without impurities. In intrinsic semiconductors, each electron that gets promoted into the conduction band leaves behind exactly one hole in the valence band. Therefore, the concentrations of electrons and holes are exactly the same.

To emphasize that the semiconductor is intrinsic, the carrier concentrations, as well as the Fermi energy, carry the subscript i. Using the results obtained in the previous section, we can figure out the concentration of charge carriers

$$n_{0i} = N_C e^{-(E_C-E_{Fi})/kT} \ , \quad p_{0i} = N_V e^{(E_V-E_{Fi})/kT}$$

and the intrinsic Fermi energy, E_{Fi}. Namely, by setting

$$n_{0i} = p_{0i} \equiv n_i \ ,$$

we obtain an equation for E_{Fi}, which is easily solved:

$$E_{Fi} = \frac{E_C + E_V}{2} + \frac{kT}{2}\ln\frac{N_V}{N_C} = \frac{E_C + E_V}{2} + \frac{3kT}{4}\ln\frac{m^*_h}{m^*_e} \ .$$

In the second equality, we used the relation between the effective densities of states and the effective masses of the two charge carriers.

At zero temperature, the intrinsic Fermi energy is located exactly in the middle of the band gap. As the temperature increases, it shifts toward the conduction band if $N_V > N_C$ ($m^*_h > m^*_e$) or toward the valence band if $N_V < N_C$ ($m^*_h < m^*_e$). This is similar to the result obtained in Sect. 3.7, which predicts that the Fermi level shifts toward the state of the lower degeneracy as the temperature goes up. Here, the role of the state degeneracy is played by the effective densities of states.

Substitution of the result for E_{Fi} into the expression for the electron (or hole) concentration, we obtain the intrinsic carrier concentration:

$$n_i = n_{0i} = p_{0i} = \sqrt{N_C N_V} e^{-E_g/(2kT)} .$$

The prefactor $\sqrt{N_C N_V}$ is typically of the order of $10^{17} - 10^{20}$ cm^{-3}. In semiconductors with E_g of about 1 eV at room temperature, $kT = 0.0259$ eV, intrinsic concentration is of the order of $10^7 - 10^{11}$ cm^{-3}.

On the other hand, in the materials even with a relatively small band gap value, say, $E_g = 3$ eV and $N_C \sim N_V \sim 10^{19}$ cm^{-3}, the intrinsic carrier concentration is about $7 \cdot 10^{-7}$ cm^{-3}. One would need a cube of about 1 m side length to have just one free electron in it. To have a measurable concentration of conduction electrons, such a material would need to be heated up above its melting temperature.

5.6 Doping and Extrinsic Semiconductors

One can control the concentration of conduction electrons and holes by adding the so-called shallow impurities that can be easily turned into positive or negative ions. This procedure is called doping. Doping can be performed in several ways, such as ion implantation or diffusion.

Donors are dopants that give electrons to the lattice. An example of donor doping is adding elements from column V of the periodic table to Si. Group-V elements, such as arsenic or phosphorus, have 5 valence electrons, whereas in Si lattice, bonds are formed by 4 valence electrons. This means that one of the valence electrons of a donor atom does not participate in chemical bonding with the neighboring Si atoms and is therefore relatively loosely attached to its host. It can be easily removed from the donor impurity and become a free conductance electron. It cannot go into the valence band because almost all Bloch states in the valence band are occupied. After losing its electron, the donor atom turns into a positively charged ion.

Acceptors are dopants that catch the host lattice electrons, thereby creating holes in the lattice. An example of acceptor doping is adding elements from the third column of the periodic table to Si. Group-III elements, such as boron or gallium, have three valence electrons, i.e., they are one electron short of forming a covalent bond with the four neighboring Si atoms. To make up for this deficiency, an acceptor impurity may take an electron from the host lattice, thereby becoming a negatively charged ion and creating a hole in the material. That electron comes from the valence band because the conduction band has much fewer electrons than the valence band.

In III–V semiconductors, e.g., GaAs, a group-IV element, such as Si or Ge, may act as either a donor or an acceptor impurity, depending on which lattice atoms it substitutes. This is called the amphoteric doping. The concentration and the type of the charge carriers in this case depend on the technological details of doping procedure.

A compensated semiconductor is the one where both donor and acceptor impurities are present.

We will assume that the concentration of the impurity atoms is at least several orders of magnitude smaller than the concentration of the host atoms. That is, the doping is too weak to change the band structure of the material. Instead, the impurity atoms will introduce additional energy levels within the band gap. They will also change the Fermi energy by affecting the concentrations of electrons and holes in the material.

Charge carriers whose concentration is the larger are called majority carriers; charge carriers with the lower concentration are called minority carriers. If the majority carriers are electrons, a semiconductor is said to be of n-type; if the majority carriers are holes, it is a p-type semiconductor:

$$\text{n-type: } n_0 > n_i > p_0 = n_i^2/n_0 \,,$$

$$\text{p-type: } p_0 > n_i > n_0 = n_i^2/p_0 \,.$$

5.7   Impurity Energy Levels

Consider donors first. All but one valence electron of a donor impurity is taking part in chemical bonding, whereas the remaining electrons are relatively loosely bound to its host. We can view the neutral donor atom as a positive ion with an electron bound to it. The binding energy of the electron to the impurity ion can be found in exact same way as the electron energy levels in hydrogen (see Sect. 1.5), but with two modifications.

First, the electron mass must be replaced by the conductivity effective mass, $m_e \to m_e^*$.

Second, calculation of hydrogen energy spectrum was performed in vacuum, whereas a donor ion and an extra electron find themselves in a semiconductor medium. The two charges interact directly via own electric field. They also polarize the medium, and the polarized medium generates an additional electric field that affects the interaction between the two charges. This additional field weakens the electron–ion interaction as compared to the interaction in vacuum. This weakening can be accounted for by modifying the Coulomb interaction energy:

$$U(\vec{r}) = -\frac{k_e}{\epsilon} \frac{e^2}{r} \,.$$

The parameter ϵ is called the dielectric constant, or the relative permittivity, of the medium. Its typical semiconductor value is of the order of 10, see Appendix A.3. Hence, our second modification is to replace $e^2 \to \frac{e^2}{\epsilon}$.

For an electron in the ground state of a hydrogen atom, ionization energy is defined as the energy necessary to promote the electron to vacuum:

$$E_i = 1\,\text{Ry} = \frac{k_e^2 m_e e^4}{2\hbar^2} = 13.6\,\text{eV} \ .$$

In a semiconductor, ionization energy is the energy necessary to promote an extra electron from the bound state with a donor atom to the bottom of the conduction band. In other words, it is the energy difference between the bottom of the conduction band and the donor energy level. By making the two replacements just described, we find the donor ionization energy in a semiconductor to be

$$E_{di} = E_C - E_d = \frac{k_e^2 m_e^* e^4}{2\hbar^2 \epsilon^2} = \frac{m_e^*/m_e}{\epsilon^2}\,\text{Ry} \ .$$

Here, E_d is the donor energy level, which lies below the conduction band edge by donor ionization energy, $E_d = E_C - E_{di}$ (Fig. 5.3).

Given that, typically, m_e^*/m_e varies from about 0.05 to 2 and ϵ is between 10 and 20, we find that the typical value of donor ionization energy is of the order of 0.02 ... 0.2 eV. For instance, for silicon, we obtain $E_{di} = 0.0258$ eV. This value is identical to the thermal energy at room temperature, $kT = 0.0259$ eV. Based on this estimate, we conclude that at room temperature, a large part of donor atoms is ionized.

It is also instructive to determine the typical distance between the extra valence electron and the donor impurity ion. In hydrogen, the Bohr radius of the electron orbit in the ground state ($n = 1$) is $r_B = \frac{\hbar^2}{k_e m_e e^2}$. With the replacements described above, we find that in a semiconductor, the radius of the first Bohr orbit is

$$r_d = \frac{\hbar^2 \epsilon}{k_e m_e^* e^2} = r_B \frac{\epsilon}{m_e^*/m_e} \approx (0.53\,\text{Å})\frac{\epsilon}{m_e^*/m_e} \ .$$

Numerically, the electron orbit radius in a semiconductor is about 10 nm, which is ca. 20 lattice constants.

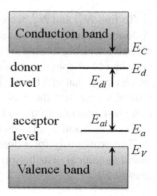

Fig. 5.3 Donor and acceptor levels relative to the conduction and valence band energies

With respect to the acceptor impurities, the results are quite similar. A neutral acceptor atom in the crystal lattice can be viewed as a bound state of a negatively charged ion and a hole orbiting around it. Therefore, the ionization energy E_{ai} of an acceptor level E_a, that is, the energy necessary to remove this hole from the atom into the valence band, is

$$E_{ai} = E_a - E_V = \frac{k_e^2 m_h^* e^4}{2\hbar^2 \epsilon^2} = \frac{m_h^*/m_e}{\epsilon^2} \, \text{Ry} \, ,$$

and the radius of the hole orbit around an acceptor atom is

$$r_a = \frac{\hbar^2 \epsilon}{k_e m_h^* e^2} = r_B \frac{\epsilon}{m_h^*/m_e} \, .$$

The calculations performed in this section are approximate. They neglect the tensorial character of the effective mass and do not distinguish between heavy and light holes. In addition, they do not account for the interaction between the extra electron (or hole) attached to the donor (or acceptor) ion with its other electrons. But, in spite of their crude nature, these calculations give the right order of magnitude of the ionization energy, which has to be measured experimentally. Some values of the ionization energies for Si, Ge, and GaAs are given in Appendix A.3.

5.8   Statistics of Donors and Acceptors

Let N_d be the concentration of donor atoms, N_d^0 the concentration of neutral donors, and N_d^+ the concentration of donors that have given an electron to the lattice and became positive ions. Obviously,

$$N_d = N_d^0 + N_d^+ \, .$$

To find N_d^+ and N_d^0, we use the grand canonical distribution derived in Sect. 3.4. A donor impurity can exist either as a positive ion or as a neutral atom with an extra electron orbiting around it. That electron has two possible spin orientations: either spin up ($\uparrow$) or spin down ($\downarrow$). It is reasonable to assume that when a donor captures an electron, that electron exists in the ground state and has the energy E_d; in other words, we neglect the excited states of the electron. Denoting the number of extra electrons attached to the donor as n, we have three possibilities: $n = 0$; $n = 1$ and $s = \uparrow$; $n = 1$ and $s = \downarrow$, where s stands for spin.

The probability for a donor to have $n = 0$ or 1 valence electrons is given by the grand canonical distribution with the chemical potential μ being identical to Fermi energy E_F:

$$P_n = \frac{1}{Z} e^{(E_F - E_d)n/kT} \;, \quad Z = 1 + g_d e^{(E_F - E_d)n/kT} \;, \quad g_d = 2 \;.$$

The donor degeneracy factor $g_d = 2$ in the expression for Z comes from two possible spin orientations.

The probability for a donor atom to be ionized, i.e., not have an extra electron, corresponds to $n = 0$. This probability is just the fraction of ionized donors, $P_0 = N_d^+ / N_d$. Therefore, the concentration of ionized donors is

$$N_d^+ = \frac{N_d}{1 + g_d e^{(E_F - E_d)/kT}} \;.$$

The concentration of donors that are not ionized is

$$N_d^0 = N_d - N_d^+ = \frac{N_d}{1 + \frac{1}{g_d} e^{(E_d - E_F)/kT}} \;.$$

In view of the relations

$$E_d = E_C - E_{di} \;, \quad n_0 = N_C e^{-(E_C - E_F)/kT} \;,$$

we can rewrite the exponential as

$$e^{(E_F - E_d)/kT} = e^{(E_F - E_C + E_{di})/kT} = \frac{n_0}{N_C} e^{E_{di}/kT} \;,$$

giving

$$N_d^+ = \frac{N_d}{1 + g_d \frac{n_0}{N_C} e^{E_{di}/kT}} \;, \quad N_d^0 = \frac{N_d}{1 + \frac{1}{g_d} \frac{N_C}{n_0} e^{-E_{di}/kT}} \;.$$

The expressions relating the concentration of ionized acceptor atoms, N_a^-, and neutral acceptor atoms, N_a^0, to the total acceptor concentration, N_a, can be obtained by making two modifications in the results for obtaining the donors.

First, donor degeneracy factor must be replaced with the acceptor counterpart, $g_d \to g_a$. Similar to the donor case, the acceptor ground state can be occupied with either a heavy or a light hole of spin $1/2$ or $-1/2$. The total number of combinations of hole spin and "heaviness" is

$$g_a = 4 \;.$$

Second, we should keep in mind that the hole energies are measured downward, with $E^{(h)} = -E + \Delta E$, see Sect. 4.5. Then, the difference between donor level and Fermi energies must be replaced with

$$E_d - E_F \rightarrow E_a^{(h)} - E_F^{(h)} = -E_a + E_F \, ,$$

where E_a is the acceptor level energy.

After making these replacements, we obtain the concentrations of negatively charged and neutral acceptor impurities:

$$N_a^- = \frac{N_a}{1 + g_a e^{(E_a - E_F)/kT}} \, , \quad N_a^0 = \frac{N_a}{1 + \frac{1}{g_a} e^{(E_F - E_a)/kT}} \, .$$

In view of the relations

$$E_a = E_V + E_{ai} \, , \quad p_0 = N_V e^{-(E_F - E_V)/kT} \, ,$$

we can rewrite these expressions in an equivalent form:

$$N_a^- = \frac{N_a}{1 + g_a \frac{p_0}{N_V} e^{E_{ai}/kT}} \, , \quad N_a^0 = \frac{N_a}{1 + \frac{1}{g_a} \frac{N_V}{p_0} e^{-E_{ai}/kT}} \, .$$

5.9 Mass Action Law

The expressions for n_0 and p_0 in Sect. 5.2 contain one unknown parameter, the Fermi energy E_F. But note that the product of the two concentrations does not depend on the Fermi energy. Indeed, multiplication of these results gives

$$n_0 p_0 = N_C N_V e^{-(E_C - E_V)/kT} = N_C N_V e^{-E_g/kT} \, .$$

In the second equality, we used the fact that $E_C - E_V = E_g$, the band gap energy.

Doping of a semiconductor is an efficient way to control the concentrations of electrons and holes. The expressions for the equilibrium concentrations derived in Sect. 5.2 still hold in doped as well as intrinsic semiconductors, with the difference between the two being in the location of the Fermi energy E_F.

The electron and hole concentrations can be written in a different form, which does not involve the effective densities of states $N_{C,V}$ and the Fermi energy. For this, we add and subtract the intrinsic Fermi energy to the arguments in the exponential functions:

$$n_0 = N_C e^{-(E_C + E_{Fi} - E_{Fi} - E_F)/kT} \, , \quad p_0 = N_V e^{(E_V - E_{Fi} + E_{Fi} - E_F)/kT} \, .$$

But $N_C e^{-(E_C - E_{Fi})/kT} = N_V e^{(E_{Fi} - E_V)/kT} = n_i$. We thus have

$$n_0 = n_i e^{(E_F - E_{Fi})/kT} \, , \quad p_0 = n_i e^{(E_{Fi} - E_F)/kT} \, .$$

With these expressions, we can rewrite the mass action law as

$$n_0 p_0 = n_i^2 \, .$$

The product of electron and hole concentrations is the intrinsic concentration squared for any level of doping.

The mass action law tells us that the concentration of one type of carriers uniquely determines carrier concentration of the other type. Because this law is derived using Boltzmann approximation, it holds provided that Fermi energy is separated from the edges of the conduction and valence bands by at least a few kT.

5.10 Charge Neutrality Equation

At a first glance, doping of a material with, say, donors should not affect the concentration of holes, p_0, which should remain equal the intrinsic concentration. Donor doping should increase the concentration of electrons, n_0, from the value n_i to the value $n_i + N_d^+$, where N_d^+ is the concentration of ionized doping.

This intuition is wrong. Donor doping provides electrons not only to the conduction band, but also fills the empty states in the valence band, whose energy is smaller than the donor level, $E_V < E_d$. As a result, not only the electron concentration increases, but also hole concentration gets reduced due to donor doping.

Same is true about acceptors. Acceptor impurities not only take electrons from the valence band, thereby creating more holes, but they also capture the conduction electrons, thereby decreasing their concentration.

The relation between the equilibrium carrier concentrations n_0 and p_0 on the one hand and between the ionized impurity concentrations N_d^+ and N_a^- on the other hand is slightly more complex. Because the material must be electrically neutral, the concentration of positive charges must be exactly balanced by the concentration of negative charges. The positive charges are mobile holes and immobile donor ions. The negative charges are mobile conduction electrons and immobile acceptor ions. We thus arrive at the charge neutrality equation:

$$n_0 + N_a^- = p_0 + N_d^+ \, ,$$

or, explicitly,

$$n_i e^{(E_F - E_{Fi})/kT} + \frac{N_a}{1 + g_a e^{(E_a - E_F)/kT}} = n_i e^{(E_{Fi} - E_F)/kT} + \frac{N_d}{1 + g_d e^{(E_F - E_d)/kT}} \, .$$

This equation contains only one unknown: the Fermi energy E_F. Once it is found, electron and hole concentrations can be determined using the expressions obtained earlier, viz., $n_0 = n_i e^{(E_F - E_{Fi})/kT}$ and $p_0 = n_i e^{(E_{Fi} - E_F)/kT}$.

An exact analytical solution of the charge neutrality equation can also be obtained, in principle, by introducing a variable $x = e^{(E_F - E_{Fi})/kT}$, which then will have to be found from a fourth-degree equation of the form $x^4 + ax^3 + bx^2 + cx + d = 0$ with the coefficients a, b, c, and d being some combinations of the parameters N_a, N_d, n_i, g_d, g_a, $e^{(E_{Fi} - E_d)/kT}$, and $e^{(E_a - E_{Fi})/kT}$.

A general solution of a fourth-degree equation exists, but it is too complex to be of practical utility; hence, it will not be considered. Another option is to solve the charge neutrality equation numerically. But the most instructive approach is to focus on a few practically relevant special limit cases, referred to as ionization regimes, in which analytical approximations for the Fermi energy and the carrier concentrations can be obtained.

5.11   Ionization Regimes

5.11.1   Complete Ionization

The simplest and the most important practically is the regime when the overwhelming majority of the impurity atoms are ionized. This regime is realized when the energy differences $E_F - E_a$ and $E_d - E_F$ exceed thermal energy, kT, by some reasonably large amount, say, by a factor of 3 or more. Then, $e^{(E_a - E_F)/kT}$ and $e^{(E_F - E_d)/kT}$ are both much smaller than 1, and we can approximate

$$N_d^+ = \frac{N_d}{1 + g_d e^{(E_F - E_d)/kT}} \approx N_d \,, \quad N_a^- = \frac{N_a}{1 + g_a e^{(E_a - E_F)/kT}} \approx N_a$$

by neglecting the exponentials in the denominators of both expressions. Our charge neutrality condition then reads

$$n_0 + N_a = p_0 + N_d \,.$$

From the mass action law, we can express

$$p_0 = \frac{n_i^2}{n_0} \,,$$

resulting in a quadratic equation for n_0:

$$n_0 + N_a = \frac{n_i^2}{n_0} + N_d \quad \Rightarrow \quad n_0^2 + (N_d - N_a)n_0 - n_i^2 = 0 \,.$$

The solution of this equation is

$$\text{n-type: } n_0 = \frac{N_d - N_a}{2} + \sqrt{\left(\frac{N_d - N_a}{2}\right)^2 + n_i^2}, \quad p_0 = \frac{n_i^2}{n_0}.$$

From the point of view of numerical accuracy, these expressions work the best when the doping is of n-type, $N_d > N_a$. But consider what happens in the extreme opposite case $N_a \gg n_i, N_d$. Then, the terms n_i and N_d are negligibly small relative to N_a, leaving us with $n_0 \approx -\frac{N_a}{2} + \frac{N_a}{2} = 0$ and $p_0 = n_i^2/n_0 \to \infty$.

Hence, in the opposite case of p-type doping, $N_a > N_d$, it is more advantageous to use a mathematically equivalent, but numerically a different set of formulae:

$$\text{p-type: } p_0 = \frac{N_a - N_d}{2} + \sqrt{\left(\frac{N_a - N_d}{2}\right)^2 + n_i^2}, \quad n_0 = \frac{n_i^2}{p_0}.$$

In particular, when doping of just one type is present with the concentration much higher than intrinsic concentration, the following simple results hold:

$$\text{for } N_d \gg n_i \gg N_a : n_0 = N_d, \quad p_0 = \frac{n_i^2}{N_d};$$

$$\text{for } N_a \gg n_i \gg N_d : p_0 = N_a, \quad n_0 = \frac{n_i^2}{N_a}.$$

The relations above are derived under the assumption that the energy differences between donor level and Fermi energy, as well as between Fermi energy and acceptor level, are large compared to kT. To check if these assumptions hold true, we need to find Fermi energy. For this, we can use the relations like $n_0 = N_C e^{-(E_C - E_F)/kT} = n_i e^{(E_F - E_{Fi})/kT}$, and $p_0 = N_V e^{(E_V - E_F)/kT} = n_i e^{(E_{Fi} - E_F)/kT}$. From these, we find several equivalent expressions for Fermi energy in a doped semiconductor:

$$E_F = E_C - kT \ln \frac{N_C}{n_0} = E_{Fi} + kT \ln \frac{n_0}{n_i} = E_V + kT \ln \frac{N_V}{p_0} = E_{Fi} - kT \ln \frac{p_0}{n_i}.$$

It follows from the second equality that for n-type doping, $n_0 > n_i > p_0 = n_i^2/n_0$, Fermi energy lies above the intrinsic Fermi energy, $E_F > E_{Fi}$. Increasing the donor concentration raises Fermi energy.

Likewise, the last equality shows that for p-type doping, Fermi energy is below the intrinsic Fermi energy, $E_F < E_{Fi}$, and increasing the acceptor concentration lowers Fermi energy.

We can now establish the condition under which all impurities are ionized. Let us focus on the n-type semiconductor for definiteness. In order for almost all donors to be ionized, $N_d^+ \approx N_d$, we must have

$$g_d \, e^{(E_F - E_d)/kT} \ll 1 \, .$$

Adding and subtracting the conduction band energy E_C in the argument of the exponential and keeping in mind that $E_C - E_d = E_{di}$, we rewrite this as

$$g_d \, e^{(E_F - E_C + E_C - E_d)/kT} = g_d \, e^{(E_F - E_C)/kT} \, e^{E_{di}/kT} \ll 1 \, .$$

But $e^{(E_F - E_C)/kT} = n_0/N_C \approx N_d/N_C$ under the complete ionization assumption and for $N_d \gg n_i$. We conclude that the complete donor ionization assumption holds true, provided that

$$g_d e^{E_{di}/kT} \frac{N_d}{N_C} \ll 1 \;\; \Rightarrow \;\; N_d \ll N_C \frac{e^{-E_{di}/kT}}{g_d} \, .$$

Similarly, for acceptor doping, the condition of complete acceptor ionization, $N_a^- \approx N_a$, is fulfilled if

$$N_a \ll N_V \frac{e^{-E_{ai}/kT}}{g_a} \, .$$

At the room temperature and at typical values of impurity ionization energies, the ratios $e^{-E_{di}/kT}/g_d$ and $e^{-E_{ai}/kT}$ are of the order of 0.1. Then, the full ionization of impurities is achieved if the impurity concentrations $N_{d,a}$ do not exceed a few percent of the effective densities of states $N_{C,V}$.

As the temperature is lowered, however, the exponential terms in the numerator, $e^{-E_{di}/kT}$ and $e^{-E_{ai}/kT}$, quickly go to zero. Also the effective densities of states decrease on cooling as $N_{C,V} \propto T^{3/2}$. This means that the complete ionization assumption breaks down at low temperatures.

5.11.2 Intrinsic Regime

Intrinsic carrier concentration, $n_i(T)$, increases with temperature, whereas the doping concentrations, N_a and N_d, are temperature-independent. Hence, when the temperature exceeds a certain value, denoted as T_i, intrinsic concentration becomes much greater than the difference of the impurity concentrations, $n_i \gg |N_d - N_a|$. Then, the difference $N_d - N_a$ can be neglected in the expression $n_0 = \frac{N_d - N_a}{2} + \sqrt{\left(\frac{N_d - N_a}{2}\right)^2 + n_i^2}$, and charge carrier concentrations become approximately equal the intrinsic concentration:

$$n_0(T) \approx p_0(T) \approx n_i(T) = \sqrt{N_C(T)N_V(T)} e^{-E_g/(2kT)} \, .$$

It increases with temperature.

Let us now establish the temperature T_i, at which the complete ionization regime ends and the intrinsic regime begins. At this temperature, intrinsic concentration should be comparable to the magnitude of the difference $N_d - N_a$, which we assume to be non-zero:

$$n_i(T_i) = |N_d - N_a|, \quad \text{or} \quad \sqrt{N_C(T_i)N_V(T_i)}e^{-E_g/(2kT_i)} = |N_d - N_a|.$$

Rearranging the terms in this equation, we find

$$kT_i = \frac{E_g}{\ln \frac{N_C(T_i)N_V(T_i)}{(N_d - N_a)^2}}.$$

Recalling that $N_{C,V}(T) = N_{C,V}(T_0)(T/T_0)^{3/2}$, where the reference temperature is usually taken to be $T_0 = 300\,\mathrm{K}$, we finally have an equation, from which T_i can be found:

$$T_i = \frac{E_g/k}{\ln \left[\frac{N_C(T_0)N_V(T_0)}{(N_d - N_a)^2} \left(\frac{T_i}{T_0} \right)^3 \right]}.$$

One can solve this equation for T_i analytically using the so-called Lambert W-function.[2] Alternatively, an iterative solution can be performed, namely, starting with $T_i = T_0$ and plugging in this value into the right-hand side of the last equation to obtain a better estimate for T_i, which is again plugged in into the right-hand side to get a new T_i. This procedure is repeated until the results converge. In practice, only two or three iterations are sufficient to find T_i to a good accuracy. For the typical semiconductors and typical doping levels, the intrinsic regime begins at T_i exceeding the room temperature by a few hundred Kelvins.

5.11.3 Carrier Concentration in a Semiconductor with One Type of Doping at Not Too High Temperatures

Consider the case when only one type of doping impurities—assume for definiteness donors—is present. Let us focus on the practically interesting case when the temperature is small enough for the intrinsic concentration to be negligible in comparison to the concentration of the ionized donors:

$$N_a = 0, \quad N_d^+(T) \gg n_i(T) \text{ at } T < T_i.$$

---

[2] Lambert's function $W(x)$ is the solution of $W(x)e^{W(x)} = x$.

We do not assume complete ionization. In view of the above inequality, the concentration of the conduction electrons is given by

$$n_0 = N_d^+ = \frac{N_d}{1 + g_d e^{(E_F - E_d)/kT}}$$

to a very good accuracy. Because $n_0 = N_C e^{(E_F - E_C)/kT}$, we obtain an equation, from which Fermi energy can be found:

$$N_C e^{(E_F - E_C)/kT} = \frac{N_d}{1 + g_d e^{(E_F - E_d)/kT}} .$$

Denoting

$$x = e^{(E_F - E_C)/kT},$$

we rewrite our equation in terms of this new variable: $N_C x = \frac{N_d}{1 + x g_d e^{E_{di}/kT}}$. Multiplying both sides by the denominator of the right-hand side, we obtain a quadratic equation for x:

$$N_C x (1 + x g_d e^{E_{di}/kT}) = N_d .$$

Its solution reads

$$x = \sqrt{\frac{e^{-2E_{di}/kT}}{4 g_d^2} + \frac{N_d}{g_d N_C} e^{-E_{di}/kT}} - \frac{e^{-E_{di}/kT}}{2 g_d} .$$

By the very definition of x, Fermi energy and electron concentration can be immediately expressed in terms of x as

$$E_F = E_C + kT \ln x , \quad n_0 = N_C x , \quad p_0 = \frac{n_i^2}{N_C x} .$$

Here, the hole concentration follows from the mass action law.

The case when only acceptor-type impurities are present with $N_a(T) \gg n_i(T)$, $N_d = 0$ can be treated analogously and is left for the reader as an exercise.

5.11.4 Electron Freeze-Out Regime

At the thermal energies kT below the impurity ionization energy, only a fraction of the impurity atoms are ionized, and this fraction decreases to zero as the temperature is decreased. This ionization regime is called the freeze-out regime.

Let us consider for simplicity the case when only donor impurities are present. Then, we can use the formula for n_0 obtained in the previous section. In that formula, further approximations can be made. First, we note that under the square root, we have

$$\frac{e^{-2E_{di}/kT}}{4g_d^2} \ll \frac{N_d}{g_d N_C} e^{-E_{di}/kT} \quad \text{at} \quad kT \ll E_{di} .$$

This is so, because the ratio $E_{di}/kT \gg 1$ is a large number, implying that $e^{-E_{di}/kT} \ll 1$, and hence $e^{-2E_{di}/kT} \ll e^{-E_{di}/kT}$. As for the prefactor $N_d/(g_d N_C)$, it is also large in view of the fact that $N_C(T) \propto T^{3/2}$ decreases on cooling, and hence, $N_d/(2N_C(T))$ increases. The expression for electron concentration then simplifies to

$$n_0(T) = N_C(T) \left(\sqrt{\frac{N_d}{g_d N_C(T)}} e^{-E_{di}/(2kT)} - \frac{e^{-E_{di}/kT}}{2g_d} \right) .$$

By the same reasoning, we can neglect the second term in this expression relative to the first one and obtain

$$n_0(T) = \sqrt{\frac{N_d N_C(T)}{g_d}} e^{-E_{di}/(2kT)} .$$

The Fermi energy in the freeze-out regime is given by

$$E_F = E_C - \frac{E_{di}}{2} + \frac{kT}{2} \ln \frac{N_d}{g_d N_C}$$

as can be obtained from the expression $n_0 = N_C e^{(E_F - E_C)/kT}$ and the approximate formula for n_0 just obtained. Note that in the limit of zero temperature, Fermi energy goes to the limit

$$E_F \to E_C - \frac{E_{di}}{2} = \frac{E_C + E_d}{2} \quad \text{as } T \to 0$$

in full agreement with our observation made earlier that at zero temperature, Fermi energy is just in between the highest occupied state (donor level) and the lowest unoccupied state (conduction band edge), see Sect. 3.7.

We can establish the temperature T_f that separates the freeze-out regime from the complete ionization regime. At this temperature, the two contributions under the square root in the expression for n_0 from the previous section must be equal:

$$\frac{e^{-2E_{di}/kT_f}}{4g_d^2} = \frac{N_d}{g_d N_C(T_f)} e^{-E_{di}/kT_f} .$$

Taking the natural logarithm of both sides and rearranging the terms, we obtain

$$kT_f = \frac{E_{di}}{\ln \frac{N_C(T_f)}{4g_d N_d}}$$

with

$$N_C(T_f) = N_C(T_0) \left(\frac{T_f}{T_0}\right)^{3/2} ,$$

$T_0 = 300\,\text{K}$ being the standard reference temperature. Thus

$$T_f = \frac{E_{di}/k}{\ln\left[\frac{N_C(T_0)}{4g_d N_d} \left(\frac{T_f}{T_0}\right)^{3/2}\right]} .$$

This equation can be solved analytically using Lambert's W-function or iteratively by first setting, e.g., $T_f = T_0$ (or any other reasonable first guess, e.g., $T_f = 0.1 E_{di}/k$), and then obtaining the next improved result for T_f according to its right-hand side until convergence is reached.

5.12 Numerical Determination of Fermi Energy and Carrier Concentrations

One possible way to find E_F, n_0, and p_0 from the charge neutrality equation, $n_0 + N_a^- = p_0 + N_d^+$, and mass action law, $p_0 = n_i^2/n_0$, is as follows. Assume for definiteness that the semiconductor is of n-type. Then, by applying this same procedure used to treat the complete ionization regime, we find the electron and hole concentrations

$$n_0 = \frac{N_d^+ - N_a^-}{2} + \sqrt{\left(\frac{N_d^+ - N_a^-}{2}\right)^2 + n_i^2} , \quad p_0 = \frac{n_i^2}{n_0}$$

and the Fermi energy

$$E_F = E_C + kT \ln\left(\frac{N_d^+ - N_a^-}{2N_C} + \sqrt{\left(\frac{N_d^+ - N_a^-}{2N_C}\right)^2 + \frac{n_i^2}{N_C^2}}\right) .$$

The expressions for N_d^+ and N_a^- have been established earlier in Sect. 5.8.

The right-hand side of the last equation contains the ionized impurity concentrations, which themselves depend on the Fermi energy. This equation can be solved iteratively. Starting with some initial trial value of Fermi energy, one can determine N_d^+ and N_a^- corresponding to this initial value of E_F; upon substitution of the ionized impurity concentrations so obtained into the last equation, one finds a new, more accurate estimate for E_F, which is again used to find better estimates for N_d^+ and N_a^-. This procedure can be repeated many times until the desired accuracy is achieved. In practice, the number of iterations is of the order of 10.

For a p-type semiconductor, the same procedure can be performed, but with the interchanged $n_0 \leftrightarrow p_0$ and $N_d^+ \leftrightarrow N_a^-$.

Shown in Fig. 5.4 are the results of numerical and analytical calculations of the Fermi energy and carrier concentrations in a hypothetical semiconductor with the parameters indicated in the figure caption. For these parameters, the intrinsic temperature can be found to be $T_i = 562\,\mathrm{K}$, and the freeze-out temperature is $T_f = 112\,\mathrm{K}$. These two temperatures are marked by vertical lines in the figure. The agreement between the analytical and numerical results at $0 < T < T_i$ is smaller than the line thickness.

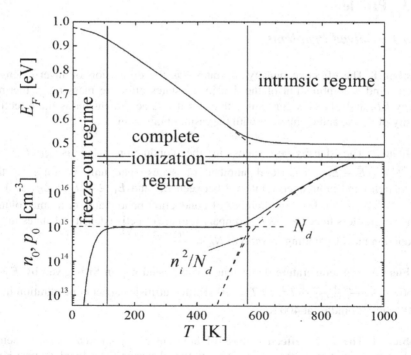

Fig. 5.4 The results of numerical calculations of the Fermi energy and electron and hole equilibrium concentrations in a hypothetical n-type semiconductor with the band gap $E_g = 1\,\mathrm{eV}$, the effective densities of states at $T_0 = 300\,\mathrm{K}$ taken to be $N_C(T_0) = N_V(T_0) = 10^{19}\,\mathrm{cm}^{-3}$, donor concentration $N_d = 10^{15}\,\mathrm{cm}^{-3}$, donor ionization energy $E_{di} = 0.05\,\mathrm{eV}$ and degeneracy factor $g_d = 2$, and no acceptors, $N_a = 0$

It is seen that Fermi energy, measured relative to the top of the valence band, monotonically decreases with temperature and finally saturates at $T > T_i$ at the level of 0.5 eV, i.e., in the middle of the band gap.

As for the charge carrier concentrations, three regimes can be distinguished. (1) At low temperatures $T < 100$ K, thermal energy is insufficient to ionize impurity atoms, and let alone to promote electrons from the valence to the conduction band. As a result, the freeze-out of conduction electrons occurs, where both electron and hole concentrations go to zero on cooling. (2) In the temperature range between about 100 K and 500 K, the majority electron concentration has a plateau at the level $n_0 = N_d = 10^{15}$ cm$^{-3} \gg n_i$, whereas the minority hole concentration is much smaller than this value. This regime corresponds to complete ionization of donors. (3) At even higher temperatures, $T > 500$ K, intrinsic carrier concentration becomes comparable to and eventually starts to dominate over N_d on heating. In this intrinsic regime, characterized by the inequality $n_i(T) \gg N_d$, we have $n_0 \approx p_0 \approx n_i$, independent of the doping level.

5.13   Problems

5.13.1   Solved Problems

Problem 1 The effective density of states has the dimension of inverse length cubed. Find a combination of the density of states effective mass $m_{e,h}^*$, thermal energy kT, and Planck's constant h that has the same dimension as the effective density of states, and compare it with the result obtained in Sect. 5.2.

Problem 2 The density of states in the conduction band is $g_C(E) = \frac{4\pi(2m_e^*)^{3/2}}{h^3}\sqrt{E - E_C}$. The total number of states (per unit volume) in the conduction band in an energy interval between E_C and $E_C + \Delta E$ is $Z_C(\Delta E) = \int_{E_C}^{E_C+\Delta E} dE\, g_C(E)$. Find the number of conduction band states in a unit volume with energies less than $E_C + kT$. Compare it to the effective density of states in the conduction band by finding its ratio to $N_C = \frac{2(2\pi\, m_e^*\, kT)^{3/2}}{h^3}$.

Problem 3 The temperature dependence of the band gap in Si is given by $E_g = 1.166\,\text{eV} - \frac{4.77 \cdot 10^{-4}\,\text{eV/K}}{T+636\,\text{K}}T^2$. At $T = 300$ K, the intrinsic carrier concentration in Si is 10^{10} cm^{-3}. Find n_i at 400 K.

Problem 4 The DOS effective mass of holes in a hypothetical intrinsic semi-conductor is ten times larger than that of the electrons. The band gap is 1 eV, independent of temperature. At $T = 300$ K, the intrinsic concentration is 10^9 cm^{-3}.

(a) Find the difference $E_{Fi} - E_V$ at 300 K.
(b) Find the effective densities of states N_C and N_V at 300 K.
(c) At what temperature is Fermi energy $0.6\,E_g$ above the valence band edge?

Problem 5 A Si sample is doped with boron (ionization energy $E_{ai} = 45$ meV, degeneracy = 4) at $N_a = 10^{16}\,\text{cm}^{-3}$. Take into account the temperature dependence of the band gap. The semiconductor parameters needed to solve the problem can be found in Appendix A.3.

Problem 6 A p-type semiconductor has the band gap $E_g = 0.8\,\text{eV}$, the electron and hole effective masses are the same, and $n_i = 2 \cdot 10^9\,\text{cm}^{-3}$. The temperature is $T = 300\,\text{K}$ and $N_d = 0$. At what concentration of acceptors will Fermi energy be above the valence band edge by $\Delta E = 0.2\,\text{eV}$?

Problem 7 A compensated semiconductor with the intrinsic carrier density $n_i(300\,\text{K}) = 10^{10}\,\text{cm}^{-3}$ and $E_g = 0.9\,\text{eV}$ is doped with $N_d = 10^{16}\,\text{cm}^{-3}$ donors and $N_a = 10^{12}\,\text{cm}^{-3}$ acceptors. Determine Fermi energy relative to the intrinsic Fermi energy at $T = 200\,\text{K}$ in the complete ionization limit.

5.13.2 Practice Problems

Problem 1 Derive the hole concentration p_0 and the respective effective density of states N_V following the same steps as for electrons in Sect. 5.2.

Problem 2 At $T = 300\,\text{K}$, the effective density of electron states in a semiconductor is $10^{20}\,\text{cm}^{-3}$. What is the effective density of states at 200 K and at 400 K?

Problem 3 Estimate electron concentration (= hole concentration) in an intrinsic semiconductor with $E_g = 0.7\,\text{eV}$ and in a dielectric with $E_g = 7\,\text{eV}$ at 300 K. Take a reasonable value for N_C and N_V, e.g., $10^{19}\,\text{cm}^{-3}$.

Problem 4 The concentration of holes in a p-type semiconductor (no donors) exceeds the concentration of electrons by 25 times. Find the ratio of the concentration of ionized acceptors to the intrinsic concentration.

Problem 5 A Si sample is doped with a donor impurity at $N_d = 10^{17}\,\text{cm}^{-3}$. At the temperature $T = 30\,\text{K}$, electron concentration in this sample was measured to be $n_0 = 10^{12}\,\text{cm}^{-3}$. Find the impurity ionization energy E_{di}. Hint: $n_0 = N_d^+ \gg n_i \gg p_0$ at 30 K.

Problem 6 A compensated semiconductor with the band gap of 0.8 eV is doped with equal concentrations of donors and acceptors. The respective ioniza-

energies are $E_{di} = 0.05\,\text{eV}$ and $E_{ai} = 0.02\,\text{eV}$. Find the Fermi energy relative to the top of the valence band, $E_F - E_V$, in this material at absolute zero temperature.

5.13.3 Solutions

Problem 1 For definiteness, let us focus on the conduction electrons; the discussion of holes is similar. To express N_C in terms of m_e^* and kT, we consider the units in which these parameters are measured (we use SI units in this problem):

$$[N_C] = \text{m}^{-3}\,,\quad [m_e^*] = \text{kg}\,,\quad [kT] = \text{J} = \frac{\text{kg} \cdot \text{m}^2}{\text{s}^2}\,.$$

We may also need Planck's constant, measured in

$$[h] = \text{J} \cdot \text{s} = \frac{\text{kg} \cdot \text{m}^2}{\text{s}}\,.$$

We express N_C as a product

$$N_C \propto (m_e^*)^\alpha\,(kT)^\beta\,h^\gamma$$

with unknown α, β, and γ and a non-dimensional prefactor. The dimension of this product is

$$[N_C] = \text{kg}^\alpha \cdot \left(\frac{\text{kg} \cdot \text{m}^2}{\text{s}^2}\right)^\beta \cdot \left(\frac{\text{kg} \cdot \text{m}^2}{\text{s}}\right)^\gamma = \text{kg}^{\alpha+\beta+\gamma}\,\text{m}^{2(\beta+\gamma)}\,\text{s}^{-(2\beta+\gamma)}\,.$$

It must be equal to m^{-3}. Hence, we must have

$$\alpha + \beta + \gamma = 0\,,\quad 2(\beta + \gamma) = -3\,,\quad 2\beta + \gamma = 0\,.$$

Subtracting the third equation from the second, we get $\gamma = -3$. Substitution of this value of γ back into the second equation gives $\beta = 3/2$. Finally, substitution of these values of β and γ into the first equation gives $\alpha = 3/2$. Hence, up to a non-dimensional constant

$$N_C \propto \frac{(m_e^* kT)^{3/2}}{h^3}\,.$$

Problem 2 Substitution of the formula for $g_C(E)$ into the definition of $Z_C(\Delta E)$ gives

$$Z_C(\Delta E) = \frac{4\pi (2m_e^*)^{3/2}}{h^3} \int_{E_C}^{E_C + \Delta E} dE \sqrt{E - E_C} = \frac{8\pi (2m_e^*)^{3/2}}{3h^3} \Delta E^{3/2} .$$

For $\Delta E = kT$, we have

$$Z_C(kT) = \frac{8\pi (2m_e^* kT)^{3/2}}{3h^3} .$$

The ratio of this parameter to N_C has the value

$$\frac{Z_C(kT)}{N_C} = \frac{8\pi \cdot 2^{3/2}/3}{2(2\pi)^{3/2}} = \frac{4}{3\sqrt{\pi}} = 0.752 .$$

This is close to 1. This result confirms that the effective density of states can be interpreted as the concentration of states in the conduction band whose energies are higher than E_C by approximately one thermal energy kT.

Problem 3 The intrinsic carrier concentration is given by

$$n_i(T) = \sqrt{N_C(T)N_V(T)} \, e^{-E_g(T)/(2kT)} ,$$

where the effective densities of states in the conduction and valence bands are proportional to $T^{3/2}$. We need to recalculate the square root of their product, $\sqrt{N_C(T)N_V(T)} \propto T^{3/2}$ from the reference temperature $T_0 = 300\,\mathrm{K}$ to a different temperature T:

$$\sqrt{N_C(T)N_V(T)} = \sqrt{N_C(T_0)N_V(T_0)} \left(\frac{T}{T_0}\right)^{3/2} .$$

Likewise, we need to recalculate the exponential in the expression for n_i:

$$e^{-E_g(T)/(2kT)} = e^{-E_g(T_0)/(2kT_0)} \, e^{E_g(T_0)/(2kT_0) - E_g(T)/(2kT)} .$$

Multiplying these two expressions together, we find

$$n_i(T) = n_i(T_0) \left(\frac{T}{T_0}\right)^{3/2} e^{E_g(T_0)/(2kT_0) - E_g(T)/(2kT)} .$$

It remains to plug in the numbers. We start with the band gap at $T_0 = 300\,\mathrm{K}$ and $T = 400\,\mathrm{K}$:

$$E_g(300\,\mathrm{K}) = 1.166 - \frac{4.77 \cdot 10^{-4}}{300 + 636} 300^2 = 1.120\,\mathrm{eV} ,$$

$$E_g(400\,\text{K}) = 1.166 - \frac{4.77 \cdot 10^{-4}}{400 + 636} 400^2 = 1.092\,\text{eV}\ .$$

Next, we find the thermal energy at $T = 400\,\text{K}$, given that $kT_0 = 0.02585\,\text{eV}$:

$$kT = 0.02585 \cdot \frac{400}{300} = 0.03447\,\text{eV}\ .$$

Finally, we determine the intrinsic concentration at 400 K:

$$n_i(400\,\text{K}) = 10^{10}\,(400/300)^{3/2}\,e^{1.120/(2\cdot0.02585)-1.092/(2\cdot0.03447)} = 5.21 \cdot 10^{12}\,\text{cm}^{-3}\ .$$

Problem 4

(a) Intrinsic Fermi energy happens to be exactly in the middle of the band gap only if the electron and hole effective masses are the same; this is not the case in most semiconductors. When the effective masses are different, we have a general expression

$$E_{Fi} = \frac{E_C + E_V}{2} + \frac{3kT}{4}\ln\frac{m_h^*}{m_e^*}\ ,$$

from which it follows that

$$E_{Fi} - E_V = \frac{E_g}{2} + \frac{3kT}{4}\ln\frac{m_h^*}{m_e^*}\ .$$

Numerically,

$$E_{Fi} - E_V = 0.5 + 0.75 \cdot 0.02585 \cdot \ln(10) = 0.545\,\text{eV}\ .$$

(b) We know that the effective density of states is proportional to the effective mass of the charge carriers in the respective band raised to the power $3/2$:

$$N_C \propto (m_e^*)^{3/2}\ ,\quad N_V \propto (m_v^*)^{3/2}\ .$$

Knowing that holes are 10 times as heavy as the electrons, we can find the ratio of the effective densities of states as

$$\frac{N_C}{N_V} = \left(\frac{m_e^*}{m_h^*}\right)^{3/2}\ .$$

On the other hand, the expression for n_i contains the product $N_C N_V$:

$$n_i^2 = N_C N_V e^{-E_g/kT} \ \Rightarrow\ N_C N_V = n_i^2 e^{E_g/kT}\ .$$

Multiplying the last two equations together, we obtain the effective density of states in the conduction band:

$$N_C^2 = \left(\frac{m_e^*}{m_h^*}\right)^{3/2} n_i^2 e^{E_g/kT} \quad \Rightarrow \quad N_C = \left(\frac{m_e^*}{m_h^*}\right)^{3/4} n_i^2 e^{E_g/(2kT)} .$$

The expression for N_V is obtained by interchanging electron and hole effective masses:

$$N_V = \left(\frac{m_h^*}{m_e^*}\right)^{3/4} n_i^2 e^{E_g/(2kT)} .$$

Numerically,

$$N_C = \frac{10^9 \cdot e^{0.5/0.02585}}{10^{3/4}} = 4.47 \cdot 10^{16} \, \text{cm}^{-3} ,$$

$$N_V = 10^{3/4} \cdot 10^9 \cdot e^{0.5/0.02585} = 1.41 \cdot 10^{18} \, \text{cm}^{-3} .$$

(c) In the second formula from the solution to part (a), we set $E_{Fi} - E_V = cE_g$, where $c = 0.6$:

$$cE_g = \frac{E_g}{2} + \frac{3kT}{4} \ln \frac{m_h^*}{m_e^*} .$$

Solving this for kT, we find the thermal energy corresponding to the temperature at which the Fermi level is higher than the valence band energy by a desired amount:

$$kT = \frac{4}{3} E_g \frac{c - 1/2}{\ln \frac{m_h^*}{m_e^*}} .$$

Numerically, this is

$$kT = \frac{4}{3} \cdot (1 \, \text{eV}) \cdot \frac{0.1}{\ln(10)} = 0.05791 \, \text{eV} .$$

To convert thermal energy into temperature, we divide it by 0.02585 eV and multiply by 300 K:

$$T = \frac{0.05791}{0.02585} \cdot 300 = 672 \, \text{K} .$$

Problem 5 The concentration of ionized acceptors is given by a general formula

$$N_a^- = \frac{N_a}{1 + g_a e^{(E_a - E_F)/kT}} \, ,$$

which can be transformed into

$$x = \frac{N_a^-}{N_a} = \frac{1}{1 + g_a e^{(E_a - E_V + E_V - E_F)/kT}} = \frac{1}{1 + g_a e^{E_{ai}/kT} e^{(E_V - E_F)/kT}} \, .$$

In the second exponential in the denominator, we recognize the concentration of holes:

$$p_0 = N_V \, e^{(E_V - E_F)/kT} \, .$$

So, we can write $e^{(E_V - E_F)/kT} = \frac{p_0}{N_V}$, giving

$$x = \frac{1}{1 + g_a e^{E_{ai}/kT} \frac{p_0}{N_V}} \, .$$

On the other hand,

$$p_0 = \frac{N_a^-}{2} + \sqrt{\left(\frac{N_a^-}{2}\right)^2 + n_i^2} \, .$$

It is reasonable to assume that at low temperatures, the concentration of ionized acceptors greatly exceeds the intrinsic concentration, $N_a^- \gg n_i$, allowing us to approximate

$$p_0 = N_a^- = x N_a$$

to a good accuracy. Then, substitution of this expression into the expression for x just obtained gives

$$x = \frac{1}{1 + g_a \frac{N_a}{N_V} e^{E_{ai}/kT} x} = \frac{1}{1 + Ax} \, , \quad A = g_a \frac{N_a}{N_V} e^{E_{ai}/kT} \, .$$

This is a quadratic equation for the fraction of ionized acceptors, x:

$$Ax^2 + x - 1 = 0 \quad \Rightarrow \quad x = -\frac{1}{2A} + \sqrt{\frac{1}{4A^2} + \frac{1}{A}} \, .$$

The possibility of a negative sign in front of the square root has been discarded because x must be positive. It remains to plug in the numbers.

(a) At $T = 200\,\mathrm{K}$, $kT = 0.02585 \cdot 200/300 = 0.01723\,\mathrm{eV}$. The effective density of states in the valence band at this temperature is

$$N_V(200\,\mathrm{K}) = N_V(300\,\mathrm{K})\left(\frac{200}{300}\right)^{3/2} = 1.116 \cdot 10^{19}\,\mathrm{cm}^{-3}\,.$$

This gives the parameter A to be

$$A = 4 \cdot \frac{10^{16}}{1.116 \cdot 10^{19}} \cdot e^{0.045/0.01723} = 0.04883\,.$$

Using this value, we find $x = 0.955$, i.e., 95.5% of acceptors are ionized.

(b) At $T = 50\,\mathrm{K}$, $kT = 0.02585 \cdot 50/300 = 0.004308\,\mathrm{eV}$. The effective density of states in the valence band at this temperature is

$$N_V(50\,\mathrm{K}) = N_V(300\,\mathrm{K})\left(\frac{50}{300}\right)^{3/2} = 1.395 \cdot 10^{18}\,\mathrm{cm}^{-3}\,.$$

This gives the parameter A to be

$$A = 4 \cdot \frac{10^{16}}{1.395 \cdot 10^{18}} \cdot e^{0.045/0.004308} = 986.3\,.$$

Using this value, we find $x = 0.0313$, i.e., 3.13% of acceptors are ionized.

Problem 6 Since $m_e^* = m_h^*$, the intrinsic Fermi level lies exactly in the middle of the band gap, i.e.,

$$E_{Fi} - E_V = \frac{E_g}{2} = 0.4\,\mathrm{eV}\,.$$

We wish to have

$$E_F - E_V = 0.2\,\mathrm{eV}\,,$$

which implies that

$$E_{Fi} - E_F = 0.2\,\mathrm{eV}\,.$$

We can figure out the concentration of holes at this value of the Fermi energy:

$$p_0 = N_V\, e^{(E_V - E_F)/kT} = N_V\, e^{(E_V - E_{Fi} + E_{Fi} - E_F)/kT} = n_i\, e^{(E_{Fi} - E_F)/kT}\,,$$

where we used the one of the previously obtained expressions for the intrinsic concentration, $n_i = N_V\, e^{(E_V - E_{Fi})/kT}$. Numerically,

$$p_0 = 2 \cdot 10^9 \cdot e^{0.2/0.02585} = 4.583 \cdot 10^{12}\,\mathrm{cm}^{-3} \ .$$

This value is much bigger than n_i. From this, we conclude that to a good accuracy $p_0 = N_a^- = N_a$, as practically all acceptors are ionized at 300 K. Hence, $N_a = 4.58 \cdot 10^{12}\,\mathrm{cm}^{-3}$.

Problem 7 First, we find the intrinsic concentration at $T = 200\,\mathrm{K}$ from its value at $T_0 = 300\,\mathrm{K}$:

$$n_i(T) = n_i(T_0) \left(\frac{T}{T_0} \right)^{3/2} e^{\frac{E_g}{2kT_0} - \frac{E_g}{2kT}} = n_i(T_0) \left(\frac{T}{T_0} \right)^{3/2} e^{\frac{E_g}{2kT_0}\left(1 - \frac{T_0}{T}\right)} \ .$$

Numerically,

$$n_i(200\,\mathrm{K}) = 10^{10} \cdot \left(\frac{200}{300} \right)^{3/2} e^{\frac{0.9}{2 \cdot 0.02585}\left(1 - \frac{300}{200}\right)} = 9.031 \cdot 10^5\,\mathrm{cm}^{-3} \ .$$

This number is much smaller than the difference $N_d - N_a$. Hence, in this n-type semiconductor, the concentration of electrons is to a very good accuracy

$$n_0 = N_d - N_a = 9.999 \cdot 10^{15}\,\mathrm{cm}^{-3} \ .$$

Electron concentration is related to the intrinsic concentration by

$$n_0 = N_C\,e^{-(E_C - E_F)/kT} = N_C\,e^{-(E_C - E_{Fi} + E_{Fi} - E_F)/kT} = n_i\,e^{(E_F - E_{Fi})/kT} \ ,$$

where we used one of the previously obtained expressions for the intrinsic concentration, $n_i = N_C\,e^{-(E_C - E_{Fi})/kT}$. We obtain

$$E_F - E_{Fi} = kT\,\ln \frac{n_0}{n_i} \ .$$

Numerically,

$$E_F - E_{Fi} = 0.02585 \cdot \frac{200}{300} \cdot \ln \frac{9.999 \cdot 10^{15}}{9.031 \cdot 10^5} = 0.399\,\mathrm{eV} \ .$$

Chapter 6
Carrier Concentration and Electric Potential

6.1 Electron and Hole Concentrations in a Non-uniform Electric Potential

If the electric potential $\phi(\vec{r})$ inside a semiconductor is non-uniform, the charge carrier concentrations will differ from the values n_0 and p_0 obtained in the previous chapter. The reason is simple: the mobile charge carriers get redistributed in the material in response to the electric field $\vec{\mathcal{E}} = -\vec{\nabla}\phi(\vec{r})$ produced by the spatial inhomogeneity of the potential. An overall motion of the charge carriers means that they are not in thermal equilibrium. In this chapter, we assume that the semiconductor is not a part of a closed circuit; hence, the electric current due to a redistribution of electrons and holes eventually stops, and the system comes to a new equilibrium state.

Electrons tend to migrate from the regions with the lower electric potential to the regions with the higher potential. Likewise, holes migrate from the high-potential to the low-potential regions. As the charge carriers move from one region to another, they leave behind the ionized impurities. As a result, an internal disbalance of positive and negative charges is created, which itself produces an internal electric field. The internal field counteracts the external field, so that the total electric field in the material eventually becomes zero and further migration of the free charge carriers stops.

This reasoning shows that the electric potential inside the semiconductor and the distribution of charge carriers affect each other, which makes finding the new equilibrium state a non-trivial task. We begin by establishing the effect of a non-uniform potential $\phi(\vec{r})$ on the carrier concentrations. We mentally divide the material into small volume elements ΔV such that the electric potential in each element is approximately constant. The local concentration of electrons and holes is defined as the ratio of their average number in each element to the volume ΔV of that element.

© The Author(s), under exclusive license to Springer Nature Switzerland AG 2022
M. Evstigneev, *Introduction to Semiconductor Physics and Devices*,
https://doi.org/10.1007/978-3-031-08458-4_6

The energy of a charged particle due to the presence of an electric potential $\phi(\vec{r})$ is $q\phi(\vec{r})$. That is, the energy of an electron near the top of the conduction band changes from E_C to $E_C - e\phi(\vec{r})$, and the energy of a hole (from the holes' perspective) near the top of the valence band changes from $E_V^{(h)}$ to $E_V^{(h)} + e\phi(\vec{r})$. We can repeat the derivation from Sect. 5.2 to obtain the local concentrations of electrons and holes at a point $\vec{r}$ with the result

$$n_0(\vec{r}) = N_C e^{-(E_C - E_F - e\phi(\vec{r}))/kT} , \quad p_0(\vec{r}) = n_i^2/n_0(\vec{r}) = N_V e^{(E_V - E_F - e\phi(\vec{r}))/kT} .$$

It should be borne in mind that the electric potential is defined up to an arbitrary constant. If the potential is uniform, $\phi(\vec{r}) = \text{const}$, this constant should be set to zero, as the carrier concentrations should be given by the expressions from Sect. 5.2 in this case.

It is expedient to cast the expressions for $n_0(\vec{r})$ and $p_0(\vec{r})$ into a form that is independent of the choice of that constant. Namely, the electron and hole concentrations at a point $\vec{r}$ are related to their values at some reference point $\vec{r}_0$ by

$$n_0(\vec{r}) = n_0(\vec{r}_0)\, e^{(\phi(\vec{r}) - \phi(\vec{r}_0))/V_{th}} , \quad p_0(\vec{r}) = p_0(\vec{r}_0)\, e^{-(\phi(\vec{r}) - \phi(\vec{r}_0))/V_{th}} .$$

Here, in order to avoid using the letter e in the meaning of the elementary charge and the base of an exponential (e.g., writing $e^{e\phi(\vec{r})/kT}$ is a bit clumsy), we have introduced the thermal voltage

$$V_{th} = \frac{kT}{e} .$$

At the room temperature $T = 300\,\text{K}$, it has the value $V_{th} = 0.02585\,\text{V}$.

6.2 Poisson's Equation

A non-uniform distribution of electric charges inside a semiconductor due to, e.g., a non-uniform doping profile, results in an onset of an internal electric field, $\mathcal{E}(\vec{r})$, related to an electrostatic potential as $\mathcal{E}(\vec{r}) = -\vec{\nabla}\phi(\vec{r})$. The electric field depends on the distribution of the charges inside the material. Poisson's equation relates the electric potential to charge density. It is the familiar Gauss law expressed in the form of a differential equation.

For the sake of simplicity, we focus here on a special case when the electric charge density depends on the x-coordinate only and is uniform in the yz-plane: $\rho = \rho(x)$. Consider a thin slab of thickness Δx and base area A perpendicular to the x-axis, see Fig. 6.1. By Gauss' theorem, the electric field flux through its surface equals the charge inside the slab divided by the product $\epsilon_0\epsilon$, where ϵ is the relative dielectric constant of the material, $\epsilon_0 = 8.854 \cdot 10^{-14}\,\text{F/m}$ is the permittivity of

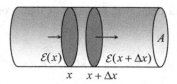

Fig. 6.1 Gaussian surface inside a semiconductor is a cylinder; the electric field flux through it arises due to a non-uniform distribution of electric charge in the x-direction

vacuum, and ϵ is the relative permittivity of the material:

$$\int d\vec{A} \cdot \vec{\mathcal{E}} = \frac{A \int_x^{x+\Delta x} dx' \, \rho(x')}{\epsilon_0 \epsilon} .$$

As the electric field has only the x-component, the flux through the slab is given by $\mathcal{E}(x + \Delta x)A - \mathcal{E}(x)A$. For very small Δx, we can neglect the x-dependence of the electric charge density inside the slab; hence, the total charge contained inside the slab is $\rho(x)A\Delta x$. We thus have

$$\mathcal{E}(x + \Delta x)A - \mathcal{E}(x)A = \frac{\rho(x)A \, \Delta x}{\epsilon_0 \epsilon} .$$

Dividing both sides by the slab volume $A \, \Delta x$ and taking the limit $\Delta x \to 0$, we have

$$\lim_{\Delta x \to 0} \frac{\mathcal{E}(x + \Delta x) - \mathcal{E}(x)}{\Delta x} = \frac{d\mathcal{E}(x)}{dx} = \frac{\rho(x)}{\epsilon_0 \epsilon} .$$

Because the electrostatic potential is non-uniform in the x-direction only, we have

$$\mathcal{E}(x) = -\frac{d\phi(x)}{dx} .$$

Hence, we arrive at Poisson's equation:

$$\frac{d^2\phi}{dx^2} = -\frac{\rho(x)}{\epsilon_0 \epsilon} .$$

In a semiconductor, the charge density is a sum of the positive charge densities of holes and ionized donors and the negative charge densities of electrons and ionized acceptors:

$$\rho(x) = e(p_0(x) + N_d^+(x) - n_0(x) - N_a^-(x)) .$$

For the sake of generality, we allowed the doping concentrations to depend on the coordinate x. In equilibrium, the local concentrations of carriers depend on the potential, as established in Sect. 6.1. Because of that, Poisson's equation is non-linear:

$$\frac{d^2\phi}{dx^2} = -\frac{e}{\epsilon_0\epsilon}\left(p_0(x_0)e^{-\phi(x)/V_{th}} + N_d^+(x) - n_0(x_0)e^{\phi(x)/V_{th}} - N_a^-(x)\right),$$

where x_0 is a reference point, where the potential has the value $\phi(x_0) = 0$. Its exact solution in a one-dimensional geometry will be presented in the end of this chapter. But also approximate solutions can be obtained when the surface potential is small compared to thermal voltage and when it is large, but not too high. These approximations are discussed below.

6.3   Approximate Solution of Poisson's Equation

6.3.1   Problem Formulation

Consider a thick semiconductor slab with a flat boundary in the yz-plane extending in the positive x-direction. The semiconductor finds itself in thermal equilibrium at the temperature sufficiently high for complete ionization of all doping impurities. For definiteness, we assume that the semiconductor is of p-type, as in Fig. 6.2a, with all acceptors ionized and no donors:

$$N_a^- \approx N_a \gg n_i , \quad N_d = 0 .$$

The behavior of the electrostatic potential is described by Poisson's equation

$$\frac{d^2\phi(x)}{dx^2} = -\frac{e}{\epsilon_0\epsilon}(p_0(x) - n_0(x) - N_a)$$

with

$$p_0(x) = N_a e^{-\phi(x)/V_{th}} , \quad n_0(x) = \frac{n_i^2}{p_0(x)} = \frac{n_i^2}{N_a}e^{\phi(x)/V_{th}} .$$

The electrostatic potential on the surface has a fixed value ϕ_S, and very far away from the surface, it is zero. The value of the surface potential is maintained by an external voltage applied to a metal electrode. In order to prevent the current from flowing, that electrode is separated from the semiconductor by a thin insulating layer, see Fig. 6.2. Note that due to the presence of this insulating layer and due to the fact that the metal and the semiconductor have different work functions, the voltage applied to the metal electrode differs from the value ϕ_S. In this chapter, we will simply assume ϕ_S to be given.

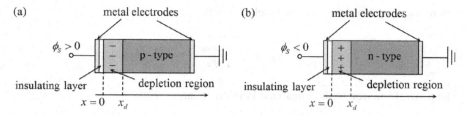

Fig. 6.2 The development of a depletion layer near when (**a**) a positive bias is applied to the surface of a p-type semiconductor and (**b**) a negative bias is applied to the surface of an n-type semiconductor

We wish to find out how the electrostatic potential $\phi(x)$ decreases from the initial value

$$\phi(0) = \phi_S$$

to the value far from the surface

$$\phi(\infty) = 0 .$$

Furthermore, because there is no electric field at infinity, we have a further condition

$$\frac{d\phi}{dx}\bigg|_{x=\infty} = 0 .$$

6.3.2   Debye Screening

First, let us focus on the situation when the surface potential is small compared to the thermal voltage. Then, we can approximate

$$p_0(x) = N_a e^{-\phi(x)/V_{th}} \approx N_a \left(1 - \frac{\phi(x)}{V_{th}}\right)$$

and neglect the electron concentration $n_0(x) \approx \frac{n_i^2}{N_a}\left(1 + \frac{\phi(x)}{V_{th}}\right)$ relative to the hole and acceptor concentrations. This allows us to linearize Poisson's equation with respect to $\phi(x)/V_{th}$ as

$$\frac{d^2\phi(x)}{dx^2} = \frac{eN_a}{\epsilon_0\epsilon V_{th}}\phi(x) = \frac{\phi(x)}{\lambda_D^2} ,$$

where we introduced the so-called Debye length

$$\lambda_D = \sqrt{\frac{\epsilon_0 \epsilon V_{th}}{e N_a}} \ .$$

The general solution of the linearized Poisson equation reads

$$\phi(x) = A e^{-x/\lambda_D} + B e^{x/\lambda_D} \ .$$

Due to the boundary condition $\phi(\infty) = 0$, we have to set B to zero. Next, considering the boundary condition at $x = 0$, we find that the parameter A is, then, the surface potential ϕ_S, yielding

$$\phi(x) = \phi_S e^{-x/\lambda_D} \ .$$

We see that the free charge carriers (holes in this case) redistribute themselves in the material in response to the external potential in such a way that the electric potential in the material quickly goes to zero. This phenomenon is called screening.

6.3.3 Depletion Approximation

A simple approximate expression for the potential $\phi(x)$ in the semiconductor can be obtained if the surface potential ϕ_S has the same sign as the majority charge carriers, i.e., $\phi_S > 0$ for a p-type semiconductor (Fig. 6.2a) and $\phi_S < 0$ for an n-type semiconductor (Fig. 6.2b). In this case, the majority carriers are pushed away from the surface. On the other hand, we assume that the magnitude of the surface potential is not large enough to attract a considerable amount of the minority carriers. As a result, a region is formed near the surface, where neither the majority nor the minority carriers compensate the uniform charge of the dopant ions. The charge density inside this region is close to the charge density of the dopants. It is negative in a p-semiconductor and positive in an n-semiconductor. This region is called the depletion region or the space-charge region (SCR).

Within the depletion approximation, the boundary of the SCR is viewed as sharp. That is, the true (but unknown) distribution of the electric charge density is approximated by a box distribution:

$$\rho(x) \approx \begin{cases} -e N_a & \text{for } x \le x_d \ , \\ 0 & \text{for } x > x_d \ , \end{cases}$$

where x_d is the yet unknown width of the depletion region, see Fig. 6.3. With this approximation for the charge density, Poisson's equation

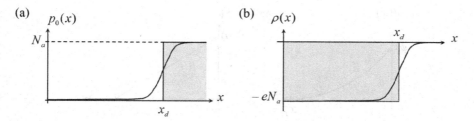

Fig. 6.3 (a) Schematic illustration of the hole concentration vs. the coordinate x (solid lines) and the depletion approximation for $p_0(x)$ (shaded area). (b) Schematic illustration of the charge density vs. coordinate (solid line) and its depletion approximation (shaded area)

$$\frac{d^2\phi(x)}{dx^2} = \begin{cases} \frac{eN_a}{\epsilon_0\epsilon} & \text{for } x \leq x_d \\ 0 & \text{for } x > x_d \end{cases}$$

is easy to solve both inside and outside the SCR. It should be borne in mind that the solutions of Poisson's equation obtained in the two regions, $x \leq x_d$ and $x > x_d$, must coincide at the boundary of the SCR, $x = x_d$.

Integrating Poisson's equation once with respect to x, we have

$$\frac{d\phi(x)}{dx} = \begin{cases} \frac{eN_a}{\epsilon_0\epsilon}(x - x_d) & \text{for } x \leq x_d \\ 0 & \text{for } x > x_d \end{cases}.$$

It is 0 at $x > x_d$ because the electric field $-d\phi/dx$ must vanish at $x \to \infty$. The integration constant in the expression for $d\phi/dx$ for $x \leq x_d$ is chosen so as to make sure that the derivative $d\phi/dx$ is continuous at $x = x_d$.

One more integration gives

$$\phi(x) = \begin{cases} \frac{eN_a}{2\epsilon_0\epsilon}(x - x_d)^2 & \text{for } x \leq x_d \text{ ,} \\ 0 & \text{for } x > x_d \text{ .} \end{cases}$$

Note that $\phi(x)$ is continuous at the SCR boundary $x = x_d$. We now can determine the width x_d of the depletion region from the surface value of the potential:

$$\phi(0) = \frac{eN_a x_d^2}{2\epsilon_0\epsilon} = \phi_s \implies x_d = \sqrt{\frac{2\epsilon_0\epsilon}{eN_a}\phi_s} \text{ .}$$

This completes the derivation.

Keeping in mind that the electric field induced in the semiconductor is $\mathcal{E}(x) = -d\phi(x)/dx$, we conclude that it decreases linearly from the maximal value

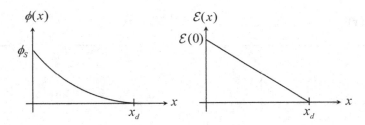

Fig. 6.4 Electric potential and electric field in the semiconductor obtained within the depletion approximation

$$\mathcal{E}(0) = \frac{eN_a x_d}{\epsilon_0 \epsilon} = \sqrt{\frac{2eN_a}{\epsilon_0 \epsilon} \phi_S}$$

at $x = 0$ to zero at $x = x_d$, see Fig. 6.4.

For an n-type semiconductor with $\phi_S < 0$, a similar analysis gives the same formulae, but with the acceptor concentration replaced by the donor concentration, $N_a \rightarrow N_d$, and the opposite sign of $\phi(x)$:

$$\phi(x) = \begin{cases} -\frac{eN_d}{2\epsilon_0 \epsilon}(x - x_d)^2 & \text{for } x \leq x_d \\ 0 & \text{for } x > x_d \end{cases} \quad, \quad x_d = \sqrt{\frac{2\epsilon_0 \epsilon}{eN_d}|\phi_S|} \; .$$

Combining the results for the p- and n-type semiconductors together, we express the electric potential in the depletion region as

$$\phi(x) = \phi_S \left(1 - \frac{x}{x_d}\right)^2 \;, \quad x_d = \sqrt{\frac{2\epsilon_0 \epsilon}{eN_{dop}}|\phi_S|} \;,$$

where $N_{dop} = N_d$ or N_a depending on the semiconductor type.

6.3.4 Validity Range of the Depletion Approximation

Let us now determine more precisely the validity range of the depletion approximation. As stated earlier, for this approximation to be valid, the concentration of both majority and minority carriers must be much smaller than the doping concentration. Assuming for definiteness the semiconductor to be p-type, we demand that

$$p_0(x) \ll N_a \quad \text{and} \quad n_0(x) \ll N_a \quad \text{for } x < x_d \; .$$

The charge carrier concentrations are related to the electric potential by

$$p_0(x) = p_0(\infty)e^{-\phi(x)/V_{th}} = N_a e^{-\phi(x)/V_{th}} \,,$$

$$n_0(x) = n_0(\infty)e^{\phi(x)/V_{th}} = \frac{n_i^2}{N_a}e^{\phi(x)/V_{th}} \,,$$

where we used the general formulas from Sect. 6.1 and the fact that $N_a \gg n_i$, which allows us to approximate to a good accuracy $p_0(\infty) = N_a$ and $n_0(\infty) = n_i^2/N_a$. We thus arrive at two conditions:

$$e^{\phi(x)/V_{th}} \gg 1 \,, \quad e^{\phi(x)/V_{th}} \ll \frac{N_a^2}{n_i^2}$$

within the depletion region $x < x_d$.

It is clear that the first condition cannot be fulfilled at all values of $x < x_d$. Indeed, as the coordinate increases from 0, the potential $\phi(x)$ decreases and eventually reaches the value of a few V_{th} at some coordinate x_0:

$$\phi(x_0) = \phi_S \left(1 - \frac{x_0}{x_d}\right)^2 = K V_{th} \,,$$

where $K \sim 2$ or 3. The depletion approximation actually becomes poor at

$$x > x_0 = x_d \left(1 - \sqrt{\frac{K V_{th}}{\phi_S}}\right) .$$

Note that $x_0 < x_d$. In order for the validity range of this approximation to be broad, x_0 must be only slightly smaller than x_d, which is only possible if

$$\phi_S \gg V_{th} \,.$$

This is the first validity criterion of the approximation. As usual, the strong inequality $\gg$ means that the surface potential must exceed the thermal voltage by at least 10 times.

However, the surface potential cannot be too large because otherwise the second condition,

$$e^{-\phi_S/V_{th}} \ll \frac{N_a^2}{n_i^2} \,,$$

will be violated. Thus, the depletion approximation is valid if the surface potential has the value within the range

$$V_{th} \ll \phi_S < 2V_{th} \ln \frac{N_a}{n_i} \,.$$

It should be much higher than the thermal voltage, but smaller than the so-called threshold potential

$$\phi_T = 2\frac{E_{Fi} - E_F}{e} = 2V_{th} \ln \frac{N_a}{n_i}$$

by at least a few multiples of the thermal voltage. For an n-type semiconductor with $\phi_S < 0$, a similar condition can be derived.

To summarize, both for n- and p-semiconductors, the depletion approximation is valid, provided that

$$V_{th} \ll |\phi_S| < 2V_{th} \ln \frac{N_{dop}}{n_i} \ ,$$

where the doping concentration N_{dop} is either N_d or N_a, depending on the semiconductor type. This double inequality can only be fulfilled if the doping level is sufficiently large. Namely, we must have

$$2V_{th} \ln \frac{N_{dop}}{n_i} \gg V_{th} \ \Rightarrow \ N_{dop} \gg n_i \ .$$

More precisely, N_{dop} must exceed the intrinsic concentration by at least a few hundred times.

If the magnitude of the surface potential exceeds the threshold potential ϕ_T, an inversion layer is formed near the semiconductor surface. In the inversion layer, the concentration of the minority carriers exceeds the doping level. As a result, the conductivity of the inversion layer is comparable to or may even exceed the conductivity of the bulk semiconductor. The depletion approximation cannot be applied to describe this effect, and one must resort to the exact solution of Poisson's equation.

6.4   Band Diagrams and Band Bending

In thermal equilibrium, Fermi level must be constant everywhere in the semiconductor. The non-uniformity of the carrier concentrations $n_0(x)$ and $p_0(x)$ in the presence of an x-dependent potential $\phi(x)$ implies that the position of the band edges E_V and E_C relative to the Fermi level E_F depends on the coordinate x. Namely, the valence band edge gets closer to the Fermi level in the direction of the hole concentration increase, and the conduction band edge gets closer to E_F in the direction of the electron concentration increase. Mathematically,

$$E_C(x) = E_C - e\phi(x) \ , \ \ E_V(x) = E_V - e\phi(x) \ ,$$

Fig. 6.5 Band bending in an n-type semiconductor when the surface potential ϕ_S is (**a**) negative and (**b**) positive

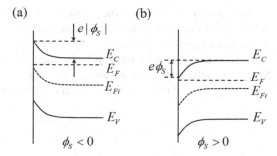

where E_C and E_V (without the explicit x-dependence) are the band energies far from the surface, where the electric potential is zero.

This is conveniently depicted in an energy band diagram, which shows how the important energy levels change with the spatial coordinate x. Figure 6.5 exemplifies the energy band diagrams of an n-type semiconductor with the negative and positive surface potential. If the surface potential is positive, the semiconductor bands bend down near the surface. If the surface potential is negative, the bands bend up. The size of band bending is $e|\phi_S|$. The direction of band bending has to do with the fact that the electron energy in the presence of the electric potential is shifted by the amount $-e\phi(x)$.

The energy band diagram does not have to be perfectly scale-accurate; however, it must show the relative arrangement of the important energies in a semiconductor. For example, if the semiconductor is of n-type, its Fermi level E_F far from the surface must be higher than the intrinsic Fermi level $E_{Fi} \approx (E_C + E_V)/2$; in other words, E_F must be closer to the conduction band edge than to the valence band edge. In a p-type semiconductor, E_F must be closer to the valence band than to the conduction band. Note that the position of the intrinsic Fermi level bends together with that band edges $E_C(x)$, $E_V(x)$. If the intrinsic Fermi level $E_{Fi}(x)$ crosses the semiconductor Fermi energy E_F, this means that the conductivity type of the semiconductor changes from the p- to the n-type at the point of crossing where $E_{Fi}(x) = E_F$.

6.5 Electric Potential in a Semiconductor from Poisson's Equation

6.5.1 Exact Solution of Poisson's Equation

Because the right-hand side of Poisson's equation strongly depends on the unknown electrostatic potential, its analytical solution may seem a formidable task even for complete ionization. Yet, Poisson's equation can be solved analytically under the complete ionization conditions.

For the sake of brevity of the formulae, we still assume that $N_a^- \approx N_a \gg n_i$ and $N_d = 0$. But the method developed here applies to arbitrary doping levels with both donor and acceptor impurities.

It is convenient to cast Poisson's equation, formulated in the beginning of Sect. 6.3, into a non-dimensional form. Let us introduce the non-dimensional potential $y(x)$ and the non-dimensional coordinate t according to

$$y(x) = \frac{\phi(x)}{V_{th}} , \quad t = \frac{x}{\lambda_D} .$$

Here, λ_D is Debye length defined in Sect. 6.3.2. It is also expedient to define the non-dimensional threshold potential from Sect. 6.3.4 as

$$y_T = \frac{\phi_T}{V_{th}} = 2\ln\frac{N_a}{n_i} = 2\frac{E_{Fi} - E_F}{kT} , \quad e^{y_T} = \frac{N_a^2}{n_i^2} .$$

With these definitions, Poisson's equation is conveniently written in the non-dimensional form:

$$\frac{d^2 y(t)}{dt^2} = -\left(e^{-y(t)} - e^{y(t)-y_T} - 1\right)$$

with the boundary conditions

$$y(0) = y_S = \frac{\phi_S}{V_{th}} , \quad y(\infty) = \frac{dy}{dt}(t = \infty) = 0 .$$

Poisson's equation is reminiscent of a Newtonian equation of motion for a unit-mass particle with position $y(t)$ in a potential $V(y)$:

$$\frac{d^2 y}{dt^2} = -\frac{dV(y)}{dy} .$$

The potential is obtained by integration:

$$\frac{dV(y)}{dy} = e^{-y} - e^{y-y_T} - 1 \Rightarrow V(y) = -e^{-y} - e^{y-y_T} - y .$$

This mechanical analogy allows us to use the law of energy conservation. Namely, multiplying both sides of the "equation of motion" with the "velocity" dy/dt, we obtain

$$\frac{dy}{dt}\frac{d^2 y}{dt^2} = -\frac{dy}{dt}\frac{dV(y)}{dy} \Rightarrow \frac{d}{dt}\left(\frac{1}{2}\left(\frac{dy}{dt}\right)^2 + V(y)\right) = 0 .$$

At any moment of time, the total "energy" of the particle

$$E = \frac{1}{2}\left(\frac{dy}{dt}\right)^2 + V(y) = \frac{1}{2}\left(\frac{dy}{dt}\right)^2 - e^{-y} - e^{y-y_T} - y$$

is a constant. The value of the "energy" E can be established from the boundary condition at $t = \infty$, where $y = 0$ and $dy/dt = 0$:

$$E = -1 - e^{-y_T} .$$

Substitution of this expression into the definition of "energy," we conclude that

$$\frac{1}{2}\left(\frac{dy}{dt}\right)^2 - e^{-y} - e^{y-y_T} - y = -1 - e^{-y_T} ,$$

and thus the velocity is given by

$$\frac{dy}{dt} = -\text{sign}(\phi_S)\sqrt{2\left(y + e^{-y} - 1 + e^{-y_T}\left(e^y - 1\right)\right)} .$$

The correct sign in front of the square root must be opposite to the sign of the surface potential ϕ_S. If $\phi_S > 0$, then the potential should be decreasing to 0 with distance from the surface, and thus $dy/dt < 0$. If $\phi_S < 0$, then the potential should be increasing with distance, and thus $dy/dt > 0$.

The last equation can be written in the equivalent form

$$dt = -\text{sign}(\phi_S)\frac{dy}{\sqrt{2\left(y + e^{-y} - 1 + e^{-y_T}\left(e^y - 1\right)\right)}} .$$

Integrating this from $t = 0$, where $y = y_S$, to some arbitrary t and the corresponding $y(t)$ gives

$$t = -\text{sign}(\phi_S)\frac{1}{\sqrt{2}}\int_{y_S}^{y(t)} \frac{dy'}{\sqrt{y' + e^{-y'} - 1 + e^{-y_T}\left(e^{y'} - 1\right)}} .$$

Finally, let us convert this result from the non-dimensional variables y and t into the dimensional ones, ϕ and x:

$$x = \frac{\lambda_D}{\sqrt{2}}\,\text{sign}(\phi_S)\int_{\phi(x)/V_{th}}^{\phi_S/V_{th}} \frac{dy}{\sqrt{y + e^{-y} - 1 + e^{-y_T}\left(e^y - 1\right)}} .$$

This completes the solution of Poisson's equation. The integral in this solution must be evaluated numerically. Instead of the x-dependent potential $\phi(x)$, we have the coordinate x at which a given value of the electrostatic potential ϕ is realized. That is not a problem: all that needs to be done is to plot the independent variable ϕ on the vertical axis and the dependent variable $x(\phi)$ on the horizontal axis.

Even though the solution obtained in this section is mathematically exact, it still relies on Boltzmann's approximation for the carrier concentrations. Therefore, it cannot be applied when the magnitude of the surface potential gets too large. More precisely, the Fermi energy must differ from the band edge energies $E_C(x) = E_C - e\phi(x)$ and $E_V(x) = E_V - e\phi(x)$ by at least a few thermal energies both inside of the material and on the surface. Assuming this to be the case far from the surface, we must have

$$E_C - e\phi_S - E_F > eV_{th} , \quad E_V - e\phi_S - E_F > eV_{th} .$$

If these conditions do not hold true, one cannot use Boltzmann's approximation for the charge carrier concentrations.

6.5.2 Numerical Results

The field effect refers to the possibility to control the concentration and the sign of the charge carriers near the surface of a semiconductor using an electric field. The field effect is at the heart of metal–oxide–semiconductor field effect transistor (MOSFET) operation, so let us discuss it in detail.

For definiteness, let us consider a p-type Si semiconductor doped with $N_a = 10^{15}\,\text{cm}^{-3}$ acceptors at room temperature $T = 300\,\text{K}$. Using the parameters from Appendix A.3, we find for this doping level the threshold potential and the Debye length

$$\phi_T = 2V_{th} \ln \frac{N_a}{n_i} = 0.603\,\text{V} , \quad \lambda_D = \sqrt{\frac{\epsilon_0 \epsilon V_{th}}{e N_a}} = 0.129\,\mu\text{m} .$$

Suppose that we can control the value of the electrostatic potential ϕ_S on its surface, and we wish to investigate how the potential $\phi(x)$ and the charge density

$$\rho(x) = e(p_0(x) - n_0(x) - N_a^-) , \quad p_0(x) = N_a e^{-\phi(x)/V_{th}} , \quad n_0(x) = \frac{n_i^2}{p_0(x)} ,$$

change with the coordinate x. This is achieved by solving Poisson's equation, as discussed in Sect. 6.5.1. The behavior of the electrostatic potential (normalized to the threshold value) and electric charge density $\rho(x)$ is shown in Fig. 6.6a–d. The upper plot of each part of this figure is the potential normalized to the threshold value, $\phi(x)/\phi_T$, vs. the distance measured in the units of Debye length, x/λ_D. The lower plot shows the charge density normalized to the magnitude of the charge density of the acceptors, $\rho(x)/(eN_a)$. Depending on the value of the surface potential, the following kinds of behaviors are observed:

(a) At the surface potential $\phi_S = 0.01\phi_T = 0.006$ V, electron concentration near the surface has the value $n_0(0) = \frac{n_i^2}{N_a}e^{\phi_S/V_{th}} \approx 1.25 \cdot 10^5$ cm^{-3}, much smaller than acceptor concentration. The hole concentration near the surface is very close to the acceptor concentration, $p_0(0) = N_a e^{-\phi_S/V_{th}} \approx 7.9 \cdot 10^{14}$ cm^{-3}. This allows using the Debye screening approximation from Sect. 6.3.2; in fact, the curve $\phi(x) = \phi_S e^{-x/\lambda_D}$ in this case is practically indistinguishable from the numerical curve from Fig. 6.6a.

(b) Increasing the surface potential to the value $\phi_S = 0.5\phi_T = 0.3$ V results in a formation of the space-charge region with the charge density of the acceptor background, $-eN_a$, and thickness of about $x_d \approx 0.623\ \mu$m $\approx 4.8\lambda_D$. Inside the depletion region with constant charge density, the potential decreases approximately quadratically. At $x > x_d$, the potential exhibits an exponential tail, where $\phi(x) \propto e^{-x/\lambda}$.

(c) At the surface potential equal to the threshold value, $\phi_S = \phi_T$, the size of the depletion region slightly increases as compared to plot (b). More importantly, right at the surface, a very thin layer appears, where the charge density is twice the charge density in the rest of the depletion region. The reason for this is clear: at this value of surface potential, electron concentration near the surface is the same as acceptor concentration, $n_0(0) = N_a$. On the other hand, hole concentration near the surface is equal to the electron concentration in the bulk, $p_0(0) = N_a e^{-\phi_T/V_{th}} = n_i^2/N_a$. Because the concentrations of electrons and holes near the surface are inverted relative to their bulk values, this thin electron-enriched layer is called the inversion layer.

(d) At even higher surface potential $\phi_S = 1.2\phi_T$, the potential $\phi(x)$ inside the inversion layer initially decreases from the initial value ϕ_S to the threshold value ϕ_T. This part of the plot corresponds to the electron-rich inversion layer, marked by a very large negative charge density $\rho(x)$. Near the point x where the potential reaches the value ϕ_T, the inversion layer ends and the depletion region begins. Its width and charge density are the same as at the threshold surface potential value from panel (c). That is, after the surface potential reaches the threshold value, the width of the depletion region saturates at the value

$$x_{dT} = \sqrt{\frac{2\epsilon_0\epsilon}{eN_a}\phi_T}\ ,$$

as obtained by solving Poisson's equation using the depletion approximation and setting ϕ_S to ϕ_T. A further increase of ϕ_S does not affect the depletion region width, which still remains equal x_{dT}.

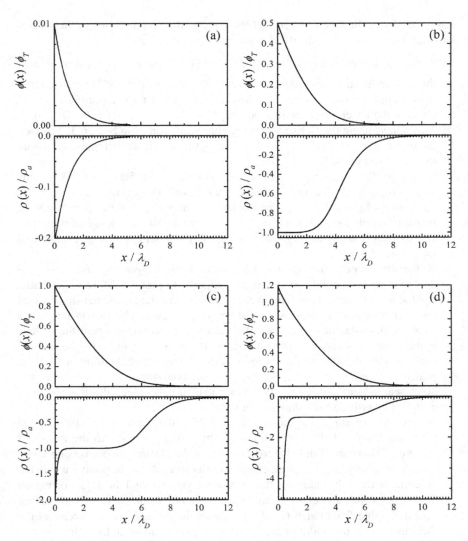

Fig. 6.6 The electric potential normalized to the threshold value $\phi_T = 0.603$ V and the charge density normalized to the magnitude of the acceptor charge density $\rho_a = eN_a = 1.602 \cdot 10^{-4}$ C/cm^3 in p-type Si doped at $N_a = 10^{15}$ cm^{-3} for different values of the surface potential: (a) $\phi_S = 0.01\phi_T$; (b) $\phi_S = 0.5\phi_T$; (c) $\phi_S = \phi_T$; (d) $\phi_S = 1.2\phi_T$

6.6 Problems

6.6.1 Solved Problems

Problem 1 The surface of a p-Si sample doped at $N_a = 10^{16}\,\text{cm}^{-3}$ contains trapped positive charges. Their surface concentration is $\eta = 10^{11}\,\text{cm}^{-2}$, i.e., the surface charge density is $e\eta = 1.602 \cdot 10^{-8}\,\text{C/cm}^2$. The potential far from the surface is zero. What is the depletion region width x_d and the value of the surface potential? Is the depletion approximation applicable to answer this question?

Problem 2 Same as in Problem 1, but this time $\eta = 10^{12}\,\text{cm}^{-2}$. Is the inversion layer formed near the surface of the sample? If yes, what is its surface charge density and the number of electrons per unit area?

Problem 3 Formulate Poisson's equation to find the electric potential in an intrinsic semiconductor with the boundary conditions $\phi(0) = \phi_S$ and $\phi(\infty) = 0$. How does this equation simplify for $|\phi_S| \ll V_{th}$? Find the Debye length of intrinsic silicon at $T = 300\,\text{K}$.

Problem 4 The value of the surface potential on an n-type Si sample with $N_d = 10^{17}\,\text{cm}^{-3}$ is $\phi_S = -0.5\,\text{V}$. At what distance x_i from the surface are electron and hole concentrations equal, $p_0(x_i) = n_0(x_i) = n_i$? Hint: Use the depletion approximation to answer this question.

6.6.2 Practice Problems

Problem 1 The surface of a p-Si sample doped at $N_a = 10^{16}\,\text{cm}^{-3}$ contains an unknown number η of positive charges per unit area. At what concentration η of those charges does an inversion layer start to form on the surface of the sample? What is the electric field value in the semiconductor right at the surface at this concentration of surface charges?

Problem 2 The potential on the surface of an n-Si sample with $N_d = 5 \cdot 10^{15}\,\text{cm}^{-3}$ is $\phi_S = -0.75\,\text{V}$. The temperature is $300\,\text{K}$. At this value of the surface potential, an inversion layer exists at the surface. Find the voltage drop across the inversion layer.

Problem 3 Consider an n-type Si sample with $N_d = 10^{16}\,\text{cm}^{-3}$. At what value of the surface potential ϕ_S are the electron and hole concentrations near the surface equal?

Problem 4 Formulate and solve Poisson's equation to find the electric potential $\phi(x)$ in an intrinsic semiconductor with the boundary conditions $\phi(0) = \phi_S$ and $\phi(\infty) = 0$. You may need the identity $\int_x^\infty \frac{dt}{\sinh(t)} = -\ln(\tanh(x/2))$.

6.6.3 Solutions

Problem 1 The threshold value of the surface potential is

$$\phi_T = 2V_{th} \ln \frac{N_a}{n_i} = 2 \cdot 0.02585 \cdot \ln \frac{10^{16}}{8.59 \cdot 10^9} = 0.7221 \, \text{V} \, .$$

The maximal surface charge density of the depletion layer can be found by assuming the surface potential to be at the threshold value:

$$-eN_a x_{dT} = eN_a \sqrt{\frac{2\epsilon_0 \epsilon}{eN_a}\phi_T} = -\sqrt{2\epsilon_0 \epsilon e N_a \phi_T}$$

$$= \sqrt{2 \cdot 8.854 \cdot 10^{-14} \cdot 11.7 \cdot 1.602 \cdot 10^{-19} \cdot 10^{16} \cdot 0.7221}$$

$$= -4.896 \cdot 10^{-8} \, \text{C/cm}^2 \, .$$

Since this number exceeds in magnitude the value of $e\eta$, we can conclude that the inversion layer is not formed at the surface, and the depletion approximation is valid.

The surface charge density (i.e., charge per unit area) of the depletion region is given by the product of the volume charge density and the depletion region width, $-eN_a x_d$. It should perfectly compensate the surface charge density due to the trapped charges. We thus have

$$eN_a x_d = e\eta \quad \Rightarrow \quad x_d = \eta/N_a = 10^{11}/10^{16} = 10^{-5} \, \text{cm} = 0.1 \, \mu\text{m} \, .$$

The depletion region is maintained by the surface potential found from the formula $x_d = \sqrt{\frac{2\epsilon_0 \epsilon}{eN_a}\phi_S}$, giving

$$\phi_S = \frac{eN_a x_d^2}{2\epsilon_0 \epsilon} = \frac{1.602 \cdot 10^{-19} \cdot 10^{16} \cdot 10^{-10}}{2 \cdot 8.854 \cdot 10^{-14} \cdot 11.7} = 0.0773 \, \text{V} \, .$$

Problem 2 As established in the solution to the previous problem, the magnitude of the maximal surface charge density in the depletion region is $eN_a x_{dT} = 4.896 \cdot 10^{-8} \, \text{C/cm}^2$. This is below the surface charge density of the trapped charges, $e\eta = 1.602 \cdot 10^{-7} \, \text{C/cm}^2$. Therefore, the depletion region alone cannot ensure the charge neutrality. The number $-eN_a x_{dT}$ found above is the surface charge density of the depletion region.

Let η_i be the concentration of electrons per unit area in the inversion layer. The surface charge density of the inversion layer, $-e\eta_i$, is found from the charge neutrality

$$e\eta - eN_a x_{dT} - e\eta_i = 0 \quad \Rightarrow \quad -e\eta_i = eN_a x_{dT} - e\eta \, ;$$

$$-e\eta_i = 4.896 \cdot 10^{-8} - 1.602 \cdot 10^{-7} = -1.112 \cdot 10^{-7} \, \text{C/cm}^{-3} \, .$$

The corresponding surface concentration of electrons is

$$\eta_i = \frac{1.112 \cdot 10^{-7}}{1.602 \cdot 10^{-19}} = 6.94 \cdot 10^{11} \, \text{cm}^{-2} \, .$$

Problem 3 In an intrinsic semiconductor, the charge density is

$$\rho(x) = e(p_0(x) + n_0(x)) \, ,$$

where the carrier concentrations in the presence of the potential $\phi(x)$ are given by

$$p_0(x) = n_i \, e^{-\phi(x)/V_{th}} \, , \quad n_0(x) = n_i \, e^{\phi(x)/V_{th}} \, .$$

Hence,

$$\rho(x) = en_i (e^{-\phi(x)/V_{th}} - e^{\phi(x)/V_{th}}) = -2en_i \sinh(\phi(x)/V_{th}) \, .$$

Poisson's equation reads

$$\frac{d^2\phi(x)}{dx^2} = \frac{2en_i}{\epsilon_0 \epsilon} \sinh(\phi(x)/V_{th}) \, .$$

At small values of its argument $|x| \ll 1$, the hyperbolic sine can be approximated as $\sinh(x) \approx x$; hence, for small potential $|\phi(x)| \ll V_{th}$, we can write

$$\frac{d^2\phi(x)}{dx^2} = \frac{2en_i}{\epsilon_0 \epsilon V_{th}} \phi(x) = \frac{\phi(x)}{\lambda_D^2} \, ,$$

where the Debye length

$$\lambda_D = \sqrt{\frac{\epsilon_0 \epsilon V_{th}}{2en_i}} \, .$$

At $T = 300 \, \text{K}$ in Si, $n_i = 8.59 \cdot 10^9 \, \text{cm}^{-3}$, so

$$\lambda_D = \sqrt{\frac{8.854 \cdot 10^{-14} \cdot 11.7 \cdot 0.02585}{2 \cdot 1.602 \cdot 10^{-19} \cdot 8.59 \cdot 10^9}} = 3.12 \cdot 10^{-3}\,\text{cm} = 31.2\,\mu\text{m}\,.$$

Problem 4 The value of the electric potential at this point is defined by the condition

$$N_d e^{\phi(x_i)/V_{th}} = \frac{n_i^2}{N_d} e^{-\phi(x_i)/V_{th}} \quad \Rightarrow \quad \phi(x_i) = V_{th} \ln \frac{n_i}{N_d}\,.$$

On the other hand, the potential depends on the coordinate x as

$$\phi(x) = \phi_S \left(1 - x/x_d\right)^2\,, \quad x_d = \sqrt{\frac{2\epsilon_0 \epsilon}{e N_d} |\phi_S|}\,.$$

At the point $x = x_i$, we must have

$$\phi_S \left(1 - \frac{x_i}{x_d}\right)^2 = V_{th} \ln \frac{n_i}{N_d} \quad \Rightarrow \quad x_i = x_d \left(1 - \sqrt{\frac{V_{th}}{\phi_S} \ln \frac{n_i}{N_d}}\right)\,.$$

Numerically,

$$x_d = \sqrt{\frac{2 \cdot 8.854 \cdot 10^{-14} \cdot 11.7 \cdot 0.5}{1.602 \cdot 10^{-19} \cdot 10^{17}}} = 8.04 \cdot 10^{-6}\,\text{cm} = 80.4\,\text{nm}\,,$$

$$x_i = 80.4 \cdot \left(1 - \sqrt{\frac{0.02585}{0.5} \cdot \ln 10^7}\right) = 7.01\,\text{nm}\,.$$

Chapter 7
Generation–Recombination Processes

7.1 Recombination Mechanisms

The electron and hole concentrations n_0 and p_0 do not change in time in the state of thermal equilibrium. Nevertheless, even in equilibrium, electrons constantly get promoted from the valence band into the conduction band and return into the valence band. Excitation of the electrons from the valence band into the conduction band results in the generation of electron–hole pairs; recombination is an elementary process, in which a conduction electron re-enters the valence band where to occupy an empty state that we call a hole. The state of thermal equilibrium is by no means static. It is a dynamic state, in which generation and recombination processes perfectly balance each other.

Domination of one of these two processes would mean an increase or a decrease of the electron and hole concentrations. Generation dominates over recombination if the initial concentrations of the charge carriers was below the equilibrium values n_0 and p_0. In the opposite case, recombination proceeds at a higher rate until equilibrium is established.

During an electron–hole recombination event, energy and momentum must be conserved. Note that the vast majority of conduction electrons occupy the states at the bottom of the conduction bands, whereas the majority of holes "live" near the top of the valence band. This means that the energy released in a recombination event is slightly greater than the band gap energy E_g by an amount of the order of the thermal energy kT, see Sect. 5.3.

The momentum released in a recombination event depends on the band structure of the semiconductor. In the direct-band gap semiconductors, the momentum at the bottom of the conduction band is the same as at the top of the valence band, so that the electron and hole have very close momentum values before the recombination. This means that very little momentum is released after they recombine. In the indirect-band gap semiconductors, where the band edges happen to be at different momenta, the momentum released has an appreciable value.

M. Evstigneev, *Introduction to Semiconductor Physics and Devices*,
https://doi.org/10.1007/978-3-031-08458-4_7

Both the energy E_g and the momentum released in a recombination event must be carried away by a third particle. This third particle can be a photon, a phonon, another electron, or a hole. It should be borne in mind that the momentum of a massless photon of energy $hf = E_g$, namely, $p_{ph} = h/\lambda = hf/c = E_g/c$, is much smaller than momentum of a massive particle of the same kinetic energy, $p_{part} = \sqrt{2mE_g}$. Indeed, the ratio of the photon-to-particle momentum is $p_{ph}/p_{part} = \sqrt{E_g/(2mc^2)}$. For the lightest massive particle, an electron, the so-called rest energy is $m_e c^2 \approx 511\,\mathrm{keV}$, whereas the typical band gap in a semiconductor $E_g \approx 1\,\mathrm{eV}$. Hence, the momentum ratio is about $p_{phot}/p_e \approx 10^{-3}$.

An important conclusion can be drawn from this estimate. If the momentum released in a recombination event is small, then photons are much more likely to participate in the recombination. If the initial state of an electron in the conduction band and its final state in the valence band differ significantly not only in energy, but also in momentum, then the probability of a photon emission is much lower.

Recombination in semiconductors can proceed according to the four main mechanisms:

Radiative recombination , in which an electron recombines with a hole with an emission of a photon. This mechanism is the dominant one in the direct-band gap semiconductors (see Fig. 7.1a), such as GaAs, the extra energy is carried away by a photon. In the indirect-band gap semiconductors, such as Si, radiative recombination is also possible, but much less likely than in the direct-band gap semiconductors. The reason is that it must be accompanied by a release of a phonon which would carry away the momentum difference between the initial electron state at the bottom of the conduction band and the final state at the top of the valence band (see Fig. 7.1b). The more particles are involved in a process, the less probable it is; hence, radiative recombination in the indirect-band gap semiconductors plays a minor role as compared to other mechanisms that follow.

Auger recombination is an example of a non-radiative mechanism, in which the energy and momentum released in a recombination event are carried away by another electron from the conduction band or by another hole from the valence band, see Fig. 7.2a and b, respectively. The charge carrier that absorbs the energy and momentum initially has high energy. It gives up very quickly its excess energy to the lattice.

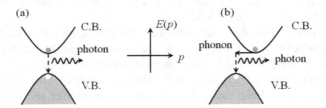

Fig. 7.1 Radiative recombination in the (**a**) direct-band gap and (**b**) indirect-band gap semiconductors. The filled circles represent electrons in the conduction band; the empty circles represent holes in the valence band

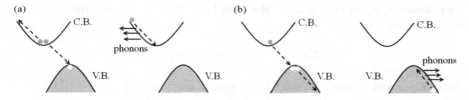

Fig. 7.2 Two types of an Auger process, in which either (**a**) an electron or (**b**) a hole absorbs the energy and momentum released in a recombination event. This particle then quickly dissipates the extra energy $\sim E_g$ by giving it away to the lattice

Trap-assisted Shockley-Read-Hall (SRH) recombination is a non-radiative recombination event assisted by the so-called deep impurities, or traps. The donors and acceptors are shallow impurities in the sense that their energies, $E_{d,a}$, are close to the band edges. In contrast, the traps are deep impurities whose energy E_t is close to the middle of the band gap. Usually, they are metal atoms of various types, such as Fe, Cu, Ni, Cr, Au, Pt, etc. The deep impurities can capture conduction electrons, which then can go into the valence band, i.e., recombine with holes. The energy E_g is released in two smaller portions, first due to electron capture by a trap (energy released = $E_C - E_t$), and then due to the recombination of the trapped electron with a hole (energy released = $E_t - E_V$). During a SRH recombination event, it is the trap atom that first absorbs the excess energy and momentum, and then dissipates it into the crystal in the form of phonons.

Surface recombination proceeds according to the SRH mechanism just described, but with the traps located on the surface of the semiconductor. Not all atoms on the surface have their valence electrons paired up with those of the neighbor atoms. Therefore, many surface atoms have unsaturated dangling bonds that can capture conduction electrons. In addition, surfaces may have a large number of defects and impurity atoms that can serve as traps as well.

Radiative and Auger recombination processes cannot be controlled externally by increasing the degree of purity of the material. They are sometimes called intrinsic recombination mechanisms. In contrast, SRH and surface recombination rates can be reduced by increasing the purity of the material, by surface passivation, and by heavy doping of the semiconductor near the surface. They are referred to as extrinsic recombination mechanisms.

7.2 Charge Carrier Dynamics

7.2.1 Generation and Recombination Rates

Under the non-equilibrium conditions, the carrier concentrations, n and p, differ from their equilibrium values, n_0 and p_0. The deviations of the respective concentrations from the equilibrium values are termed the excess electron and hole concentrations:

$$\Delta n = n - n_0, \quad \Delta p = p - p_0.$$

Normally, a semiconductor is electrically neutral, implying that

$$\Delta n = \Delta p.$$

Apart from thermal activation of electrons from the valence to the conduction band, excess charge carriers can be produced by an external agent, e.g., by illuminating the semiconductor with light. After an external generation process is switched off, the excess concentration Δn will decay to zero. In this section, we would like to describe the process of generation and equilibration of the excess carriers mathematically. For the sake of simplicity, we will assume the semiconductor to be uniform, i.e., the excess concentration Δn to be position-independent.

Let us denote the generation rate of electrons and holes as G. It is the number of electrons and holes generated in a unit volume per unit time; hence, it is measured in

$$[G] = \mathrm{cm}^{-3}\mathrm{s}^{-1}.$$

Because electrons and holes are produced in pairs, their generation rates are the same, and we do not need to introduce separate generation rates for electrons and holes.

The total generation rate or electron–hole pairs, G_{tot}, consists of two contributions: the thermal generation, G_{th}, which is acting in thermal equilibrium, and excess non-thermal generation, G, which is due to an external perturbation, such as light:

$$G_{tot} = G_{th} + G.$$

The thermal generation at the rate G_{th} originates from the energy exchange between the valence band electrons with the lattice vibrations and electromagnetic waves that exist within the semiconductor. It cannot be switched on or off. In contrast, the non-thermal generation is usually induced and can be controlled by an experimentalist.

In principle, the generation rate may depend on the excess concentration Δn, but only if Δn becomes very large, e.g., comparable to the effective densities of states in the valence and conduction band. Indeed, at the excess concentration this high, most of the Bloch states near the top of the valence band are empty, and most states near the bottom of the conduction band are occupied by electrons. Both of these factors make further generation of excess electron–hole pairs more difficult.

Here, we will focus on the usual case when $\Delta n \ll N_C, N_V$. Then, the generation rate G_{tot} is independent of the excess carrier concentration Δn.

We also introduce the recombination rate R as the number of recombination events that happen in a unit volume per unit time, i.e., it is measured in

$$[R] = \text{cm}^{-3}\text{s}^{-1},$$

just like the generation rate. In contrast to the generation rate, the recombination rate R does depend on the excess concentration Δn. The more carriers are present, the more recombination events should take place in a unit volume per unit time. Hence, the recombination rate $R(\Delta n)$ increases with the excess concentration.

Assuming a uniform distribution of excess carriers in the semiconductor, the rate of change of the excess carrier concentration is given by the difference of the generation and the recombination rates:

$$\frac{d\Delta n}{dt} = G_{tot} - R(\Delta n) = G + G_{th} - R(\Delta n).$$

In thermal equilibrium, charge carriers are generated only due to the thermal excitation. Furthermore, there are no excess carriers. Hence, thermal equilibrium is defined by the conditions $G = 0$ and $\Delta n = 0$. Furthermore, in thermal equilibrium, the concentration of carriers does not change in time. Hence, the time derivative

$$\left.\frac{d\Delta n}{dt}\right|_{G=0,\Delta n=0} = G_{th} - R(0) = 0.$$

The thermal generation rate is exactly equal the recombination rate at zero excess concentration:

$$G_{th} = R(0).$$

In the out-of-equilibrium case, it is convenient to combine the thermal generation rate and the recombination rate into a single parameter, called the net recombination rate:

$$U(\Delta n) = R(\Delta n) - G_{th} = R(\Delta n) - R(0).$$

The net recombination rate is positive if the recombination processes dominate over thermal generation at $\Delta n > 0$. It is negative if $\Delta n < 0$, in which case thermal generation dominates. In thermal equilibrium, $\Delta n = 0$ and $U(0) = 0$.

The recombination rate is actually a sum of several contributions due to different recombination mechanisms operative in the volume of a semiconductor. As discussed above, these are the radiative recombination (subscript r), the Auger recombination (subscript A), and the Shockley-Read-Hall recombination mechanism (subscript SRH). The net recombination rate for each of these recombination channels is given by the difference

$$U_r(\Delta n) = R_r(\Delta n) - R_r(0), \quad U_A(\Delta n) = R_A(\Delta n) - R_A(0),$$

$$U_{SRH}(\Delta n) = R_{SRH}(\Delta n) - R_{SRH}(0).$$

7.2.2 Recombination Time Approximation

Consider a situation when some non-thermal generation mechanism was present initially, but was switched off at $t = 0$. Let us find out how the non-equilibrium excess concentration $\Delta n(t)$ decays to zero at $t \geq 0$. Since at $t \geq 0$ we have $G = 0$, the equation that governs the equilibration of charge carriers is

$$\frac{d\Delta n}{dt} = G_{th} - R(\Delta n) = -U(\Delta n).$$

Let us assume that the excess concentration is so small that we can linearize the net recombination rate near the point $U(0) = 0$ as

$$U(\Delta n) = U'(0)\Delta n + \ldots \approx \frac{\Delta n}{\tau} \text{ for } \Delta n \ll n_0 + p_0$$

and neglect the higher-order terms. Of course, when we say that Δn is small, we should also answer the question: small compared to what? As will be shown below in this chapter, the excess concentration should be much smaller than the majority carrier concentration.

The parameter

$$\tau = \frac{1}{U'(0)} \equiv \left(\frac{dU(\Delta n)}{d\Delta n} \bigg|_{\Delta n=0} \right)^{-1}$$

is called the recombination time, or the lifetime of the charge carriers. Within the relaxation time approximation, the excess carrier concentration is an exponentially decaying function of time: Within this approximation, we can write

$$\frac{d\Delta n}{dt} = -\frac{\Delta n}{\tau} \quad \Rightarrow \quad \Delta n(t) = \Delta n(0)e^{-t/\tau}.$$

Typically, several recombination channels act simultaneously in the material: the radiative, SRH, and Auger recombination. Then, the net recombination rate is the sum of the contributions from all recombination channels,

$$U(\Delta n) = U_r(\Delta n) + U_A(\Delta n) + U_{SRH}(\Delta n),$$

all of which depend on Δn. Taking the derivatives of all three recombination rates at $\Delta n = 0$, we obtain the expression for the total recombination time

$$\frac{1}{\tau} = \frac{1}{\tau_r} + \frac{1}{\tau_A} + \frac{1}{\tau_{SRH}},$$

where τ_r, τ_A, and τ_{SRH} are the recombination times due to each individual channel. The net lifetime τ is smaller than the smallest recombination time present.

If an external non-thermal generation mechanism is acting, e.g., due to the illumination of the material with light, then $G > 0$, and the equation for the excess concentration assumes the form

$$\frac{d\Delta n}{dt} = -\frac{\Delta n}{\tau} + G,$$

provided that Δn is not too big to invalidate the recombination time approximation for $U(\Delta n)$. Starting from some initial value $\Delta n(0)$, the excess concentration will then evolve according to

$$\Delta n(t) = \Delta n(0)e^{-t/\tau} + G\tau \left(1 - e^{-t/\tau}\right),$$

until it reaches a steady state with

$$\frac{d\Delta n}{dt} = 0, \quad \Delta n = G\tau,$$

after which it will not change. The electron and hole concentrations in the steady state are given by

$$n = n_0 + G\tau, \quad p = p_0 + G\tau.$$

7.3 Radiative Recombination

A single radiative recombination event involves an electron and a hole and does not require a third particle to take place. Therefore, the number of radiative recombination events that happen in a unit volume per unit time is proportional to the electron and hole concentrations:

$$R_r = Bnp.$$

The proportionality constant B is called the radiative recombination coefficient. It is measured in

$$[B] = cm^3/s.$$

In Si (an indirect-band gap semiconductor), it has the value of $B_{Si} \approx 4.7 \cdot 10^{-15}\,cm^3/s$ at 300 K, whereas in GaAs (a direct-band gap semiconductor), it is five orders of magnitude higher, $B_{GaAs} \approx 7 \cdot 10^{-10}\,cm^3/s$.

By writing $n = n_0 + \Delta n$ and $p = p_0 + \Delta n$, we can express the radiative recombination rate in terms of the excess carrier concentration Δn:

$$R_r(\Delta n) = B(n_0 + \Delta n)(p_0 + \Delta n) = Bn_0 p_0 + B(n_0 + p_0)\Delta n + B\Delta n^2.$$

The net radiative recombination rate is

$$U_r(\Delta n) = R_r(\Delta n) - R_r(0) = B(n_0 + p_0)\Delta n + B\Delta n^2.$$

By taking its derivative with respect to Δn at $\Delta n = 0$, we find the radiative recombination time

$$\tau_r = \frac{1}{U_r'(0)} = \frac{1}{B(n_0 + p_0)}.$$

Observing that the net recombination rate can be written as $U_r(\Delta n) = \frac{\Delta n}{\tau_r} + B\Delta n^2$, we conclude that the radiative recombination time description is a good approximation provided that the first term is much bigger than the second term, i.e.,

$$\Delta n \ll n_0 + p_0.$$

In the opposite limit, $\Delta n \gg n_0 + p_0$, the net radiative recombination rate increases as the second power of Δn: $U_r = B\Delta n^2$.

7.4   Auger Recombination

When an electron and a hole recombine, the energy and momentum released can be transferred to either another electron or to another hole. Hence, Auger recombination involves three particles and may proceed via two channels, depending on whether it is an electron or a hole that carries away the energy. Let us denote the electron-assisted Auger recombination rate as $R_{A,e}$ and the hole-assisted rate as $R_{A,h}$. The Auger recombination rate is

$$R_A = R_{A,e} + R_{A,h}.$$

The rate of an Auger process involving two electrons and a hole is proportional to the hole concentration and to the square of electron concentration:

$$R_{A,e} = C_e n^2 p.$$

Similarly,

$$R_{A,h} = C_h n p^2.$$

The proportionality constants C_e and C_h are called the Auger recombination coefficients. They are measured in

$$[C_e] = [C_h] = cm^6/s$$

and their typical value is of the order of 10^{-30} cm^6/s in a number of semiconductors, including Si and GaAs.

As before, we write $n = n_0 + \Delta n$ and $p = p_0 + \Delta n$ to express the Auger recombination rate in terms of the excess carrier concentration Δn:

$$R_{A,e}(\Delta n) = C_e(n_0 + \Delta n)^2(p_0 + \Delta n)$$
$$= C_e n_0^2 p_0 + C_e(n_0^2 + 2n_0 p_0)\Delta n + C_e p_0 \Delta n^2 + C_e \Delta n^3,$$
$$R_{A,h}(\Delta n) = C_h p_0^2 n_0 + C_h(p_0^2 + 2n_0 p_0)\Delta n + C_h n_0 \Delta n^2 + C_h \Delta n^3.$$

The net recombination rates $U_A(\Delta n)$ are obtained by subtracting the values $R_{A,e}(0) = C_e n_0^2 p_0$ and $R_{A,h}(0) = C_h p_0^2 n_0$ from the respective expressions:

$$U_A(\Delta n) = U_{A,e}(\Delta n) + U_{A,h}(\Delta n),$$
$$U_{A,e}(\Delta n) = C_e(n_0^2 + 2n_0 p_0)\Delta n + C_e p_0 \Delta n^2 + C_e \Delta n^3,$$
$$U_{A,h}(\Delta n) = C_h(p_0^2 + 2n_0 p_0)\Delta n + C_h n_0 \Delta n^2 + C_h \Delta n^3.$$

Taking the derivative of the sum of these two expressions at $\Delta n = 0$, we find the Auger recombination time to be

$$\tau_A = \frac{1}{C_e(n_0^2 + 2n_0 p_0) + C_h(p_0^2 + 2n_0 p_0)} = \frac{1}{C_e n_0^2 + C_h p_0^2 + 2(C_e + C_h)n_i^2}.$$

If the material is of n-type, the electron-mediated Auger recombination process dominates, and if it is of p-type, the hole-assisted Auger mechanism is more efficient. Correspondingly,

$$\tau_A \approx \begin{cases} \frac{1}{C_e n_0^2} & \text{if } n_0 \gg n_i \gg p_0, \\ \frac{1}{C_h p_0^2} & \text{if } p_0 \gg n_i \gg n_0. \end{cases}$$

The recombination time approximation is valid under the condition that the linear term in the expression for $U_A(\Delta n)$ is much higher than the quadratic and cubic terms, i.e.,

$$C_e(n_0^2 + 2n_0 p_0) + C_h(p_0^2 + 2n_0 p_0) \gg (C_e p_0 + C_h n_0)\Delta n + (C_e + C_h)\Delta n^2.$$

The right-hand side of this strong inequality monotonically increases with Δn. Hence, this inequality is fulfilled for Δn much smaller than the value at which the two sides of the inequality become equal. We thus arrive at the condition

$$\Delta n \ll \sqrt{\frac{(C_e p_0 + C_h n_0)^2}{4(C_e + C_h)^2} + \frac{C_e n_0^2 + C_h p_0^2 + 2n_0 p_0(C_e + C_h)}{C_e + C_h}} - \frac{C_e p_0 + C_h n_0}{2(C_e + C_h)}.$$

In an n-type semiconductor, we can neglect all terms in the right-hand side that contain p_0 and obtain

$$\Delta n \ll n_0 \frac{C_h}{C_e + C_h} \left(\sqrt{1 + 4\frac{C_e(C_e + C_h)}{C_h^2}} - 1 \right) \approx \begin{cases} 2\frac{C_e}{C_h} n_0 & \text{if } C_h \gg C_e ; \\ 2n_0 & \text{if } C_e \gg C_h ; \\ n_0 & \text{if } C_e \approx C_h. \end{cases}$$

For a p-type semiconductor, the condition is similar, but with n_0 replaced by p_0 and the coefficients C_e and C_h interchanged:

$$\Delta n \ll p_0 \frac{C_e}{C_e + C_h} \left(\sqrt{1 + 4\frac{C_h(C_e + C_h)}{C_e^2}} - 1 \right) \approx \begin{cases} 2\frac{C_h}{C_e} p_0 & \text{if } C_e \gg C_h ; \\ 2p_0 & \text{if } C_h \gg C_e ; \\ p_0 & \text{if } C_e \approx C_h. \end{cases}$$

For an intrinsic semiconductor, the recombination time approximation is justified for $\Delta n \ll n_i$, irrespective of the relation between the Auger coefficients C_e and C_h.

7.5 Shockley-Read-Hall (SRH) Recombination

7.5.1 Electron and Hole Capture and Emission by the Traps

Suppose the material contains deep impurity atoms, or traps, with the energy E_t and concentration N_t. Let us denote the concentration of those traps that are filled

Fig. 7.3 An illustration of electron capture, electron emission, hole capture, and hole emission by a trap

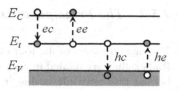

with an electron as N_t'. We focus on the case when the trap concentration is too small to affect the charge neutrality. We furthermore assume that a trap atom can accommodate not more than just one electron.

The four possible exchange processes may happen between a trap and the valence and conduction bands. They are schematically shown in Fig. 7.3, in which the filled circles represent electrons and empty circles represent holes:

Electron capture (ec) is a transition of an electron from the conduction band to the trap level. The rate of this process is proportional to the concentration of empty traps, $N_t - N_t'$, and to the non-equilibrium electron concentration n. Denoting the proportionality constant as c_{ec}, the rate of electron capture is

$$R_{ec} = c_{ec}(N_t - N_t')n.$$

Electron emission (ee) is a back-transition of an electron from a trap into the conduction band. The rate of this process is proportional to the concentration of filled traps, N_t', with the proportionality constant c_{ee}:

$$R_{ee} = c_{ee}N_t'.$$

Hole capture (hc) is a transition of an electron from a trap into the valence band. The rate of hole capture is proportional to the concentration of holes and the concentration of traps filled with an electron:

$$R_{hc} = c_{hc}N_t'p.$$

Hole emission (he) is a transition of an electron from the valence band to a trap. Its rate is proportional to the concentration of empty traps:

$$R_{he} = c_{he}(N_t - N_t').$$

The emission processes can be interpreted as generation and the capture processes as recombination of charge carriers.

7.5.2 The Principle of Detailed Balance

The rate constants of the four processes, c_{ec}, c_{ee}, c_{hc}, and c_{he}, are not independent. Rather, they are related to each other by the principle of detailed balance. The idea behind this principle is that in thermal equilibrium, the rate of each elementary process is equal to the rate of the reverse process. That is, electron capture rate exactly equals electron emission rate, and hole capture and emission processes proceed at the same rate as well.

In thermal equilibrium (subscript 0), the probability for a trap to be occupied by an electron is given by the Fermi-Dirac distribution:

$$f_t = \frac{1}{1 + e^{(E_t - E_F)/kT}}, \quad N'_{t0} = N_t f_t.$$

Strictly speaking, the exponential factor should be multiplied by the trap degeneracy factor $g_t = 2$ due to the two possible spin orientations of the captured electron. We absorb this factor into the definition of the trap energy, which we write as

$$E_t = E_{t0} + kT \ln g_t,$$

where E_{t0} is the true energy of the trap.

Because in equilibrium the rates of direct and reverse processes are the same,

$$R_{ec0} = R_{ee0}$$

with $n = n_0$, $p = p_0$, we have:

$$c_{ec} N_t (1 - f_t) n_0 = c_{ee} N_t f_t.$$

Dividing both sides by $f_t N_t$, we can express the electron emission rate constant as

$$c_{ee} = c_{ec}(f_t^{-1} - 1)n_0 = c_{ec} e^{(E_t - E_F)/kT} n_i e^{(E_F - E_{Fi})/kT} = c_{ec} n_i e^{(E_t - E_{Fi})/kT}.$$

In the same manner, from the condition $R_{hc0} = R_{he0}$ we find

$$c_{he} = c_{hc} n_i e^{(E_{Fi} - E_t)/kT}.$$

Hence, instead of four rate constants, we need only two capture constants: $c_e \equiv c_{ec}$ and $c_h \equiv c_{hc}$, where we dropped the subscript c as superfluous. The out-of-equilibrium capture and emission rates for electrons and holes are:

$$R_{ec} = c_e (N_t - N'_t) n,$$
$$R_{ee} = c_e n_i e^{(E_t - E_{Fi})/kT} N'_t,$$

$$R_{hc} = c_h N_t' p,$$

$$R_{he} = c_h n_i e^{(E_{Fi} - E_t)/kT} (N_t - N_t').$$

The electron and hole capture coefficients have the dimension

$$[c_{e,h}] = cm^3/s.$$

They must be proportional to the thermal velocity of the charge carriers, see Sect. 5.4.3,

$$v_{e,h|th} = \sqrt{\frac{3kT}{m_{e,h}^*}},$$

as the faster the charge carriers move, the more often they collide with the traps. Hence,

$$c_{e,h} = \sigma_{e,h} v_{e,h|th}.$$

The proportionality constants $\sigma_{e,h}$ are called the capture cross section of electrons and holes. The capture cross sections depend on the probability that a collision between a respective charge carrier and a trap will result in the carrier capture. The capture cross section has the dimension of area and its typical value ranges from 10^{-18} to 10^{-14} cm^2.

7.5.3   The Net SRH Recombination Rate

The concentration of filled traps, N_t', is related to the non-equilibrium electron and hole concentrations, n and p. To establish this relation, we note that, due to electric neutrality, the net electron SRH recombination rate,

$$U_e = R_{ec} - R_{ee},$$

must be the same as the net hole recombination rate,

$$U_h = R_{hc} - R_{he}.$$

Using the expressions for the rates R_{ec}, R_{ee}, R_{hc}, and R_{he} from the previous section, we have

$$c_e \left[(N_t - N_t')n - n_i e^{(E_t - E_{Fi})/kT} N_t' \right] = c_h \left[N_t' p - n_i e^{(E_{Fi} - E_t)/kT} (N_t - N_t') \right].$$

Solving this equation for N_t', we find:

$$N_t' = N_t \frac{c_e n + c_h n_i e^{(E_{Fi}-E_t)/kT}}{c_e(n + n_i e^{(E_t-E_{Fi})/kT}) + c_h(p + n_i e^{(E_{Fi}-E_t)/kT})}.$$

Finally, substitution of this result into the net recombination rate, we find after some simple algebra:

$$U_{SRH} = U_e = U_h = N_t \frac{c_e c_h (np - n_i^2)}{c_e(n + n_i e^{(E_t-E_{Fi})/kT}) + c_h(p + n_i e^{(E_{Fi}-E_t)/kT})}.$$

This formula can be written in a more compact form by introducing the characteristic concentrations

$$n_1 = n_i e^{(E_t-E_{Fi})/kT}, \quad p_1 = n_i e^{(E_{Fi}-E_t)/kT}.$$

It is convenient to divide the numerator and the denominator by the product $N_t c_e c_h$ and to introduce the electron and hole capture times

$$\tau_e = \frac{1}{N_t c_e} = \frac{1}{N_t \sigma_e v_{th\,e}}, \quad \tau_h = \frac{1}{N_t c_h} = \frac{1}{N_t \sigma_h v_{th\,h}},$$

where we used the expressions for the capture coefficients in terms of the capture cross sections. It is seen that the recombination times decrease with temperature as $1/\sqrt{T}$. The net SRH recombination rate is

$$U_{SRH} = \frac{np - n_i^2}{\tau_h(n + n_1) + \tau_e(p + p_1)}.$$

Recalling that $n = n_0 + \Delta n$ and $p = p_0 + \Delta n$, as well as keeping in mind the mass action law, $n_0 p_0 = n_i^2$, the SRH recombination rate is written in an equivalent form

$$U_{SRH} = \frac{\Delta n(n_0 + p_0 + \Delta n)}{\tau_h(n_0 + n_1) + \tau_e(p_0 + p_1) + (\tau_e + \tau_h)\Delta n}.$$

7.5.4   SRH Recombination Time

At small excess concentrations,

$$\Delta n \ll n_0 + p_0$$

the excess recombination rate is proportional to Δn:

$$U_{SRH} \approx \frac{\Delta n}{\tau_{SRH}},$$

and the SRH recombination time is given by

$$\tau_{SRH} = \frac{\tau_h(n_0 + n_1) + \tau_e(p_0 + p_1)}{n_0 + p_0}.$$

Consider what happens if one type of charge carriers, say, electrons, dominates. Note that the parameters n_1 and p_1 are of the similar order of magnitude as the intrinsic concentration, because E_t is close to the middle of the band gap. Then, we can neglect the small concentrations, which are defined by the inequality

$$n_0 \gg n_1 \sim p_1 \sim n_i \gg p_0 = n_i^2/n_0.$$

This gives the SRH recombination time $\tau_{SRH} \approx \tau_h$. More generally,

$$\tau_{SRH} \approx \begin{cases} \tau_h & \text{if } n_0 \gg n_i \gg p_0 \\ \tau_e & \text{if } p_0 \gg n_i \gg n_0 \end{cases}.$$

The SRH recombination time tends to the value of the capture time of the minority carriers: holes in an n-semiconductor and electrons in an p-semiconductor. This may seem a bit counter-intuitive, but is relatively easy to understand as follows. Suppose, the semiconductor is of n-type, and so most traps are occupied by electrons. A trapped electron is likely to return back to the conduction band in an n-type semiconductor, because its valence band does not have many holes. Since recombination means transition of an electron from a trap to the valence band (or, equivalently, hole capture by an occupied trap), recombination time is determined by the hole capture process. A similar argument applies if a semiconductor is of p-type.

Another important observation is that at the excess concentration higher than all other concentrations found in the general formula for U_{SRH}, we have

$$U_{SRH} \approx \frac{\Delta n}{\tau_e + \tau_h} \quad \text{for } \Delta n \gg n_0, p_0, n_i, n_1, p_1.$$

This has the familiar form of the excess concentration to the recombination time ratio. We see that, in contrast to the radiative and Auger mechanisms, the recombination time approximation recovers its validity in the case of SRH recombination also at large excess concentrations, but with a different recombination time $\tau_{SRH} = \tau_e + \tau_h$. Its value is of the same order of magnitude as τ_{SRH} in the small-Δn limit. It is for this reason that the SRH recombination is often described within the relaxation time approximation, viz. $U_{SRH} \approx \Delta n/\tau_{SRH}$, both at low and high excess concentrations, with a constant τ_{SRH} independent of Δn.

7.6   Surface Recombination

So far, we assumed an infinite semiconductor, whereas a real-life semiconductor device has a surface. As a result of an abrupt termination of the material, new electron states appear on the surface, called the surface states. They have energies within the band gap. Hence, they act as the centers where recombination can proceed according to the SRH mechanism. In addition, a semiconductor surface usually contains defects and impurities, which likewise act as recombination centers.

Surface recombination proceeds via the SRH mechanism. Let us assume for simplicity that surface traps are uniformly distributed in a thin surface layer of thickness δx with the concentration N_t, which differs from the trap concentration far from the surface. The number of recombination events that happen inside this layer per unit time is

$$\left(\frac{d\Delta N}{dt}\right)_{surf} = U A \,\delta x,$$

where A is the surface area, ΔN is the total number of the excess charge carriers inside the surface layer, and the net recombination rate U is given by the SRH expression. Hence,

$$\left(\frac{d\Delta N}{dt}\right)_{surf} = N_t \frac{c_e c_h (np - n_i^2)}{c_e(n + n_1) + c_h(p + p_1)} A \,\delta x$$

with

$$n_1 = n_i e^{(E_t - E_{Fi})/kT}, \quad p_1 = n_i e^{(E_{Fi} - E_t)/kT},$$

as before. Denoting the surface concentration of traps as

$$\eta_t = N_t \,\delta x, \quad [\eta_t] = \text{cm}^{-2},$$

and keeping in mind that

$$np - n_i^2 = (n_0 + \Delta n)(p_0 + \Delta n) - n_0 p_0 = (n_0 + p_0 + \Delta n)\Delta n,$$

we write the rate of change of the excess carrier concentration as

$$\left(\frac{d\Delta N}{dt}\right)_{surf} = A S,$$

where we introduced the so-called surface recombination velocity

$$S = \eta_t \frac{c_e c_h (n_0 + p_0 + \Delta n)}{c_e(n_0 + n_1) + c_h(p_0 + p_1) + (c_e + c_h)\Delta n}.$$

It is the same for both electrons and holes. In the limits of small and large excess concentrations Δn, surface recombination velocity is

$$S(\Delta n \to 0) = \eta_t \frac{c_e c_h (n_0 + p_0)}{c_e(n_0 + n_1) + c_h(p_0 + p_1)}, \quad S(\Delta n \to \infty) = \eta_t \frac{c_e c_h}{c_e + c_h}.$$

The two limit values are of the same order of magnitude, allowing us to conclude that the surface recombination velocity depends relatively weakly on the excess concentration Δn near the surface. It is often an acceptable approximation in modeling various semiconductor devices to take the surface recombination velocity as a constant independent of the excess concentration.

An implicit assumption made in this discussion is that the surface traps are characterized by a single energy level. In reality, this is not the case: surface trap energies are distributed close to the middle of the band gap. Because it is difficult to control the physico-chemical process on the surface, the surface recombination velocity has to be measured experimentally rather than calculated theoretically.

Surface recombination is an undesirable effect in the technological applications. Reduction of the surface recombination rate is achieved by surface passivation, i.e., deposition of a thin passive layer on the semiconductor surface. For example, Si surface is usually passivated by Si_3N_4 or SiO_2. Surface passivation is an important step in semiconductor technology, especially in the design of solar cells.

Surface passivation cannot be performed underneath the metal contacts, because a non-conducting passivating layer would prevent the current to be collected from the semiconductor device. In that case, an alternative method of reducing the surface recombination can be used. The idea is that recombination requires both electrons and holes; hence, prevention of one type of charge carriers from coming to the surface would drastically decrease the surface recombination rate. Doping the surface region with the same type of impurities as the bulk, but at a much larger concentration, achieves just that: it prevents the minority carriers from coming close to the surface and to recombine with the majority carriers.

7.7   Quasi-Fermi Energies

Suppose that the non-equilibrium conditions are created in a semiconductor by an external perturbation, such as illumination of the material by light. The material then strives to restore equilibrium by two main mechanisms: (1) scattering of electrons and holes by the lattice vibrations, defects, and each other, and (2) the generation-recombination processes.

Each of these mechanisms is characterized by its own characteristic rate. It turns out that, usually, equilibration within the electron and hole subsystems proceeds much faster than equilibration between them. This is so, because the recombination time is typically much longer than the scattering time. As a result, electron and hole subsystems are close to equilibrium within themselves, but they are far from equilibrium with each other.

This means that the occupation probabilities of the momentum states for electrons and holes are given by the equilibrium Fermi-Dirac distributions. However, because the concentration of particles is intimately related to their Fermi energy, we conclude that electron and hole subsystems must be characterized by different Fermi energies, denoted as E_{Fe} and E_{Fh}. That is, electron and hole occupation probabilities of some momentum state $\vec{p}$ are given by

$$f_e(E(\vec{p})) = \frac{1}{1 + e^{(E(\vec{p})-E_{Fe})/kT}}, \quad f_h(E(\vec{p})) = \frac{1}{1 + e^{(E_{Fh}-E(\vec{p}))/kT}},$$

where the energy $E(\vec{p})$ belongs to the conduction band in the electron case, and in the case of holes, it belongs to the valence band. To emphasize the fact that the system is not in equilibrium, the individual electron and hole Fermi energies are called quasi-Fermi energies, or imrefs.[1]

The electron and hole quasi-Fermi energies are related to their non-equilibrium concentrations by the general formulae from Sect. 5.1, but with n_0 and p_0 replaced with n and p:

$$n = \int_{E_C}^{\infty} dE\, g_C(E)\, f_e(E), \quad p = \int_{-\infty}^{E_V} dE\, g_V(E)\, f_h(E).$$

If Boltzmann approximation is valid, then these expressions simplify to

$$n = N_C e^{-(E_C-E_{Fe})/kT} = n_i e^{(E_{Fe}-E_{Fi})/kT},$$

$$p = N_V e^{(E_V-E_{Fh})/kT} = n_i e^{(E_{Fi}-E_{Fh})/kT},$$

where E_{Fi} is the intrinsic Fermi energy. The mass action law generalizes to

$$np = n_i^2 e^{(E_{Fe}-E_{Fh})/kT}.$$

After switching off the external perturbation, a semiconductor material will tend towards the equilibrium state, where electron and hole imrefs are equal to each other. In that state, the earlier obtained equilibrium version of the mass action law, $n_0 p_0 = n_i^2$, is restored.

---

[1] Singular: imref. This word is derived from the name Fermi spelled backwards, and is sometimes interpreted as "imaginary reference."

7.8 Problems

7.8.1 Solved Problems

Problem 1 Consider a pure semiconductor with no traps, so that the Shockley-Read-Hall recombination mechanism is inoperative in this material. The excess carrier concentration has the initial value $\Delta n(0)$ sufficiently low for the Auger recombination to be negligible relative to the radiative recombination. The excess concentration $\Delta n(t)$ is allowed to decay to zero at $t \geq 0$ as the material comes to equilibrium. Assume the radiative recombination coefficient B and the equilibrium concentrations n_0 and p_0 to be known. Derive the time-dependent excess concentration $\Delta n(t)$ by properly taking into account the radiative recombination. Plot the ratio $\Delta n(t)/\Delta n(0)$ vs. the ratio t/τ_r, where τ_r is the radiative recombination time, for several values of $\Delta n(0)/(n_0 + p_0) = 0.1, 0.5, 1, 10$, and compare your curves with the exponentially decaying function e^{-t/τ_r} obtained within the relaxation time approximation.

Problem 2 Shockley-Read-Hall lifetime in a semiconductor is $\tau_{SRH} = 1\,\mathrm{ms}$. Its radiative recombination coefficient is $B = 10^{-12}\,\mathrm{cm}^3/\mathrm{s}$, and Auger recombination coefficients $C_e = C_h = 10^{-29}\,\mathrm{cm}^6/\mathrm{s}$. It is doped with $10^{15}\,\mathrm{cm}^{-3}$ donors, and its intrinsic concentration is $10^9\,\mathrm{cm}^{-3}$. The semiconductor is illuminated by light, which generates excess charge carriers at the rate $G = 10^{15}\,\mathrm{cm}^{-3}\,\mathrm{s}^{-1}$. Find the excess carrier concentration in the steady state and $1\,\mathrm{ms}$ after illumination was turned off. Find the difference between the electron and hole quasi-Fermi energies in the presence of illumination.

Problem 3 How does the difference between the electron and hole quasi-Fermi energies depend on time as the semiconductor returns to equilibrium from some non-equilibrium initial excess concentration $\Delta n(0)$? To answer this question, assume the recombination time approximation to be valid and the recombination time τ to be known. Derive an expression for the time-dependent difference $\Delta E_F(t) = E_{Fe}(t) - E_{Fh}(t)$ vs. t for

(a) an intrinsic semiconductor;
(b) an n-type semiconductor with $n_0 \gg n_i \gg p_0$.

Plot the ratio $\Delta E_F(t)/\Delta E_F(0)$ vs. t/τ for several values of $\Delta n(0)/n_i$ in part (a) and for several values of $\Delta n(0)/p_0$ in part (b).

Problem 4 Assuming that the electron and hole capture times, τ_e and τ_h, are known, at what trap energy E_t will the SHR recombination rate maximal?

Problem 5 The excess charge carriers are uniformly distributed in a thin flat semiconductor slab with the concentration $\Delta n(t) = \Delta n(0)e^{-t/\tau_{eff}}$ that decays to zero with the recombination time τ_{eff}. The recombination time depends on the thickness W of the slab. At $W = 1\,\mu m$, it is measured to be $0.2\,\mu s$, whereas at $W = 10\,\mu m$, it is $0.5\,\mu s$. The effective recombination time depends on the sample thickness due to the surface recombination. Based on this information, express the effective recombination time τ_{eff} in terms of the bulk recombination time τ and the surface recombination velocity S. Find τ and S from the data provided.

7.8.2   Practice Problems

Problem 1 It is possible to use the recombination time approach not only at low but also at high excess carrier concentration Δn by formulating its dynamics as

$$\frac{d\Delta n}{dt} = -\frac{\Delta n}{\tau(\Delta n)}$$

with the recombination time $\tau(\Delta n)$ that depends on Δn. Write down the expressions for $\tau(\Delta n)$ for the cases of radiative and Auger recombination channels.

Problem 2 At $T = 300\,K$, the radiative recombination coefficient in Si is $B = 4.7 \cdot 10^{-15}\,cm^3/s$, and Auger recombination coefficients are $C_e = 1.1 \cdot 10^{-30}\,cm^6/s$ and $C_h = 0.3 \cdot 10^{-30}\,cm^6/s$. Its intrinsic concentration is $n_i = 8.59 \cdot 10^9\,cm^{-3}$.

(a) At what excess carrier concentration will the net Auger recombination rate be the same as the net radiative recombination rate in intrinsic silicon?
(b) At what excess carrier concentration will the net Auger recombination rate be the same as the net radiative recombination rate in p-Si doped at $10^{15}\,cm^{-3}$ acceptors?

Problem 3 The excess charge carriers are generated by light at a rate G in a doped semiconductor with given values of the carrier concentrations n_0 and p_0, a constant SRH lifetime τ_{SRH}, and a given radiative recombination coefficient B.

(a) Neglecting Auger recombination, express the excess carrier concentration Δn in the steady state in terms of these parameters.
(b) Same as in (a), but using the recombination time approximation in the radiative recombination term.
(c) In order for the recombination time approximation to be valid, the excess generation rate must be much smaller than a certain value. Establish this value by comparing your results from parts (a) and (b).

7.8.3 Solutions

Problem 1 The excess concentration changes in time according to

$$\frac{d\Delta n(t)}{dt} = U_r(\Delta n) = -B(n_0 + p_0 + \Delta n(t))\Delta n(t).$$

To solve this equation, we first rewrite it as

$$\frac{d\Delta n(t)}{(n_0 + p_0 + \Delta n(t))\Delta n(t)} = -B\,dt$$

and integrate the right-hand side over time from 0 to t. This corresponds to the integration of the left-hand side from the initial excess concentration $\Delta n(0)$ to the value $\Delta n(t)$:

$$\int_{\Delta n(0)}^{\Delta n(t)} \frac{d\Delta n'}{(n_0 + p_0 + \Delta n')\Delta n'} = -B\,t.$$

To perform the integration, we use an identity $\frac{1}{(a+x)x} = \frac{1}{a}\left(\frac{1}{x} - \frac{1}{a+x}\right)$, which allows us to rewrite the integral as a sum of two simpler integrals

$$\frac{1}{n_0 + p_0}\left(\int_{\Delta n_0}^{\Delta n(t)} \frac{d\Delta n'}{\Delta n'} - \int_{\Delta n_0}^{\Delta n(t)} \frac{d\Delta n'}{n_0 + p_0 + \Delta n'}\right) = -B\,t.$$

Each of the two integrals can be evaluated using the standard formula for the antiderivative $\int dx/)x + a) = \ln(x + a)$:

$$\frac{1}{n_0 + p_0}\left(\ln\frac{\Delta n(0)}{\Delta n(t)} - \ln\frac{n_0 + p_0 + \Delta n(0)}{n_0 + p_0 + \Delta n(t)}\right) = -B\,t.$$

By multiplying both sides by $n_0 + p_0$ and then exponentiating, we obtain

$$\frac{\Delta n(0)}{\Delta n(t)}\frac{n_0 + p_0 + \Delta n(t)}{n_0 + p_0 + \Delta n(0)} = e^{t/\tau_r}, \quad \tau_r = \frac{1}{B(n_0 + p_0)}.$$

Solving this for the excess concentration $\Delta n(t)$, we get:

$$\frac{\Delta n(t)}{\Delta n(0)} = \left(e^{t/\tau_r} + \frac{\Delta n(0)}{n_0 + p_0}\left(e^{t/\tau_r} - 1\right)\right)^{-1}.$$

The time dependence of the so-obtained excess concentration is shown in Fig. 7.4a. We see that at a finite $\Delta n(0)$ value, the relaxation time approximation

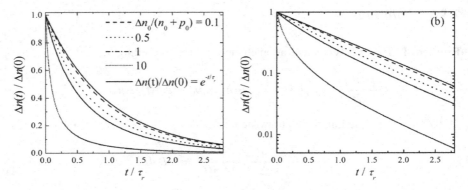

Fig. 7.4 The time dependence of the excess concentration in a semiconductor where only the radiative recombination mechanism is operative with (**a**) linear scale and (**b**) logarithmic scale chosen on the vertical axis

overestimates the values of $\Delta n(t)$ at $t > 0$. The lower the initial excess concentration $\Delta n(0)$, the closer the curves are to the one obtained within the relaxation time approximation. Increasing $\Delta n(0)$ results in a faster decay of $\Delta n(t)$ to zero than the single-exponential curve $\Delta n(t)/\Delta n(0) = e^{-t/\tau_r}$.

Nevertheless, at large times, the logarithm of the excess concentration decreases with time with the same slope equal to $1/\tau_r$ for all initial concentration values, see Fig. 7.4b.

Problem 2 We begin by assuming that the recombination time approximation is valid with the recombination time:

$$\tau = \left(\frac{1}{\tau_{SRH}} + \frac{1}{\tau_r} + \frac{1}{\tau_A} \right)^{-1}$$

with the SRH recombination time $\tau_{SRH} = 1$ ms. The radiative recombination time

$$\tau_r = \frac{1}{BN_d} = \frac{1}{10^{-12} \cdot 10^{15}} = 10^{-3}\,\text{s} = 1\,\text{ms}.$$

Because $n_0 = N_d = 10^{15}\,\text{cm}^{-3} \gg p_0$, the Auger recombination is mediated by the electrons. The respective recombination time

$$\tau_A = \frac{1}{C_e N_d^2} = \frac{1}{10^{-29} \cdot 10^{30}} = 0.1\,\text{s} = 100\,\text{ms}.$$

The hole Auger recombination process can be neglected.

The net lifetime is, then

$$\tau = (1 + 1 + 0.01)^{-1} = 0.498\,\text{ms}.$$

To find the excess carrier concentration, we equate the generation rate to the recombination rate, approximated as

$$\frac{\Delta n}{\tau} = G \quad \Rightarrow \quad \Delta n = G\tau = 0.498 \cdot 10^{-3} \cdot 10^{15} = 4.98 \cdot 10^{11}\,\text{cm}^{-3}.$$

This number is much smaller than the doping level of $10^{15}\,\text{cm}^{-3}$, which means that the relaxation time approximation is indeed valid. The difference between the electron and hole quasi-Fermi energies is found from

$$(N_d + \Delta n)(p_0 + \Delta n) \approx N_d \Delta n = n_i^2 e^{-(E_{Fe} - E_{Fh})/kT},$$

giving

$$E_{Fe} - E_{Fh} = kT \ln \frac{N_d \Delta n}{n_i^2} = 0.02585 \cdot \ln(4.98 \cdot 10^6) = 0.398\,\text{eV}.$$

The excess carrier concentration at time $t = 1\,\text{ms}$ after switching off the illumination is

$$\Delta n(t) = \Delta n(0)\,e^{-t/\tau} = 4.98 \cdot 10^{11}\,e^{-1/0.498} = 6.69 \cdot 10^9\,\text{cm}^{-3}.$$

Problem 3 We begin by writing the generalization of the mass action law

$$(n_0 + \Delta n(t))(p_0 + \Delta n(t)) = n_i^2\,e^{\Delta E_F(t)/kT},$$

from which the difference of the quasi-Fermi energies is obtained:

$$\frac{\Delta E_F(t)}{kT} = \ln \frac{(n_0 + \Delta n(t))(p_0 + \Delta n(t))}{n_i^2}$$

$$= \ln\left(1 + \frac{(n_0 + p_0)\Delta n(0)}{n_i^2}e^{-t/\tau} + \frac{\Delta n^2(0)}{n_i^2}e^{-2t/\tau}\right).$$

Here, we used the identity $n_0 p_0 = n_i^2$ and employed the expression $\Delta n(t) = \Delta n(0)e^{-t/\tau}$ obtained within the relaxation time approximation.

In the case (a), we have $n_0 = p_0 = n_i$, giving

$$\frac{\Delta E_F(t)}{kT} = \ln\left(1 + 2\frac{\Delta n(0)}{n_i}e^{-t/\tau} + \frac{\Delta n^2(0)}{n_i^2}e^{-2t/\tau}\right) = 2\ln\left(1 + \frac{\Delta n(0)}{n_i}e^{-t/\tau}\right).$$

In the case (b), we neglect the terms proportional to p_0 and $\Delta n^2(0)$ in the natural logarithm, giving

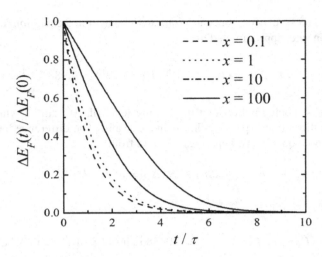

Fig. 7.5 The time dependence of the ratio $\Delta E_F(t)/\Delta E_F(0)$ for several values of the parameter x, see main text

$$\frac{\Delta E_F(t)}{kT} = \ln\left(1 + \frac{n_0 \Delta n(0)}{n_i^2} e^{-t/\tau}\right) = \ln\left(1 + \frac{\Delta n(0)}{p_0} e^{-t/\tau}\right),$$

where we used the mass action law in the second equality.

Let us denote the ratio of the initial excess carrier concentration to n_i (case (a)) or to p_0 (case (b)) as x:

$$x = \begin{cases} \Delta n(0)/n_i & \text{case (a)}, \\ \Delta n(0)/p_0 & \text{case (b)}. \end{cases}$$

Then, in both cases,

$$\frac{\Delta E_F(t)}{\Delta E_F(0)} = \frac{\ln(1 + x\, e^{-t/\tau})}{\ln(1 + x)}.$$

This dependence for several values of x is shown in Fig. 7.5. For $x \ll 1$, the term $x e^{-t/\tau} \ll 1$ for all values of t. Using the approximation $\ln(1 + \epsilon) \approx \epsilon$ for $|\epsilon| \ll 1$, we have

$$\frac{\Delta E_F(t)}{\Delta E_F(0)} \approx \frac{1 + x e^{-t/\tau}}{1 + x} \quad \text{for } x \ll 1.$$

In the opposite limit $x \gg 1$, the difference of the quasi-Fermi levels initially decreases linearly as

$$\frac{\Delta E_F(t)}{\Delta E_F(0)} \approx \frac{\ln(xe^{-t/\tau})}{\ln x} = 1 - \frac{t}{\tau \ln x} \quad \text{for } x \gg 1 \text{ and } t < \tau \ln x,$$

and then enters the exponential decrease.

Problem 4 The SRH recombination rate

$$U_{SRH} = \frac{np - n_i^2}{\tau_h(n + n_1) + \tau_e(p + p_1)}$$

with $n_1 = n_i e^{(E_t - E_{Fi})/kT}$ and $p_1 = n_i e^{(E_{Fi} - E_t)/kT}$ is maximal when the denominator is minimal. The denominator can be written as

$$\tau_h(n + n_1) + \tau_e(p + p_1) = \tau_h n + \tau_e p + \tau_h n_i e^{(E_t - E_{Fi})/kT} + \tau_e n_i e^{(E_{Fi} - E_t)/kT}$$

$$= \tau_h n + \tau_e p + n_i \sqrt{\tau_e \tau_h} \left(\sqrt{\frac{\tau_h}{\tau_e}} e^{(E_t - E_{Fi})/kT} + \sqrt{\frac{\tau_e}{\tau_h}} e^{-(E_t - E_{Fi})/kT} \right)$$

$$= \tau_h n + \tau_e p + 2n_i \sqrt{\tau_e \tau_h} \cosh\left(\frac{E_t - E_{Fi}}{kT} + \frac{1}{2} \ln \frac{\tau_h}{\tau_e} \right).$$

In the last step, we expressed $\sqrt{\tau_h/\tau_e}$ as $e^{\frac{1}{2} \ln \frac{\tau_h}{\tau_e}}$ and used the definition of a hyperbolic cosine, $\cosh(x) = (e^x + e^{-x})/2$. The denominator is minimal when the hyperbolic cosine has the smallest value of 1 at

$$E_t = E_{Fi} + \frac{kT}{2} \ln \frac{\tau_h}{\tau_e}.$$

Because $\cosh(x)$ increases exponentially fast when the magnitude of its argument $|x|$, the most active recombination centers are the ones whose energy is close to the intrinsic Fermi level.

Problem 5 The excess carriers are lost due to the recombination in the bulk and on the surface of the sample. Let

$$\Delta N = AW \Delta n$$

be the total number of excess charge carriers in the sample of surface area A and thickness W. Within the relaxation time approximation, it decreases according to

$$\frac{d\Delta N(t)}{dt} = -\frac{\Delta N(t)}{\tau} - 2SA\Delta n,$$

where the factor of 2 takes care of the fact that the sample has two surfaces. So,

$$AW\frac{d\Delta n}{dt} = -AW\frac{\Delta n}{\tau} - 2SA\Delta n = -\frac{\Delta n}{\tau_{eff}},$$

from which equation we express the effective recombination time in terms of the bulk recombination time and the surface recombination velocity:

$$\frac{1}{\tau_{eff}} = \frac{1}{\tau} + 2\frac{S}{W}.$$

From the numerical data, namely, $\tau_{eff}(W = 1\,\mu m) = 0.2\,\mu s$ and $\tau_{eff}(W = 10\,\mu m) = 0.5\,\mu s$, we obtain a system of two equations,

$$5\,\mu s^{-1} = \frac{1}{\tau} + (2\,\mu m^{-1})S, \quad 2\,\mu s^{-1} = \frac{1}{\tau} + (0.2\,\mu m^{-1})S.$$

We find:

$$S = 1.67\,\mu m/\mu s = 167\,cm/s, \quad \tau = 0.6\,\mu s.$$

Chapter 8
Carrier Transport

8.1 Flux and Electric Current Density

It is of fundamental importance in semiconductor device theory to know how much electric charge is transferred from one part of the device to another by electrons and holes. In the following discussion, we will adopt the particle-based rather than the wave-based picture, i.e., we will treat electrons and holes as classical particles with the conductivity effective masses m_e^* and m_h^* and charges $-e$ and e, respectively.

Consider electrons in a physically small volume ΔV centered at a position $\vec{r}$, see Fig. 8.1. By "physically small," we mean that its linear size is much smaller than all other characteristic dimensions of the system, but still large enough to contain many electrons at the local concentration $n(\vec{r}, t)$, which is practically constant inside ΔV. Note that we dropped the subscript 0 in the electron and hole concentrations, as now these systems are not in equilibrium.

Each electron inside ΔV has its own group velocity, $\vec{v}_j$. We can perform an arithmetic averaging over all velocities of the electrons contained in ΔV and thus obtain the average velocity:

$$\vec{v}_e(\vec{r}, t) = \frac{1}{n(\vec{r}, t)\Delta V} \sum_j \vec{v}_j .$$

Here, summation is performed over all electrons in ΔV, $n(\vec{r}, t)\Delta V$ is the number of electrons in ΔV, and $\vec{v}_j$ is the group velocity vector of the jth electron. In general, electron concentration and average velocity depend on time and spatial coordinates.

We choose ΔV to be a small cylinder with the cross-sectional area ΔA and length ΔL inside a semiconductor, see Fig. 8.1. We orient the axis of this cylinder along the mean velocity $\vec{v}_e$ and define a small time interval Δt according to

$$\Delta L = |\vec{v}_e|\Delta t .$$

© The Author(s), under exclusive license to Springer Nature Switzerland AG 2022
M. Evstigneev, *Introduction to Semiconductor Physics and Devices*,
https://doi.org/10.1007/978-3-031-08458-4_8

Fig. 8.1 Flow of electrons in
a semiconductor

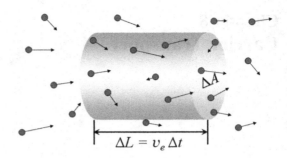

The cylinder contains $\Delta N = n\Delta V = n\,\Delta A\,|\vec{v}_e|\Delta t$ electrons. Each of those electrons moves with its own individual velocity, but on average, all electrons move toward the cylinder's base with the velocity $\vec{v}_e$. Then, the number of electrons that traverse the area ΔA in time Δt will be just ΔN. When divided by the product $\Delta A \cdot \Delta t$, this gives us the electron flux density

$$\vec{j}_e(\vec{r}, t) = n(\vec{r}, t)\vec{v}_e(\vec{r}, t) = \frac{1}{\Delta V}\sum_j \vec{v}_{ej} \, .$$

It is a vector whose magnitude is the number of electrons that traverse a unit surface area oriented perpendicular to the particles' flow per unit time. Its direction coincides with the direction of the average electron flow at the point $\vec{r}$ at time t.

The hole flux density,

$$\vec{j}_h(\vec{r}, t) = p(\vec{r}, t)\vec{v}_h(\vec{r}, t) \, ,$$

is defined in exact same manner, where $p(\vec{r}, t)$ is the local hole concentration at the position $\vec{r}$ and time t, and $\vec{v}_h(\vec{r}, t)$ is the average velocity of a hole.

The product of the flux density by the charge of the respective carriers,

$$\vec{J}_e(\vec{r}, t) = -e\vec{j}_e(\vec{r}, t) = -en(\vec{r}, t)\vec{v}_e(\vec{r}, t) \, , \quad \vec{J}_h(\vec{r}, t) = e\vec{j}_h(\vec{r}, t) = ep(\vec{r}, t)\vec{v}_h(\vec{r}, t),$$

is the electric current density of electrons and holes. The net current density,

$$\vec{J} = \vec{J}_h + \vec{J}_e = e(\vec{j}_h - \vec{j}_e) \, ,$$

is a vector pointing in the direction of the overall electric charge transfer. Its magnitude gives the amount of charge transferred per unit time through a unit area oriented perpendicular to the direction of the overall charge flow.

8.2 Diffusion Current

If electrons are distributed non-uniformly in a semiconductor material, they will tend to move from a region of higher concentration to a region of lower concentration. The degree of non-uniformity of the concentration is measured by the spatial derivatives of $n(\vec{r}, t)$, namely, $\partial n/\partial x$, $\partial n/\partial y$, and $\partial n/\partial z$. All three partial derivatives can be combined into a single vector, the gradient, which was already introduced in the discussion of Schrödinger's equation in Sect. 1.8.1:

$$\vec{\nabla} n(\vec{r}) = \frac{\partial n}{\partial x}\vec{e}_x + \frac{\partial n}{\partial y}\vec{e}_y + \frac{\partial n}{\partial z}\vec{e}_z .$$

The gradient points in the direction of the largest increase of electron concentration.

Diffusion is described by the so-called Fick law. It says that the particles flow in the direction opposite to $\vec{\nabla} n$, i.e., in the direction of the largest concentration decrease. The magnitude of the diffusive flux density is proportional to the concentration gradient. This law applies to any type of particles, including electrons and holes in a semiconductor. Hence, the diffusion flux densities of electrons and holes are

$$\vec{j}_{e|diff} = -D_e \vec{\nabla} n , \quad \vec{j}_{h|diff} = -D_h \vec{\nabla} p .$$

The proportionality constants, D_e and D_h, are called the electron and hole diffusion coefficients. The diffusion coefficient is measured in the units of

$$[D_{e,h}] = \text{cm}^2/\text{s} .$$

The corresponding electric current density of electrons and holes is

$$\vec{J}_{e|diff} = e D_e \vec{\nabla} n , \quad \vec{J}_{h|diff} = e D_h \vec{\nabla} p .$$

8.3 Drift Current

Under the action of an electric field $\vec{\mathcal{E}}$, a free electron in vacuum would be moving with the acceleration $\vec{a}_e = -e\vec{\mathcal{E}}/m_e$. In an ideal defect-free crystal at absolute zero temperature, an electron would undergo Bloch oscillations, as explained in Sect. 4.2.4. Consider now what happens when an electric field $\vec{\mathcal{E}}$ acts in a real semiconductor.

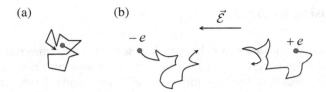

Fig. 8.2 (**a**) Motion of a charge carrier with no electric field in the sample. (**b**) Electron and hole drift in an electric field

An electron in a real material constantly experiences collisions with lattice vibrations (phonons), defects, impurity atoms, and other electrons and holes. These collisions result in sudden changes in the electron's direction of motion, but the average velocity of an electron is zero in the absence of an electric field, see Fig. 8.2a. When an electric field $\vec{\mathcal{E}}$ is applied, the electron gets accelerated by it. However, as a result of collisions, the electron loses the energy acquired from the electric field. Effectively, the field is accelerating the electron only within a short time interval between two successive collisions, see Fig. 8.2b. Same applies to holes. We denote the average value of the collision time for the two types of charge carriers as τ_e and τ_h.

Due to its acceleration by the field between the collisions, a charge carrier acquires a regular non-zero velocity component, called the drift velocity. It is given by the product of the electron's acceleration due to the electric field and the characteristic time between the collisions of an electron or a hole:

$$\vec{v}_{e|drf} = \vec{a}_e \tau_e = -\frac{e}{m_e^*} \vec{\mathcal{E}} \tau_e = -\mu_e \vec{\mathcal{E}} \,, \quad \mu_e = e\tau_e/m_e^* \,,$$

$$\vec{v}_{h|drf} = \vec{a}_h \tau_h = \frac{e}{m_h^*} \vec{\mathcal{E}} \tau_h = -\mu_h \vec{\mathcal{E}} \,, \quad \mu_h = e\tau_h/m_h^* \,.$$

Note that $\vec{v}_{e|drf}$ is oriented against the electric field because the charge of an electron is negative. Hole drift velocity $\vec{v}_{h|drf}$ is oriented along the electric field.

The proportionality constants μ_e and μ_h that relate the drift velocity and the electric field are called electron and hole mobilities. Mobility is measured in the units of

$$[\mu_{e,h}] = cm^2/(V \cdot s) \,.$$

Electron and hole drift flux densities are

$$\vec{j}_{e|drf} = -n\mu_e \vec{\mathcal{E}} \,, \quad \vec{j}_{h|drf} = p\mu_h \vec{\mathcal{E}} \,,$$

and the respective electric current densities are

$$\vec{J}_{e|drf} = e\mu_e n \vec{\mathcal{E}} , \quad \vec{J}_{h|drf} = e\mu_h p \vec{\mathcal{E}} .$$

8.4 Conductivity and Resistivity

If a voltage V is applied to a semiconductor slab of length L and cross-sectional area $A = wd$, see Fig. 8.3a, current I will be flowing through it. The ratio of the applied voltage to the current is known as resistance; according to Ohm's law, it is a voltage-independent constant:

$$R = V/I .$$

The voltage is related to the electric field according to

$$V = L\mathcal{E} .$$

On the other hand, the current is given by the product of the electric current density and the cross-sectional area of the sample:

$$I = A J_{drf} ,$$

where the drift current density is

$$J_{drf} = J_{e|drf} + J_{h|drf} = e(\mu_e n + \mu_h p)\mathcal{E} ,$$

see Sect. 8.3. Combining these expressions, we arrive at the resistance formula

$$R = \frac{L}{A(\mu_e n + \mu_h p)} = \frac{L}{A\sigma} = \frac{L}{A}\rho ,$$

which clearly shows that the resistance depends on the geometry of the sample through the ratio L/A, and on its material properties via the parameters

$$\sigma = e(\mu_e n + \mu_h p) , \quad \rho = \sigma^{-1} = \frac{1}{e(\mu_e n + \mu_h p)} ,$$

known as the conductivity and resistivity of the material. If excess carriers are not injected into the semiconductor, one can replace in these expressions the non-equilibrium concentrations n and p with the equilibrium counterparts n_0 and p_0.

8.5   Current–Voltage Measurements

8.5.1   Photoconductivity

The measurements of the current–voltage curves provide a great deal of information about the concentration and mobilities of the charge carriers. They can also be used to measure their recombination rates.

This can be done by first injecting excess carriers into the semiconductor by, e.g., illumination, and then observing its return to equilibrium after switching off the excess carrier generation process. A practical way to measure the net recombination time is to measure the time-dependent conductivity of the semiconductor sample of interest:

$$\sigma(t) = e(n(t)\mu_e + p(t)\mu_h) \, ,$$

where μ_e, μ_h are the electron and hole mobilities. Since $n(t) = n_0 + \Delta n(t)$ and $p(t) = p_0 + \Delta n(t)$, the total conductivity has the time-dependent and the time-independent parts:

$$\sigma(t) = \sigma_0 + \Delta\sigma(t) \, ,$$

where

$$\sigma_0 = e(n_0\mu_e + p_0\mu_h) \, , \quad \Delta\sigma(t) = e\Delta n(0)(\mu_e + \mu_h)e^{-t/\tau} \, .$$

If the non-thermal generation mechanism is photogeneration, i.e., the generation of charge carriers by light, then the parameter $\Delta\sigma$ is called photoconductivity. Taking the natural logarithm of the ratio $\Delta\sigma(t)$ to the initial value $\Delta\sigma(0) = \Delta n(0)(\mu_e + \mu_h)$, we immediately find:

$$\ln\frac{\Delta\sigma(t)}{\Delta\sigma(0)} = -\frac{t}{\tau} \, .$$

That is, the natural logarithm of the photoconductivity as a function of time is a straight line with the slope $-1/\tau$.

8.5.2   Hall Effect

Let us focus for the sake of simplicity on a situation when the charge carriers of one type dominate in the sample; that is, either electron or hole contribution to the conductivity can be neglected. Consider what happens when, in addition to the electric field in the x-direction, $\vec{\mathcal{E}} = \vec{e}_x V/L$, a magnetic field $\vec{\mathcal{B}} = \mathcal{B}\vec{e}_z$ is applied

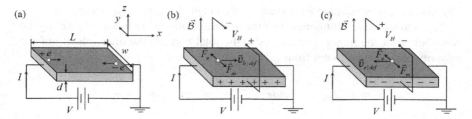

Fig. 8.3 (a) Motion of electrons and holes due to the electric field generated by the applied voltage. (b) Same as in (a), but in the presence of a magnetic field $\vec{B}$ applied perpendicular to the sample, which is assumed to be a p-type semiconductor. (c) Same as in (b), but for an n-type semiconductor sample

perpendicular to the sample in the z-direction, see Fig. 8.3b and c. A magnetic field creates a force that acts on a charge $q = e$ or $-e$ moving with the drift velocity $\vec{v}_{drf} = \vec{v}_{e|drf}$ or $v_{h|drf}$:

$$\vec{F}_m = q\,\vec{v}_{drf} \times \vec{B}.$$

If the majority charge carriers are holes, $q = +e$, then their flow coincides with the direction of the electric current I, Fig. 8.3b. If the majority carriers are electrons, $q = -e$, then their flow is against the electric current, Fig. 8.3c. In either case, the magnetic force $\vec{F}_m$ acting on a majority carrier points in the negative y-direction. As a result, the majority carriers get deflected toward the same side of the sample. This side accumulates positive or negative charge if the semiconductor is of p- or n-type, respectively, see Fig. 8.3b and c. The opposite side gets depleted of the majority carriers and therefore also gets charged up. The charge on the opposite sign is due to the ionized impurities—negative acceptors in a p-type and positive donors in an n-type semiconductor.

The process of charge build-up stops when the accumulated charges generate an electric field whose force on the majority carriers exactly balances the magnetic force,

$$F_e = F_m.$$

As a result, a potential difference develops between the two opposite sides of the sample. It is called the Hall voltage. Its polarity immediately signals the sign of the majority charge carriers, see Fig. 8.3b and c.

The magnitude of the Hall voltage allows finding their concentration. Indeed, the electric field in the y-direction and the force produced by this field on a majority carrier are

$$|\mathcal{E}_y| = |V_H|/w, \quad |F_e| = e|V_H|/w.$$

This force must be equal in magnitude of the magnetic force $|F_m| = e|v_{drf}|\mathcal{B}$.

Let us assume for definiteness that the majority carriers are holes. Neglecting the contribution of the minority electrons to the total current, the drift velocity of the majority holes can be found from

$$I = wd J_{h|drf} = wdev_{h|drf} p \quad \Rightarrow \quad v_{h|drf} = \frac{I}{wdep} \, ,$$

and thus the magnetic force is

$$|F_m| = \frac{I\mathcal{B}}{wdp} \, .$$

The condition $|F_m| = |F_e|$ gives the concentration of holes:

$$p = \frac{I\mathcal{B}}{ed|V_H|} \, .$$

If the sign of the Hall voltage reveals that the majority carriers are electrons, then their concentration is given by an exact same formula

$$n = \frac{I\mathcal{B}}{ed|V_H|} \, .$$

8.6 Temperature and Doping Level Dependence of Mobility

Collisions are responsible for the electrons' transitions from one momentum state to another. Usually, several collision mechanisms are operative, namely, scattering of electrons and holes by the lattice vibrations (phonons), scattering by the impurities, and scattering by the lattice defects. Note that all these mechanisms result in the deviation of the atomic arrangement in a semiconductor from a perfect periodic order.

The frequency of phonon scattering, v_{phonon}, increases with the temperature. The respective scattering time $\tau_{phonon} = 1/v_{phonon}$ is decreasing with T. Thus, if only phonon scattering mechanism were present, the mobility $\mu \propto \tau_{phonon}$ would be decreasing with temperature. Experimental measurements performed on semiconductor samples where the phonon scattering is the dominant one (i.e., the impurity concentration is low) suggest that $\tau_{phonon} \propto T^{-\alpha}$, where the exponent α is close to 2; e.g., $\alpha \approx 2.2$ in silicon.

Impurity scattering frequency, $v_{impurity}$, increases with the concentration of the ionized impurities,

$$N_I = N_d^+ + N_a^- \, ,$$

Table 8.1 Electron and hole mobility parameters of silicon as a function of temperature, measured in Kelvins

Parameter	Electrons	Holes
μ_0 [cm^2/(V·s)]	$88\,(300\,\mathrm{K}/T)^{0.57}$	$54.3\,(300\,\mathrm{K}/T)^{0.57}$
μ_1 [cm^2/(V·s)]	$1252\,(300\,\mathrm{K}/T)^{2.33}$	$407\,(300\,\mathrm{K}/T)^{2.23}$
N_0 [cm^{-3}]	$1.26 \cdot 10^{17}\,(T/300\,\mathrm{K})^{2.4}$	$2.35 \cdot 10^{17}\,(T/300\,\mathrm{K})^{2.4}$
α	$0.88\,(300\,\mathrm{K}/T)^{0.146}$	$0.88\,(300\,\mathrm{K}/T)^{0.146}$

and decreases with temperature. The reason for the temperature decrease is that the higher the temperature, the higher the thermal velocity of electrons and holes and the less time they spend interacting with the ionized impurities. Therefore, the probability that this interaction will result in a change of the charge carrier's momentum decreases with T.

When several scattering mechanisms are present, the total scattering frequency is the sum of the contributions from each mechanism, e.g., $\nu = \nu_{phonon} + \nu_{impurity}$ for thermal and impurity scattering. As for the time of scattering, we have $\tau^{-1} = \tau_{phonon}^{-1} + \tau_{impurity}^{-1}$. The mobility is, then, $\mu = \frac{e}{m^*(\nu_{phonon}+\nu_{impurity})}$ for electrons and holes. It decreases both with N_I and with T.

Various empirical fitting expressions for electron and hole mobilities exist. Here, we reproduce the one that has been suggested for silicon and describes the electron and hole mobilities as a function of both impurity concentration and temperature:[1]

$$\mu = \mu_0 + \frac{\mu_1}{1 + (N_I/N_0)^\alpha}.$$

The temperature dependence of the parameters μ_0, μ_1, N_0, and α is indicated in Table 8.1.

8.7 Einstein's Relation

The diffusion coefficient is a parameter that tells us how quickly an initially non-uniform distribution of charge carriers spreads out through the sample. Mobility, on the other hand, relates the average velocity of a charge carrier to the applied electric field. At a first glance, $D_{e,h}$ and $\mu_{e,h}$ are independent of each other because they describe completely different processes. Yet, there is a simple relation between them, which was originally discovered by Einstein in 1905.

As in Sect. 6.1, we place a piece of semiconductor in a non-uniform electric potential $\phi(\vec{r})$, which causes a redistribution of charge carriers inside the material.

---

[1] N.D. Arora, J.R. Hauser, and D.J. Roulston, "Electron and hole mobilities in silicon as a function of concentration and temperature", IEEE Trans. Electron Dev., ED-29, 292–295 (1982).

A non-uniform potential produces an electric field, $\vec{\mathcal{E}} = -\vec{\nabla}\phi$, which, in turn, results in a drift of charge carriers with the respective flux densities $\vec{j}_{e|drf} = -\mu_e\, n_0(\vec{r})\, \vec{\mathcal{E}}$ and $\vec{j}_{h|drf} = \mu_h\, p_0(\vec{r})\, \vec{\mathcal{E}}$. On the other hand, the non-uniformity of electron and hole concentrations results in the diffusion of these charge carriers with the flux densities $\vec{j}_{e|diff} = -D_e\vec{\nabla}n_0(\vec{r})$ and $\vec{j}_{h|diff} = -D_h\vec{\nabla}p_0(\vec{r})$. The total flux densities

$$\vec{j}_e(\vec{r}) = \vec{j}_{e|drf} + \vec{j}_{e|diff} = -\mu_e\vec{\mathcal{E}}n - D_e\vec{\nabla}n_0 = \mu_e n_0\vec{\nabla}\phi(\vec{r}) - D_e\vec{\nabla}n_0(\vec{r})\,,$$

$$\vec{j}_h(\vec{r}) = -\mu_h p_0\vec{\nabla}\phi(\vec{r}) - D_h\vec{\nabla}p_0(\vec{r})$$

must be zero in equilibrium. This is only possible if

$$\mu_e n_0(\vec{r})\vec{\nabla}\phi(\vec{r}) = D_e\vec{\nabla}n_0(\vec{r})\,, \quad -\mu_h p_0(\vec{r})\vec{\nabla}\phi(\vec{r}) = D_h\vec{\nabla}p_0(\vec{r})\,.$$

Note that

$$\frac{\vec{\nabla}n_0(\vec{r})}{n_0(\vec{r})} = \vec{\nabla}\ln n_0(\vec{r}) = \frac{\vec{\nabla}\phi(\vec{r})}{V_{th}}\,, \quad \frac{\vec{\nabla}p_0(\vec{r})}{p_0(\vec{r})} = \vec{\nabla}\ln p_0(\vec{r}) = -\frac{\vec{\nabla}\phi(\vec{r})}{V_{th}}\,,$$

as follows from the expressions for $n_0(\vec{r})$ and $p_0(\vec{r})$ from Sect. 6.1. Combining the last two equations, we obtain Einstein's relation:

$$\mu_e = \frac{D_e}{V_{th}}\,, \quad \mu_h = \frac{D_h}{V_{th}}\,.$$

While mobility is relatively easy to measure experimentally, the diffusion coefficient is not. Einstein's relation allows finding the diffusion coefficient from the knowledge of mobility.

Einstein's relation is approximate because it relies on Boltzmann's approximation for the charge carriers concentrations. It breaks down if the Fermi energy differs from the conduction or valence band edge by less than a few thermal energies kT.

8.8   Continuity Equation

As electrons and holes move inside the semiconductor under the action of an electric field $\mathcal{E}(\vec{r})$, they get generated and recombine. Hence, the respective concentrations $n(\vec{r}, t)$ and $p(\vec{r}, t)$ change both in time and in space. Continuity equation is the balance equation that relates the rate of change of the concentration to the flow of particles, as well as their generation and recombination.

Here, we will focus, for the sake of simplicity, on the one-dimensional flow of electrons in the x-direction. We assume that generation and recombination rates depend on the x-coordinate only and are constant in the yz-plane at a given x. The derivation for holes is completely analogous. Consider a small cylinder of length

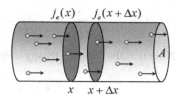

Fig. 8.4 Electron flow in a semiconductor

Δx and the base area A, see Fig. 8.4. The cylinder is oriented in the x-direction, which is the direction of the average particle flow. Its volume is $\Delta V = A\Delta x$, and it contains $\Delta N = n\Delta V$ electrons. We assume that Δx is sufficiently small for the concentration $n(x)$ to be practically constant inside the cylinder.

The number of electrons inside this cylinder changes in time for two reasons: (i) electrons flow into and out of the cylinder through its two bases, and their inflow and the outflow do not necessarily balance each other; (iii) electrons get generated and recombine inside the cylinder with the generation and recombination rates not necessarily equal each other. Mathematically, the balance of these factors can be expressed as

$$\frac{d\Delta N_e}{dt} = \left(\frac{d\Delta N_e}{dt}\right)_{in} - \left(\frac{d\Delta N_e}{dt}\right)_{out} + \left(\frac{d\Delta N_e}{dt}\right)_{gen} - \left(\frac{d\Delta N_e}{dt}\right)_{rec}.$$

The generation and recombination contributions are

$$\left(\frac{d\Delta N_e}{dt}\right)_{gen} = G\,A\,\Delta x\,, \quad \left(\frac{d\Delta N_e}{dt}\right)_{rec} = R\,A\,\Delta x\,.$$

As for the in- and outflow of the electrons, it is related to the respective current densities:

$$\left(\frac{d\Delta N_e}{dt}\right)_{in} = j_e(x)A\,, \quad \left(\frac{d\Delta N_e}{dt}\right)_{out} = j_e(x + \Delta x)A\,.$$

Note that, in contrast to the concentration, we cannot assume that electron current density is constant inside the cylinder. Had we set $j_e(x) = j_e(x + \Delta x)$, then the in- and outflow contributions would have mutually canceled.

Collecting all terms, we rewrite the balance equation as

$$\frac{\partial n(x, t)}{\partial t}A\Delta x = j_e(x)A - j_e(x + \Delta x)A + G(x)A\Delta x - R(x)A\Delta x\,.$$

Dividing both sides by $A\Delta x$ and recalling that $\frac{j_e(x+\Delta x)-j_e(x)}{\Delta x} = \frac{\partial j_e(x)}{\partial x}$ in the limit $\Delta x \to 0$, we arrive at the continuity equation for electron concentration; the hole counterpart is obtained analogously:

$$\frac{\partial n}{\partial t} = -\frac{\partial j_e}{\partial x} + G - R \, , \quad \frac{\partial p}{\partial t} = -\frac{\partial j_h}{\partial x} + G - R \, .$$

Note that the current density consists of drift and diffusion contributions:

$$j_{e,h} = j_{e,h|drf} + j_{e,h|diff} \, ,$$

$$j_{e|drf} = -\mu_e n \mathcal{E} \, , \quad j_{h|drf} = \mu_h p \mathcal{E} \, ,$$

$$j_{e|diff} = -D_e \frac{\partial n}{\partial x} \, , \quad j_{h|diff} = -D_h \frac{\partial p}{\partial x} \, .$$

8.9 Problems

8.9.1 Solved Problems

Problem 1 An n-type silicon sample at $T = 300$ K has the conductivity of $16\,(\Omega \cdot$ cm$)^{-1}$.

(a) Determine the donor concentration and the electron mobility using the empirical expressions for mobility from Sect. 8.6.
(b) What is the diffusion coefficient of electrons and holes in this sample?

Problem 2 Consider a semiconductor with donor doping concentration

$$N_d(x) = (10^{14} \, \text{cm}^{-3}) \, e^{x/(1\,\mu\text{m})}$$

that increases with x. Find the electric field and electrostatic potential distribution inside this semiconductor.

Problem 3 When a Si sample is illuminated by a lamp, its conductivity is measured as $66.6\,\Omega^{-1}\,\text{cm}^{-1}$. After illumination is switched off, the conductivity becomes $33.3\,\Omega^{-1}\,\text{cm}^{-1}$ after $0.1\,\mu\text{s}$ and $9.1\,\Omega^{-1}\,\text{cm}^{-1}$ after $2\,\mu\text{s}$. Determine the following parameters: the net recombination time in the sample.

8.9.2 Practice Problems

Problem 1 A Si sample is doped with donors at $N_d = 10^{17}\,\text{cm}^{-3}$ and acceptors at $N_a = 10^{18}\,\text{cm}^{-3}$. The temperature is 300 K. Using the mobility vs. impurity concentration formula from Sect. 8.6, determine the conductivity and the resistivity of this sample.

Problem 2 A silicon sample is acceptor-doped at $N_a = 10^{17}\,\mathrm{cm}^{-3}$. The photoconductivity of this sample is found to decay as $\sigma(t) \propto e^{-t/(0.1\,\mathrm{ms})}$. Find its Shockley–Read–Hall recombination time.

Problem 3 The geometric parameters of a semiconductor sample at $T = 300\,\mathrm{K}$ are: $d = 20\,\mu\mathrm{m}$, $w = 300\,\mu\mathrm{m}$, and $L = 1\,\mathrm{mm}$. When the voltage of 0.1 V is applied, the current of 0.3 mA flows through the sample. A magnetic field of $B = 0.1\,\mathrm{T}$ is applied in the z-direction, and the measured Hall voltage is $V_H = 20\,\mathrm{mV}$ such that the resulting Hall electric field points in the positive-y-direction. Determine the conductivity type, the majority carrier concentration, and the majority carrier mobility in this sample.

8.9.3 Solutions

Problem 1

(a) Since it is an n-type semiconductor, it is reasonable to assume that

$$n_0 \approx N_d \gg n_i \gg p_0 \,,$$

and hence, the dominant contribution to the conductivity comes from the electrons. Hence, we approximate

$$\sigma = e(\mu_e n_0 + \mu_h p_0) = e\mu_e N_d \,.$$

Using the expression from Sect. 8.6,

$$\left(\mu_0 + \frac{\mu_1}{1 + (N_d/N_0)^\alpha}\right) N_d = \frac{\sigma}{e} = \frac{16}{1.6 \cdot 10^{-19}} \,.$$

We solve this equation iteratively. First, we rewrite it as

$$N_d = \frac{\sigma/e}{\mu_0 + \frac{\mu_1}{1+(N_d/N_0)^\alpha}} = \frac{10^{20}}{88 + \frac{1252}{1+(N_d/1.26\cdot10^{17})^{0.88}}} \,.$$

Then, starting with some initial value of N_d, say, $N_d = 10^{17}\,\mathrm{cm}^{-3}$, we use the right-hand side expression to calculate an improved estimate, which is again plugged in into the right-hand side. This procedure is repeated until convergence is reached. The final result turns out to be $N_d = 1.5 \cdot 10^{17}\,\mathrm{cm}^{-3}$. Electron mobility is then found to be $\mu_e = 666\,\mathrm{cm}^2/(\mathrm{V}\cdot\mathrm{s})$.

(b) Electron diffusion coefficient is found using Einstein's relation:

$$D_e = \mu_e V_{th} = 666 \cdot 0.02585 = 17.2\,\text{cm}^2/\text{s} .$$

Hole mobility at $N_d = 1.5 \cdot 10^{17}\,\text{cm}^{-3}$ is

$$\mu_h = 54.3 + \frac{407}{1 + (1.5 \cdot 10^{17}/2.35 \cdot 10^{17})^{0.88}} = 297.5\,\text{cm}^2/(\text{V} \cdot \text{s}) .$$

Hole diffusion coefficient is, then,

$$D_h = \mu_h V_{th} = 297.5 \cdot 0.02585 = 7.69\,\text{cm}^2/\text{s} .$$

Problem 2 Since $N_d(x) \gg n_i$ at $x > 0$, the electron concentration

$$n_0(x) = N_d(x)$$

increases with x. This means that the electrons should diffuse in the negative-x-direction with the diffusion particle current density

$$j_{e|diff} = -D_e \frac{dn_0(x)}{dx} = -D_e \frac{dN_d(x)}{dx} .$$

But a small displacement of the negative electrons relative to the positive donor ions results in an onset of an electric field in the negative-x-direction. This field pulls the electrons to the right, so that there is no electron flow in the material. The drift current produced by this field,

$$j_{e|drf} = -\mathcal{E}\mu_e n_0(x) = -\mathcal{E}\mu_e N_d(x),$$

perfectly compensates the diffusion current:

$$j_{e|drf} + j_{e|diff} = -\mathcal{E}\mu_e N_d(x) - D_e \frac{dN_d(x)}{dx} = 0 .$$

From this, we find the electric field:

$$\mathcal{E} = -\frac{D_e}{\mu_e} \frac{1}{N_d} \frac{dN_d}{dx} = -V_{th} \frac{d\ln N_d}{dx} ,$$

where we used Einstein's relation, $D_e = \mu_e V_{th}$, and the fact that $d\ln N_d/dx = N_d^{-1} dN_d/dx$. Since $d\ln N_d/dx = 1\,\mu\text{m}^{-1} = 10^4\,\text{cm}^{-1}$, the electric field is

$$\mathcal{E} = -0.02585 \cdot 10^4\,\text{V/cm} = -259\,\text{V/cm} .$$

Since the electric field is related to the potential as

$$\mathcal{E} = -\frac{d\phi(x)}{dx} \; ,$$

we can conclude that the potential increases with x as

$$\phi(x) = \left(0.02585 \, \frac{V}{\mu m}\right) \cdot x = \left(258.5 \, \frac{V}{cm}\right) \cdot x \; .$$

Problem 3 In order to find N_d and τ, we thus need to fit the experimental $\sigma(t)$ curve with an exponential function:

$$\sigma(t) = \sigma_0 + \Delta\sigma(0) \, e^{-t/\tau} \; , \quad \sigma_0 = e\mu_e N_d \; , \quad \Delta\sigma(0) = e(\mu_e + \mu_h)\Delta n(0) \; ,$$

where σ_0 is the conductivity in equilibrium (no radiation), and $\Delta\sigma(0)$ is the photoconductivity.

From the experiment, we know that $\sigma(t)$ decreases by a factor of 2 in the first $0.1 \, \mu s$. This implies that τ is of the order of $0.1 \, \mu s$ as well. After $2 \, \mu s$, the exponential factor $e^{-t/\tau}$ should be negligibly small (note that $e^{-2/0.1} = e^{-20} = 2 \cdot 10^{-9}$), so we can set to a good accuracy

$$\sigma(2 \, \mu s) = \sigma_0 = 9.1 \, \Omega^{-1} \, cm^{-1} \; .$$

From the experimental data, we find that at $t = 0$, $\sigma(0) = \sigma_{eq} + \Delta\sigma_0 = 66.6 \, \Omega^{-1} \, cm^{-1}$, from which we find the photoconductivity

$$\Delta\sigma(0) = \sigma(0) - \sigma_{eq} = 66.6 - 9.1 = 57.5 \, \Omega^{-1} \, cm^{-1} \; .$$

We finally find τ by using the conductivity at $t = 0.1 \, \mu s$:

$$\tau = -\frac{t}{\ln \frac{\sigma(t) - \sigma_{eq}}{\Delta\sigma_0}} = -\frac{0.1}{\ln \frac{33.3 - 9.1}{57.5}} = 0.1155 \, \mu s \; .$$

Part III
Semiconductor Devices

Chapter 9
Metal–Semiconductor Contact

9.1 Reasons to Study

Metal–semiconductor contact is worth studying for several reasons. First, to collect a signal from a semiconductor device, one needs to attach a metal wire to it; in this case, one is interested in having a good ohmic contact with low contact resistance for both bias directions. Second, as we shall see later, metal–semiconductor contact can also act as a diode, i.e., let the current flow in only one direction. Such diodes, called Schottky diodes, can be preferred for some applications over the semiconductor pn junction diodes discussed later on in this book. The third reason is that understanding the physics of a metal–semiconductor contact, especially the energy band diagrams, will prepare us to analyze a more complicated system, the metal–oxide–semiconductor field effect transistor.

9.2 Energy Band Diagram

The minimal energy necessary to remove an electron from a piece of metal is called the metal work function. It is the difference between the energy of an electron at rest just outside the metal—the so-called vacuum level—and the Fermi energy of a metal. We will denote the work function as $e\psi_m$, a product of the elementary charge e and the parameter ψ_m that is measured in Volts:

$$e\psi_m = E_{vac} - E_F .$$

Its typical value is about 4–6 eV, depending on the material. We define the work function in a semiconductor in the same manner, i.e., as the difference between the vacuum and the Fermi energy,

© The Author(s), under exclusive license to Springer Nature Switzerland AG 2022
M. Evstigneev, *Introduction to Semiconductor Physics and Devices*,
https://doi.org/10.1007/978-3-031-08458-4_9

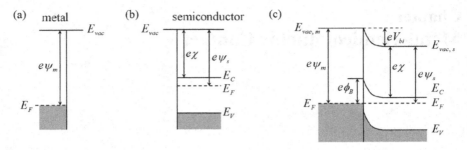

Fig. 9.1 The energy band diagram of (**a**) a metal, (**b**) a semiconductor, and (**c**) a metal–semiconductor contact

$$e\psi_s = E_{vac} - E_F \, ,$$

see Fig. 9.1a and b showing the energy band diagrams of a metal and a semiconductor that are not in contact with each other.

In addition, we define the electron affinity $e\chi$ as the energy difference between the vacuum level and the conduction band edge,

$$e\chi = E_{vac} - E_C \, .$$

Electron affinity ranges between 3 and 4.5 eV, depending on the material.

A metal and a semiconductor in contact will exchange their electrons so as to bring their Fermi energies to a common level. The Fermi level in a metal lies inside its conduction band, and therefore, the number of free electrons in a metal is in many orders of magnitude higher than the number of conduction electrons and holes in a semiconductor. For this reason, the change of the Fermi level in the metal will be negligibly small as compared to the change of the Fermi level in the semiconductor. In other words, semiconductor Fermi level will line up with the metal Fermi level, whereas the metal Fermi energy will practically remain the same.

The vacuum level of the metal and the semiconductor must be the same at the point of contact. At the same time, electron affinity and the band gap of the semiconductor must be the same both in the bulk and near the semiconductor–metal interface. Indeed, these parameters depend on the band structure of the semiconductor and do not depend on band filling.

These three observations—common Fermi level, the same vacuum level at the contact, and the constant values of the semiconductor electron affinity and band gap—necessarily imply that semiconductor bands are bent near the interface with the metal, see Fig. 9.1c showing band bending for $\psi_s < \psi_m$. Generally speaking, bands can bend both up and down, depending on the mutual relation between metal work function and semiconductor work function in the bulk. If semiconductor work function is higher than the metal work function, semiconductor bands bend down; otherwise, they bend up.

While the electron affinity and band gap are the same in the semiconductor bulk and near the interface, the semiconductor work function is position-dependent due to band bending. So is the energy difference between the Fermi level and the band edges of the semiconductor. This leads to a change of the majority carrier concentration in the semiconductor bulk and near the interface.

Even the type of the majority carriers may differ in the bulk and near the contact plane. For instance, in the case shown in Fig. 9.1c, the semiconductor is of the n-type in the bulk, but the concentration of holes exceeds the electron concentrations near the surface. This can be understood as follows. Because the semiconductor work function is smaller than the metal work function, it is easier for electrons to move over from the semiconductor into the metal than from the metal into the semiconductor. The semiconductor near the metal will then be depleted of the conduction electrons, resulting in the appearance of a space-charge region created by the positively charged donor ions near the contact.

The electrons that entered the metal from the semiconductor side, in turn, will create excess negative charge in the metal. This access charge exists as a surface charge at the metal–semiconductor interface. Strictly speaking, it also results in band bending in the metal, but the size of this bending is negligibly small as compared to the semiconductor counterpart. Band bending in the semiconductor results in an onset of the built-in voltage, which we define as the magnitude of the difference of work functions of the two materials:

$$V_{bi} = |\psi_m - \psi_s| .$$

Assuming the bulk of the semiconductor to be grounded, the value of the electric potential on the semiconductor surface is $\phi_S = -V_{bi}$ if $\psi_m > \psi_s$, see Fig. 9.1c.

An electron on the metal side with the energy equal to E_F will "see" a sharp potential barrier that prevents it from entering the semiconductor conduction band. This barrier is called the Schottky barrier. By inspecting the diagram in Fig. 9.1c, we see immediately that its height is given by

$$e\phi_B = e\psi_m - e\chi .$$

This expression is sometimes called the Schottky–Mott rule.

The metal work function ψ_m and the semiconductor electron affinity χ are bulk properties of the materials, whereas ϕ_B is the property of the metal–semiconductor interface. In real life, Schottky barrier height is given by the Schottky–Mott rule only approximately. It is also affected by other surface properties of the two materials in contact, in particular, by surface chemistry and the presence of surface defects. Therefore, in practice, Schottky barrier has to be measured experimentally.

9.3   SCR Capacitance

The potential distribution in the semiconductor is described by Poisson's equation, see Sect. 6.2:

$$\frac{d^2\phi(x)}{dx^2} = -\frac{e}{\epsilon_0\epsilon}(p_0(x) - n_0(x) + N_d) .$$

We solve it using the depletion approximation explained in Sect. 6.3.3. That is, we approximate the charge density as zero outside the depletion region of width x_d and as eN_d inside the depletion region $0 < x < x_d$. The solution of Poisson equation reads

$$\phi(x) = \begin{cases} -\frac{eN_d}{2\epsilon_0\epsilon}(x - x_d)^2 & \text{for } 0 < x < x_d , \\ 0 & \text{for } x > x_d . \end{cases}$$

Employing the boundary condition, $\phi(0) = -V_{bi}$, we immediately find the depletion region width

$$x_d = \sqrt{\frac{2\epsilon_0\epsilon V_{bi}}{eN_d}} .$$

If an external bias V is applied to the metal relative to the semiconductor, we have to replace V_{bi} with $V_{bi} - V$ in this expression:

$$x_d = \sqrt{\frac{2\epsilon_0\epsilon(V_{bi} - V)}{eN_d}} .$$

A change of the voltage V by an amount dV results in a change of the SCR width by

$$dx_d = \frac{dx_d}{dV}dV = -\sqrt{\frac{\epsilon_0\epsilon}{2eN_d(V_{bi} - V)}}\,dV .$$

This results in a change of the SCR charge by

$$dQ = -eN_d\,A\,dx_d = \sqrt{\frac{\epsilon_0\epsilon\,eN_d}{2(V_{bi} - V)}}\,A\,dV .$$

The capacitance per unit area is

Fig. 9.2 The inverse square of the capacitance per unit area of a SCR formed in an n-type semiconductor in contact with a metal with $\psi_m > \psi_s$ as a function of the applied voltage

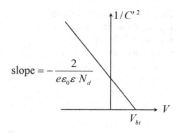

$$C' = \frac{1}{A}\frac{dQ}{dV} = \sqrt{\frac{e\epsilon_0\epsilon N_d}{2(V_{bi} - V)}} = \frac{\epsilon_0\epsilon}{x_d}.$$

This expression can also be obtained from a formal definition of a parallel-plate capacitance of thickness x_d, namely, $C = \frac{\epsilon_0\epsilon A}{x_d}$.

This result offers a practical method to determine the built-in voltage V_{bi} experimentally. Indeed, we note that the inverse capacitance squared,

$$\frac{1}{C'^2} = \frac{2}{e\epsilon_0\epsilon N_d}(V_{bi} - V),$$

depends linearly on the applied voltage V. The slope of this straight line is $-\frac{2}{e\epsilon_0\epsilon N_d}$. The built-in voltage can be found as the intercept of this straight line with the horizontal axis, see Fig. 9.2.

9.4 Ohmic Contact

For signal collection from a semiconductor device, it is desirable that the contact between the semiconductor and the metal wire had a small resistance for both directions of bias. Such contacts are called ohmic. There are two ways to produce an ohmic contact: (1) using metal–semiconductor combination that would have little or no Schottky barrier and (2) employing the tunneling effect.

Option 1 If the semiconductor is of n-type, then one can use a metal with the work function smaller than the semiconductors, $\psi_m < \psi_s$. This will cause the semiconductor bands to bend down. When no voltage is applied, the energy band diagram looks like the one in Fig. 9.3a. The band bending occurs because some electrons leave the metal into the semiconductor, creating a surface layer of an increased conduction electron concentration on the semiconductor side. Application of a positive voltage to the semiconductor will make it easier for even more electrons to be transferred from the metal into the semiconductor. Application of a positive voltage to the metal will pull electrons from the semiconductor into the metal

Fig. 9.3 An ohmic contact between a metal and (**a**) an n-type and (**b**) a p-type semiconductor

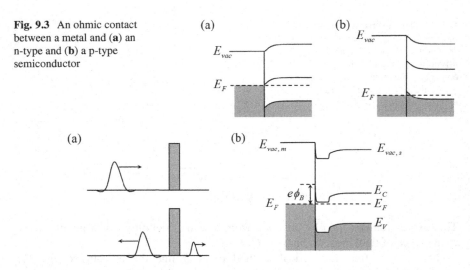

Fig. 9.4 (**a**) Tunneling of a wave packet through a potential barrier of finite width and height. (**b**) Realization of a tunneling barrier by heavy donor doping of an n-type semiconductor near the interface with a metal

because there is no potential barrier at the metal–semiconductor interface. Thus, current will flow with little resistance for both bias polarities.

If the semiconductor is of p-type, then one can use a metal whose work function is higher than that of the semiconductor, $\psi_m > \psi_s$. This will cause the semiconductor bands to bend up, creating a thin hole-enriched layer near the interface with the metal, see Fig. 9.3b. Application of a positive bias to the semiconductor will cause the majority holes to move toward the metal, and, at the same time, it will cause the electrons from the metal to flow into the semiconductor. After entering the semiconductor, those electrons will recombine with the holes, with the overall effect of electric current flowing from the semiconductor to the metal with little resistance. Application of a positive bias to the metal, on the other hand, will attract electrons from the semiconductor valence band into the metal; this is equivalent to the flow of holes away from the metal.

Option 2 When a massive classical object of some kinetic energy E hits a potential barrier of height $V_B > E$, it bounces back. In quantum mechanics, however, this is not necessarily the case. At the most fundamental level, all objects are waves described by their wave function. When such a wave meets an obstacle, the wave function cannot instantaneously become zero at the position of the barrier. It smoothly decays to zero inside the barrier but remains finite even on the other side of the barrier. This is so even when the barrier height exceeds the particle's kinetic energy. This means that a particle, such as an electron, has a small, but finite probability to tunnel through the barrier to the other side, see Fig. 9.4a. Tunneling probability is the larger the thinner the barrier and the smaller its height.

Now, a strong doping of a thin layer of an n-type semiconductor with donors will create additional band bending near the surface. The width of the SCR is inversely proportional to the square root of the doping concentration and can be made, therefore, very small. This makes it possible for electrons to tunnel from metal to semiconductor and back, see Fig. 9.4b. That is, electrons will be able to move both ways between the metal and the semiconductor via tunneling mechanism.

9.5 Rectification in a Metal–Semiconductor Contact

9.5.1 Metal/n-Type Semiconductor Junction

Qualitative Considerations

Let us derive the current–voltage relation of a junction formed by an n-type semiconductor with a metal of a higher work function, $\psi_m > \psi_s$, see Fig. 9.1c. This contact is rectifying, as the following reasoning shows. Application of a positive bias to the metal attracts the majority electrons from the semiconductor into the metal, which results in a large current in the forward direction. On the hand, application of a negative bias to the metal prevents the majority electrons to enter the metal from the semiconductor. Furthermore, the electrons of the metal likewise are not able to enter the semiconductor because of the high and sharp Schottky barrier they encounter at the metal–semiconductor boundary. Hence, the current in the reverse direction is small. It is not exactly zero because a small fraction of the metal electrons do have energy exceeding $e\phi_B$.

Assume zero bias. Electrons attempting to go from the metal into the semiconductor meet a high and sharp Schottky barrier $e\phi_B$, whereas the conduction electrons attempting to go in the opposite direction have to overcome a much shallower barrier eV_{bi} due to the built-in voltage, see Fig. 9.1c. Electron fluxes from the metal into the semiconductor and from the semiconductor into the metal must perfectly balance each other in equilibrium at zero applied voltage:

$$j_{m \to s}(0) + j_{s \to m}(0) = 0 .$$

The reason why the two fluxes balance each other in spite of the big difference between the sizes of the two barriers, $e\phi_B$ and eV_{bi}, is because the concentration of electrons in the metal greatly exceeds the majority electron concentration in the semiconductor.

Our goal is to find the total electron flux and the electric current density

$$j(V) = j_{m \to s}(V) + j_{s \to m}(V) , \quad J(V) = -ej(V)$$

when a voltage V is applied to the metal side, whereas the semiconductor is grounded.

Qualitatively, the rectification effect of the metal–semiconductor contact results from the different dependence on the applied voltage V of the potential barrier that electrons need to overcome to enter the semiconductor from the metal side and vice versa. Namely, the external voltage V has practically no effect on the height of Schottky barrier $e\phi_B$. This, in turn, implies that the electron flux density from the metal into the semiconductor is practically independent of V:

$$j_{m \to s}(V) \approx j_{m \to s}(0) = -j_{s \to m}(0) = j_S .$$

Here we denoted the so-called saturation flux density as $j_S = -j_{s \to m}(0)$.

As for the semiconductor side, application of a positive bias to the metal lowers the barrier by eV, see Fig. 9.5a, thereby exponentially increasing the number of electrons that can enter the metal from the semiconductor per unit time. This concentration gets enhanced by a factor $e^{V/V_{th}}$. We therefore expect that the electron flux into the metal should be

$$j_{s \to m}(V) = j_{s \to m}(0)e^{V/V_{th}} = -j_S e^{V/V_{th}} .$$

The current–voltage (or $J - V$) curve is thus expected to have the form

$$J(V) = -ej_{s \to m}(V) - ej_{m \to s}(V) = ej_S e^{V/V_{th}} - ej_S = J_S(e^{V/V_{th}} - 1) ,$$

where $J_S = ej_S$ is the reverse saturation current density, calculated below. The graph of this function is shown in Fig. 9.5b. A passive[1] element, in which the current to flow for one polarity of bias with very little current for the reverse polarity, is called

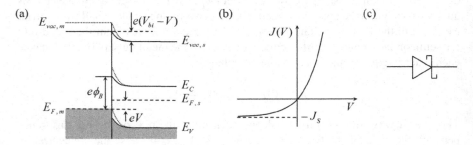

Fig. 9.5 (a) The energy band diagram of a metal contact with an n-type semiconductor when a positive voltage is applied to the metal. (b) Qualitative behavior of the current–voltage curve of a rectifying contact between a metal and an n-type semiconductor. (c) Electronic symbol of a Schottky diode

---

[1] A passive element is the one that does not require an external power source.

a diode. Specifically, a diode based on a metal–semiconductor contact is called the Schottky diode. Its electronic symbol is depicted in Fig. 9.5c.

9.5.2 Reverse Saturation Current Density of a Schottky Diode

To obtain the reverse saturation current density, we need to estimate the number of electrons that enter the metal from the semiconductor side per unit area per unit time when a voltage V is applied. Let us first consider a simpler case of a classical gas in a solid container. The gas has density n, atomic mass m, and temperature T, corresponding to thermal energy kT. The number of particles j incident on the wall of the container per unit surface area per unit time should be expressible in terms of m, kT, and n. We can figure out this expression by analyzing the dimensions of the parameters involved, which are

$$[n] = L^{-3} ; \quad [kT] = M \cdot L^2/T^2 ; \quad [m] = M .$$

Here, L, T, and M stand for the dimensions of length, time, and mass, respectively. We need to combine the parameters n, kT, and m so as to get the right dimension of j, namely,

$$[j] = 1/(L^2 \cdot T).$$

The only such combination is

$$j = bn\sqrt{\frac{kT}{m}} ,$$

where the non-dimensional prefactor b cannot be established by dimensional analysis. The kinetic theory of gases gives the value

$$b = \frac{1}{\sqrt{2\pi}} \quad \Rightarrow \quad j = n\sqrt{\frac{kT}{2\pi m}} .$$

We can use this result to obtain the semiconductor-to-metal particle current density by replacing the particle mass m with electron conductivity effective mass m_e^* and particle concentration n with the concentration n^* of those electrons that are capable of overcoming the energy barrier $e(V_{bi} - V)$ that stands on their way into the metal. In order to be able to get out of the semiconductor, a conduction electron must have the energy not less than

$$E_B = E_C + e(V_{bi} - V) = (E_C + eV_{bi}) - eV = E_F + e\phi_B - eV .$$

We used the fact that

$$E_C + eV_{bi} = E_F + e\phi_B \, ,$$

as follows from the energy band diagram at zero bias.

The concentration of conduction electrons with the energy above E_B is then

$$n^* = \int_{E_B}^{\infty} dE \, g(E) \, f(E) \approx N_C e^{-(e\phi_B - eV)/kT} = N_C e^{-(\phi_B - V)/V_{th}} \, ,$$

where $g(E)$ and $f(E)$ are electron density of states and Fermi–Dirac distribution; we used Boltzmann approximation to evaluate the integral. As obtained earlier (see Sect. 5.2), the effective density of states in the conduction band is

$$N_C = \frac{2(2\pi m_e^* kT)^{3/2}}{h^3} \, .$$

Modifying our expression for the particle flux density, j, we obtain the electron flux density from the semiconductor to the metal as

$$j_{s \to m}(V) = n^* \sqrt{\frac{kT}{2\pi m_e^*}} = N_C \sqrt{\frac{kT}{2\pi m_e^*}} \, e^{(\phi_B - V)/V_{th}} \, .$$

The saturation electron flux density is therefore

$$j_S = \frac{4\pi m_e^* k^2 T^2}{h^3} \, e^{-\phi_B/V_{th}} \, .$$

As for electric current density, $J(V) = J_S(e^{V/V_{th}} - 1)$, we get the reverse saturation current density

$$J_S = AT^2 e^{-\phi_B/V_{th}} \, ,$$

where

$$A = \frac{4\pi m_e^* k^2 e}{h^3} = (120.2 \, \text{A} \cdot \text{cm}^{-2} \cdot \text{K}^{-2}) \cdot \frac{m_e^*}{m_e}$$

is the so-called Richardson constant.[2] In semiconductors, the typical value of Richardson's constant A is about $100 \, \text{A}/(\text{K}^2 \cdot \text{cm}^2)$. With the typical Schottky barrier height of ca. $0.7 \, \text{eV}$, we find that at room temperature, the reverse saturation current density J_S is of the order of $10^{-5} \, \text{A/cm}^2$.

---

[2] It is named after Owen Richardson, who developed this theory to describe thermionic emission of electrons from metals into vacuum in the early 1900s (Nobel Prize 1928).

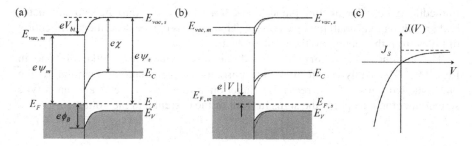

Fig. 9.6 A junction formed by a metal and a p-type semiconductor of the larger work function, $e\psi_s > e\psi_m$, (**a**) at zero applied voltage and (**b**) when a negative bias is applied to the metal. (**c**) The $J - V$ curve of this junction

9.5.3 Metal/p-Type Semiconductor Junction

The energy band diagram of a junction formed by a metal and a p-type semiconductor with a higher work function at zero applied voltage is shown in Fig. 9.6a. Due to the work function difference,

$$\psi_m < \psi_s \ ,$$

the semiconductor bands are bent down near the interface. The built-in voltage is

$$V_{bi} = \psi_s - \psi_m \ .$$

The Schottky barrier is now measured from the metal Fermi level to the valence band edge in the junction plane. For an ideal diode, it is given by

$$e\phi_B = e\chi + E_g - e\psi_m \ ,$$

as can be verified by inspecting the energy band diagram.

Application of a negative bias to the metal, Fig. 9.6b, pushes the electrons from the metal into the semiconductor and pulls the holes from the semiconductor into the metal; the two types of charge carriers meet and recombine at the metal–semiconductor boundary. This results in an appreciable current through the structure in the negative-x direction. This current should increase in magnitude with increasing the negative voltage.

On the other hand, application of a positive bias to the metal side prevents electrons from leaving the metal into the semiconductor and prevents the majority holes from coming to the metal from the semiconductor. The current, however, is not exactly zero, even when the positive voltage applied to the metal is large. A large positive voltage pulls the electrons from the metal–semiconductor interface into the volume of the metal, simultaneously generating holes on the semiconductor side immediately near the junction plane. Those holes then move into the semiconductor bulk, resulting in a small positive current. This process can be described as injection of holes from the metal into the semiconductor.

The current–voltage curve of this diode looks qualitatively like the one in Fig. 9.6c. The analytical form of the current–voltage relation is similar to the one obtained earlier for a rectifying junction formed by a metal and an n-type semiconductor, except for the signs in front of the current and the voltage:

$$J(V) = J_S(1 - e^{-V/V_{th}}) .$$

The expression for the saturation current is similar to the one obtained earlier, but with the electron conductivity effective mass replaced by the hole counterpart:

$$J_S = AT^2 e^{-\phi_B/V_{th}} , \quad A = \frac{4\pi m_h^* k^2 e}{h^3} = (120.2\, \text{A} \cdot \text{cm}^{-2} \cdot \text{K}^{-2}) \cdot \frac{m_h^*}{m_e} .$$

9.6   Non-ideality Effects

The $J - V$ relation obtained in Sect. 9.5.2 describes an ideal Schottky diode and does not account for some important effects that take place in the real metal–semiconductor contacts. One such effect overlooked by this theory is that the barrier height "seen" by electrons impinging on the metal from the semiconductor side is actually lower than the value used in the calculations. To understand this, let us consider the junction formed by a metal and an n-type semiconductor, see Fig. 9.5a, under a forward bias.

Consider a conduction electron inside the space-charge region near the metal–semiconductor interface. The ionized donors produce the electric potential in the space-charge region (see Sect. 6.3.3):

$$\phi_{SCR}(x) = (V_{bi} - V)\left(1 - \frac{x}{x_d}\right)^2 , \quad x_d = \sqrt{\frac{2\epsilon_0 \epsilon (V_{bi} - V)}{eN_d}} .$$

In addition, the electron also interacts with the electrons of the metal. This interaction can be effectively described by the method of image charges. Namely, the force produced on the electron by the metal is the same as the force between this electron of charge $-e$ and its image charge $+e$ located at a distance x on the other side of the metal–semiconductor interface, see Fig. 9.7a:

$$F_{im}(x) = -\frac{1}{4\pi \epsilon_0 \epsilon} \frac{e^2}{(2x)^2} = -\frac{e^2}{16\pi \epsilon_0 \epsilon x^2} .$$

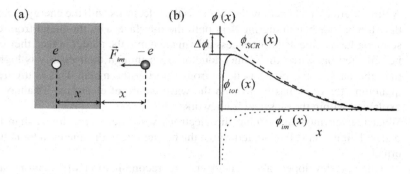

Fig. 9.7 (a) An electron (charge $-e$) at a distance x from a metal surface interacts with the metal so as if there were an image charge e at a distance x on the other side of the semiconductor–metal interface. (b) The total electric potential is a sum of the SCR-generated contribution, $\phi_{SCR}(x)$, and the image charge potential $\phi_{im}(x)$

The image-charge force can be expressed as the negative derivative of the image-charge electric potential multiplied by the electron charge, $-e$:

$$F_{im}(x) = e\frac{d\phi_{im}(x)}{dx}, \quad \phi_{im}(x) = -\frac{e}{16\pi\epsilon_0 x}.$$

The total potential in which the electron finds itself is, thus,

$$\phi_{tot}(x) = \phi_{SCR}(x) + \phi_{im}(x),$$

see Fig. 9.7b. It is seen that it has a maximum, whose height is smaller than the value $V_{bi} - V$ obtained without taking the image charge into account. The barrier height reduction $\Delta\phi$ can be obtained by approximating $\phi_{SCR}(x) \approx (V_{bi} - V)(1 - 2x/x_d)$; this approximation is valid provided that the maximum of the total potential $\phi_{tot}(x)$ is not too far from the surface. The result of this calculation is

$$\Delta\phi(V) = \left(\frac{e^3 N_d(V_{bi} - V)}{8\pi^2(\epsilon_0\epsilon)^3}\right)^{1/4}.$$

We see that the barrier height decrease depends on the applied voltage. The result is that the reverse saturation current density is increased relative to its ideal-junction value by a factor $e^{\Delta\phi(V)/V_{th}}$:

$$J_S = AT^2 e^{-\phi_B/V_{th}} e^{\Delta\phi(V)/V_{th}}.$$

Actually, this value of the saturation current density is also not quite accurate for several reasons:

(i) A further effect neglected by the theory is that electrons with the energy greater than the barrier height are incident onto the interface with the metal from the semiconductor side does not necessarily make it into the metal. In fact, they can be reflected back into the semiconductor, even though their energy is higher than the barrier that prevents them from entering the metal. This is a purely quantum effect that has to do with the wave nature of electrons; it cannot be explained within the classical framework.

(ii) What is even more important, the electrons with the energy lower than the barrier height can still make it through the barrier due to the quantum tunneling effect.

(iii) The theory developed above neglects the recombination of electrons and holes in the space-charge region and also the surface recombination on the semiconductor boundary.

(iv) There is a thin dielectric layer that separates the metal and the semiconductor. While the electrons can tunnel through this layer, there is also a finite potential difference across this layer, which affects the value of the Schottky barrier height.

All these non-ideality effects result in a deviation of the experimental current–voltage curve from the ideal one obtained here. The ideal $J - V$ curve describes the real-life devices only qualitatively correctly. In order to fit the experimental $J - V$ curves of a real Schottky diode, the ideal-diode relation is modified by introducing the so-called ideality factor n:

$$J(V) = J_S(e^{V/(nV_{th})} - 1) .$$

The ideality factor n is usually of the order of $1 - 2$, but it can also be outside of this range.

9.7 Problems

9.7.1 Solved Problems

Problem 1 Consider a junction formed by p-Si doped at $N_a = 10^{15}\,\mathrm{cm}^{-3}$ and some metal at $T = 300\,\mathrm{K}$. The semiconductor is grounded. Silicon parameters can be found in Appendix A.3. What should the metal work function be in order for the voltage applied to the metal electrode to be the same as the surface potential on the semiconductor? Draw the energy band diagram at zero voltage V, $V > 0$, and $V < 0$.

Problem 2 A junction is formed by silicon and a golden electrode with the work function $e\psi_m = 5.1\,\mathrm{eV}$. Determine the type (donors or acceptors) and the

(a)   (b)

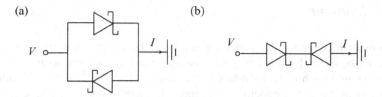

Fig. 9.8 Two Schottky diodes connected (**a**) in parallel and (**b**) in series

minimal concentration of the dopants in order for an inversion layer to form in the semiconductor near the surface.

Problem 3 Two identical ideal Schottky diodes are connected in parallel, with one in forward and the other in reverse direction, see Fig. 9.8a. Assume the reverse saturation current I_S of each diode to be known. Derive and sketch the current–voltage relation, $I(V)/I_S$ vs. V/V_{th}, of this circuit.

9.7.2 Practice Problems

Problem 1 Derive the value of the barrier lowering $\Delta\phi$ from Sect. 9.5.2.

Problem 2 When a positive bias of 0.5 V is applied to the metal side of a metal–silicon junction, the junction has the capacitance of 0.25 pF. A bias of 1.5 V results in a capacitance of 0.05 pF. The junction area $A = 10^{-5}\,\mathrm{cm^2}$, and the temperature $T = 300\,\mathrm{K}$.

(a) Is the semiconductor of p- or n-type? What is the smallest value of the applied voltage at which the space-charge region in the semiconductor starts to appear? Is the metal work function higher or lower than the semiconductors?
(b) What is the doping level of the semiconductor?
(c) What is the value of the metal work function?

Problem 3 When a positive forward bias of 0.1 V is applied to an ideal Schottky diode at $T = 350\,\mathrm{K}$, a current density $J = 5.32 \cdot 10^{-10}\,\mathrm{A/cm^2}$ is measured. The bias of 0.5 V results in the current density of $3.18 \cdot 10^{-4}\,\mathrm{A/cm^2}$.

(a) What is the reverse saturation current density of this junction?
(b) Determine the height of Schottky barrier.

Problem 4 Two identical ideal Schottky diodes are connected in series, one in the forward and the other in the reverse direction, see Fig. 9.8b. Assume the reverse saturation current, I_S, of each diode to be known. Derive the current–voltage curve $I(V)$ of this circuit. Plot the ratio I/I_S vs. the ratio V/V_{th}.

9.7.3 Solutions

Problem 1 In order for the surface potential to be exactly equal to the voltage applied to the metal, band bending must be absent in the semiconductor. The size of band bending, which is the built-in voltage, is zero only if the metal work function is the same as the semiconductor work function. Therefore, we must have

$$e\psi_m = e\psi_s = E_{vac} - E_F = E_{vac} - E_C + E_C - E_F = e\chi + E_C - E_F .$$

The energy difference $E_C - E_F$ can be found from the electron concentration:

$$n_0 = \frac{n_i^2}{N_a} = N_C\, e^{-(E_C-E_F)/kT} ,$$

giving

$$E_C - E_F = kT\,\ln\frac{N_a N_C}{n_i^2} = 0.02585\,\ln\frac{10^{15}\cdot 3.2\cdot 10^{19}}{(8.59\cdot 10^9)^2} = 0.871\,\text{eV} .$$

The metal work function must be

$$e\psi_m = 4.05 + 0.871 = 4.92\,\text{eV} .$$

The energy band diagrams of this contact for $V = 0$, $V > 0$, and $V < 0$ are shown in Fig. 9.9.

Problem 2 Let us assume that the majority carriers in the semiconductor are electrons. Referring to Fig. 9.1c, we can express the difference $E_{vac,m} - E_F$ in two ways:

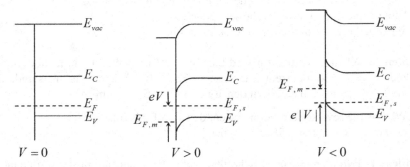

Fig. 9.9 The energy band diagram of a contact formed by a p-type silicon and a metal with $\psi_m = \psi_s$ at zero, positive, and negative voltage applied to the metal

$$\psi_m = V_{bi} + \chi + \frac{E_C - E_F}{e}.$$

The inversion layer starts forming when the concentration of holes near the surface equals the concentration of electrons far from the surface. Hole concentration near the surface is related to its value in the bulk via

$$p_{0,surf} = p_0 e^{V_{bi}/V_{th}} = \frac{n_i^2}{n_0} e^{V_{bi}/V_{th}}.$$

From the condition $p_{0,surf} = n_0$, we obtain (see also Sect. 6.3.4)

$$V_{bi} = 2V_{th} \ln \frac{n_0}{n_i}.$$

Next, the difference between the conduction band energy and the Fermi energy is related to the electron concentration by (see Sect. 5.2)

$$\frac{E_C - E_F}{e} = V_{th} \ln \frac{N_C}{n_0}.$$

Substitution of the expressions for V_{bi} and $(E_C - E_F)/e$ gives

$$\psi_m = 2V_{th} \ln \frac{n_0}{n_i} + \chi + V_{th} \ln \frac{N_C}{n_0} = \chi + V_{th} \ln \frac{n_0 N_C}{n_i^2}.$$

From this, we find the electron concentration:

$$n_0 = \frac{n_i^2}{N_C} e^{(\psi_m - \chi)/V_{th}}.$$

Taking the parameters from Appendix A.3, we find its numerical value

$$n_0 = \frac{(8.59 \cdot 10^9)^2}{3.2 \cdot 10^{19}} e^{(5.1-4.05)/0.02585} = 1.01 \cdot 10^{18} \, \text{cm}^{-3}.$$

We see that $n_0 \gg n_i$, which is consistent with our assumption that the semiconductor is of n-type.

If we assumed that it were of p-type, then we would have had to use the energy band diagram from Fig. 9.6a; a similar reasoning would have brought us to a conclusion that $n_0 \gg n_i$, which is inconsistent with the assumption about the conductivity type of the semiconductor.

Hence, the semiconductor must be doped with donors at $N_d = 1.01 \cdot 10^{18} \, \text{cm}^{-3}$.

Problem 3 The total current I is the sum of currents through the upper, I_1, and the lower, I_2, diodes,

Fig. 9.10 The current–voltage curve of a circuit, in which two diodes are connected in parallel, see Fig. 9.8a

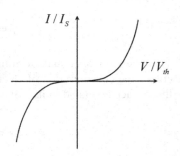

$$I = I_1 + I_2 \,.$$

Since the diodes are in parallel, the same voltage V is applied to both of them. The upper diode is in the forward direction; hence,

$$I_1(V) = I_S(e^{V/V_{th}} - 1) \,.$$

The lower diode is reverse-biased: if V is large and positive, the current through the lower diode is I_S, and if V is large and negative, $I_2 \propto -I_S e^{-V/V_{th}}$. More generally,

$$I_2(V) = -I_S(e^{-V/V_{th}} - 1) \,.$$

Adding the two currents together,

$$I(V) = I_S(e^{V/V_{th}} - e^{-V/V_{th}}) = 2I_S \sinh \frac{V}{V_{th}} \quad \Rightarrow \quad \frac{I(V)}{I_S} = 2 \sinh \frac{V}{V_{th}} \,.$$

This relation is plotted in Fig. 9.10.

Chapter 10
Metal–Oxide–Semiconductor Field Effect Transistor (MOSFET)

10.1 MOSFET Schematics and Operation Principle

A MOSFET is a workhorse of integrated electronics and is at the heart of every digital circuit. It is a building block of a charge coupled device (CCD). To name but a few of its other applications, it is used to store information and perform logical operations in digital electronics, to amplify and generate ac-signals in analog circuits, and to convert analog signals into digital ones.

A schematics of an enhancement-mode MOSFET with an n-channel, or n-MOSFET, is shown in Fig. 10.1a. It is based on a p-type substrate, with the terminal attached to it called the body (B). The body is usually grounded. In the upper part of the substrate, there are two heavily doped n^+-regions, with the source (S) and drain (D) terminals. They are separated by an insulating oxide layer, on top of which there is a metal electrode, called the gate (G). In Si-based circuits, the insulating layer is usually formed by SiO_2 (silicon dioxide), hence the name of the device. The electrodes can be made of a metal, but also they can be made of heavily doped semiconductor.

The idea of an n-MOSFET operation is that by applying a positive voltage to the gate, one affects the value of the electric potential on the semiconductor surface underneath the oxide layer. Once the gate voltage reaches a certain threshold value, V_T, the electric potential on the semiconductor surface becomes equal to the threshold potential $\phi_T = (E_{Fi} - E_F)/(2e)$, at which an electron-rich inversion layer is formed, as discussed in the end of Sect. 6.3.4. The inversion layer connects the source and the drain, allowing electric current to flow between these two electrodes.

Likewise, one may use an n-type substrate with the p^+-type source and drain regions. Then, application of a sufficiently large negative voltage to the gate will result in a formation of a hole-enriched inversion layer. This kind of a MOSFET is called the p-MOSFET in the enhancement mode. The electronic symbols of an enhancement-mode n- and p-MOSFETS are shown in Fig. 10.1a.

© The Author(s), under exclusive license to Springer Nature Switzerland AG 2022
M. Evstigneev, *Introduction to Semiconductor Physics and Devices*,
https://doi.org/10.1007/978-3-031-08458-4_10

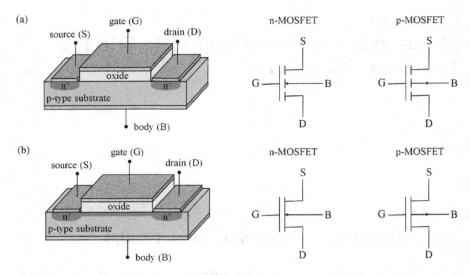

Fig. 10.1 (**a**) The structure of an enhancement-mode n-MOSFET and its electronic symbol; in a p-MOSFET, whose electronic symbol is shown to the right, the substrate is of n-type, and the heavily doped regions under the source and drain electrodes are of the p^+-type. (**b**) The sketch of a depletion-mode n-MOSFET and its electronic symbol. The depletion-mode p-MOSFET has the same structure, but with the n-type substrate, the p^+-type source and drain regions, and a hole-enriched inversion layer underneath the oxide

The action of a MOSFET is thus reminiscent of a faucet, where a small variation in the turning angle of the valve has a large effect on the water flow. Note that the oxide layer is absolutely necessary for MOSFET operation. Indeed, if one had a metal gate electrode deposited directly on the semiconductor surface, it would have been impossible to control surface conductivity by the gate voltage simply because the surface would already have been connecting the source and the drain at all voltages.

A great advantage of MOSFET technology is its flexibility. It is possible to have the parameters of the device such that the inversion layer is present even at zero gate voltage; then, one would need to apply a certain threshold voltage to the gate in order to switch the inversion layer off. A MOSFET with the pre-existing electron-rich inversion layer at zero gate voltage is termed an n-MOSFET in the depletion mode, see Fig. 10.1b. One can also build a depletion-mode p-MOSFET if one uses an n-type body and two p^+-type source and drain electrodes. The electronic symbols of depletion-mode n- and p-MOSFETs are shown in the same figure.

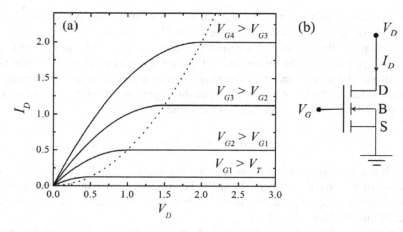

Fig. 10.2 (**a**) The current–voltage curves of an enhancement-mode n-MOSFET, connected as shown in panel (**b**), for different values of the gate voltages. The dotted curve separates the region where I_D increases with V_D from the region where I_D is constant

10.2 Qualitative Description of MOSFET $I - V$ Curve

Consider an enhancement-mode n-MOSFET with the body and the source electrodes grounded, a voltage V_D applied to the drain and a current I_D measured. Its $I - V$ curve for different values of the gate voltage V_G is shown in Fig. 10.2.

At a fixed gate voltage exceeding the threshold value, $V_G > V_T$, the drain current I_D initially increases linearly with the drain voltage. The initial slope (the conductance) dI_D/dV_D of the I_D vs. V_D curve increases as the gate voltage V_G is increased. The reason for this is clear: the higher the gate voltage, the thicker the inversion layer, and hence the lower its resistance. Interestingly, when the drain voltage attains a certain value V_D^{sat}, the drain current reaches saturation. It remains constant upon further increase of the drain voltage above V_D^{sat}-value. The saturation drain voltage increases with the gate voltage.

Let us try and understand the current saturation. In order for an inversion layer to form, it is necessary for the potential difference ΔV between the semiconductor surface underneath the oxide layer and the gate to be greater than some threshold voltage V_T. But when a voltage V_D is applied to the drain, the value of the local potential at a given point underneath the oxide will depend on the distance x from the source. It will change from $V(0) = 0$ at $x = 0$ to $V(L) = V_D$ at $x = L$, where L is the drain-to-source distance, see Fig. 10.3.

The potential difference between the gate and a point x in the inversion channel is then given by

$$\Delta V(x) = V_G - V(x).$$

If the inversion channel connects the drain and the source, then ΔV changes from the highest value V_G at $x = 0$ to the lowest value $V_G - V_D$ at $x = L$.

Increasing the drain voltage V_D results in a decrease of $\Delta V(L) = V_G - V_D$. As long as $\Delta V(L)$ is above the threshold level, the inversion channel is still connecting the drain and the source, and the current is still increasing upon further increase of the drain voltage, albeit not as sharply as in the beginning of the $I - V$ curve. But at some point, the drain voltage reaches the value

$$V_D^{sat} = V_G - V_T,$$

at which the local potential difference at $x = L$ equals $\Delta V(L) = V_T$. Then, the existence of the inversion channel cannot be supported near the drain anymore, although it still can exist at some distance away from it. This causes the phenomenon called the pinch-off, see Fig. 10.3.

At pinch-off, the inversion channel ends at some distance ΔL from the drain. The distance ΔL increases with V_D starting from the value $\Delta L = 0$ at $V_D = V_D^{sat}$.

It might come as a surprise that the drain current does not go to zero after the pinch-off, even though the inversion channel is not in contact with the drain anymore. Rather, the current saturates and does not change upon further increase of the drain voltage. This is so, because, after the pinch-off, the potential difference between one end of the inversion layer at $x = 0$ and the other end at $x = L - \Delta L$ remains constant and equals V_D^{sat}. For V_D greater than the saturation value V_D^{sat} by not too much, the distance ΔL should be much smaller than L, which implies that the inversion channel resistance R is approximately independent of V_D. Hence, also the current in the inversion layer should be equal to $I_D^{sat} = V_D^{sat}/R$ independent of V_D. The current $I_D = I_D^{sat}$ at $V_D > V_D^{sat}$ is carried by the electrons that move from the rightmost edge of the inversion layer to the drain through the space-charge region around the drain.

Our next task is to find the threshold gate voltage V_T at which the inversion layer develops and to derive an analytic expression for the $I - V$ curve of a MOSFET.

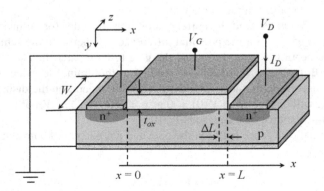

Fig. 10.3 Inversion layer at $V_D > V_D^{sat}$

10.3 Quantitative Description of a MOSFET $I - V$ Curve

Let us consider the situation when a thin inversion channel is formed, i.e., the gate voltage exceeds the threshold value, $V_G > V_T$, and the drain voltage is below the pinch-off value, $V_D < V_D^{sat}$. As we go along the inversion channel in the x-direction, the potential difference $V(x)$ between a point x in the channel and the source at the origin increases from $V(x = 0) = 0$ to $V(x = L) = V_D$. The local electric field in the inversion layer,

$$\mathcal{E}(x) = -\frac{dV(x)}{dx},$$

acts in the negative-x-direction. It pulls the electrons from the source to the drain. On average, the electrons in the inversion layer move with the drift velocity, which depends on x:

$$v_{e|drf}(x) = -\mu_e \mathcal{E}(x) = \mu_e \frac{dV(x)}{dx},$$

where μ_e is the electron mobility in the inversion channel. Note that it is appreciably smaller than electron mobility far away from the surface because the scattering of the electrons by the surface is an additional mechanism that slows them down. This mechanism is not operative in the semiconductor bulk.

Let $\eta_i(x)$ be the surface concentration of the electrons in the inversion layer. It depends on the distance x from the source, being maximal at $x = 0$ and gradually decreasing to $\eta_i(L)$ as the inversion layer gets thinner. The corresponding surface-charge density is $-e\eta_i(x)$. It is supported by the local potential difference across the oxide layer above the threshold voltage V_T,

$$\Delta V(x) - V_T = V_G - V(x) - V_T.$$

The threshold voltage must be subtracted from $\Delta V(x)$ because there is no inversion layer at $\Delta V(x) < V_T$; the inversion layer only starts to be formed at $\Delta V(x) = V_T$.

The surface-charge density $-e\eta_i(x)$ is given by

$$-e\eta_i(x) = -C'_{ox}(V_G - V_T - V(x)),$$

where

$$C'_{ox} = \frac{\epsilon_0 \epsilon_{ox}}{t_{ox}}$$

is the unit area capacitance of the insulating oxide layer between the gate electrode and the MOSFET body. The thickness of this oxide layer is t_{ox}, and its relative dielectric constant is ϵ_{ox}.

The value of the electric current at a position x is given by

$$I(x) = -eW\eta_i(x)\, v_{e|drf} = We\eta_i(x)\, \mu_e \frac{dV}{dx} = WC'_{ox}(V_G - V_T - V(x))\, \mu_e \frac{dV}{dx}.$$

Indeed, $dq = -eW\eta_i(x)dx$ is the total electron charge contained between x and $x + dx$. This charge is moving with the average velocity $v_{e|drf} = dx/dt$, resulting in the current $I(x) = dq/dt$ above. Note that for a positive drain voltage V_D, the electrons are drifting in the positive x-direction, so that the current $I(x)$ is negative.

Now, due to charge conservation, the current must be constant at all values of x. Its value is equal to the drain current, $I_D = -I(x)$ for all x. We choose the minus sign so as to make the drain current have the same sign as the drain voltage. Thus,

$$I_D = WC'_{ox}(V_G - V_T - V(x))\mu_e \frac{dV}{dx}.$$

We multiply both sides of this equation by dx and integrate its left-hand side over x from $x = 0$ to $x = L$, giving $I_D L$. Its right-hand side is integrated over the voltage from the value $V(x = 0) = 0$ to $V(x = L) = V_D$, giving $WC'_{ox}((V_G - V_T)V_D - V_D^2/2)$. We thus obtain the current–voltage curve before saturation is reached:

$$I_D(V_D) = K_n \left(2(V_G - V_T)V_D - V_D^2\right) \quad \text{for } V_D < V_D^{sat},$$

where the combination

$$K_n = \frac{W\mu_e C'_{ox}}{2L}$$

is called the conduction parameter of an n-MOSFET.

Note that $I_D = 0$ at $V_D = 0$, as one would expect. This $I_D(V_D)$ curve represents a parabola with a maximum at $V_D = V_D^{sat} = V_G - V_T$, as can be verified by differentiation.

If $V_D > V_D^{sat}$, this relation would predict a decrease of I_D with V_D. But the analytical considerations presented above are invalid if V_D exceeds the saturation value V_D^{sat} because we are now in the pinch-off regime, in which the inversion layer does not connect the source and the drain anymore. Rather than starting to decrease, the current saturates at the value

$$I_D^{sat} = I_D(V_D^{sat}) = K_n(V_G - V_T)^2 \quad \text{for } V_D > V_D^{sat}.$$

The current–voltage curve of a p-MOSFET is the same as of an n-MOSFET, with the only difference being that the conduction parameter K_n has to be replaced with

$$K_p = \frac{W\mu_h C'_{ox}}{2L},$$

where μ_h is the hole mobility near the surface.

10.4 Determination of the Threshold Voltage

10.4.1 Energy Band Diagram of a MOS Structure at Zero Gate Voltage

The inversion layer in the p-semiconductor is formed when the potential on its surface underneath the oxide reaches the threshold value

$$\phi_T = 2V_{th} \ln \frac{N_a}{n_i} = 2\frac{E_{Fi} - E_F}{e},$$

at which electron concentration near the surface becomes equal to hole concentration in the bulk. Semiconductor surface potential is controlled by the voltage on the gate electrode separated from the semiconductor by a thin dielectric (e.g., SiO_2) layer. We need to figure out the threshold gate voltage V_T, at which the surface potential attains the value ϕ_T. Solving this problem requires a careful analysis of the energy band diagram of the metal–oxide–semiconductor system when a voltage V_G is applied to the gate, while the source, the drain, and the body are grounded.

Let us forget about the source and the drain for the moment and consider the energy bands in the p-type substrate when zero voltage is applied to the gate, see Fig. 10.4, showing the metal–oxide–semiconductor (MOS) structure in panel (a) and its energy band diagram at zero bias in panel (b). Drawing the energy band diagram of a MOS structure is based on the same principles as in the simpler metal–semiconductor case.

In particular, metal and semiconductor in equilibrium must have the same Fermi energy. This is so because maintaining zero voltage between the gate and the body of a MOS structure requires connecting the two objects by a wire through which the gate and the body can exchange electrons. This electron exchange renders the Fermi energies in the gate and the substrate to be the same.

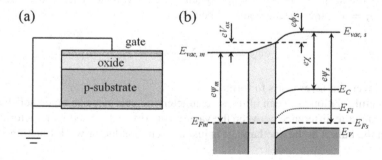

Fig. 10.4 (a) A metal–oxide–semiconductor (MOS) capacitor with zero voltage between the substrate and the gate and (b) the energy band diagram of this structure

Next, the vacuum energy level must be continuous throughout the MOS structure. Just as in the case of a metal–semiconductor contact, this implies band bending in the semiconductor and oxide layers. More precisely, because there are no free charge carriers in the oxide, a potential difference V_{ox} develops across it due to the band bending in the semiconductor and also due to the surface charges that are trapped on the oxide surface (we will discuss them later). Band bending in the semiconductor results in the onset of a surface potential ϕ_S.

The energy difference between the vacuum level and Fermi energy in the metal is the metal work function, $e\psi_m$. Electron affinity in the semiconductor bulk—that is, the energy difference between the semiconductor conduction band edge and the vacuum energy—is denoted as $e\chi$. The work function in the semiconductor is then

$$e\psi_s = E_{vac,s} - E_{Fs} = e\chi + E_C - E_F.$$

10.4.2   Energy Band Diagram of a MOS Structure for Non-zero Gate Voltage

Consider now what happens when a positive bias V_G is applied to the gate. This leads to a shift of all energy levels in the gate down by the amount eV_G, e.g., the new Fermi level in the metal is now $E_{Fm} = E_{Fs} - eV_G$. At the same time, the energy differences between the metal Fermi level E_{Fm} and the vacuum level, $e\psi_m$, remain the same as in the zero-bias case.

Because the semiconductor is grounded, all the bulk energy levels and energy differences (E_{Fs}, E_{Fi}, E_C, E_V, $E_{vac,s}$, and $e\chi$) remain the same as in the zero-bias case. But at the semiconductor–oxide interface, the size of the band bending will be affected by the voltage V_G applied to the gate. The size of the semiconductor band bending near the surface is quantified by the surface potential $e\phi_S$, which equals the difference between the characteristic energies (such as $E_{vac,s}$ or E_{Fi}) in the bulk and near the surface, see Fig. 10.5b. When the gate voltage reaches the threshold value, $V_G = V_T$, the surface potential becomes

$$\phi_S = \phi_T = 2\frac{E_{Fi} - E_F}{e},$$

and an inversion layer starts to form.

A useful relation between all these characteristic energies can be established by considering the difference between the vacuum energy and the semiconductor Fermi level in the bulk. On the one hand, it is just the semiconductor work function

$$e\psi_s = E_C - E_F + e\chi,$$

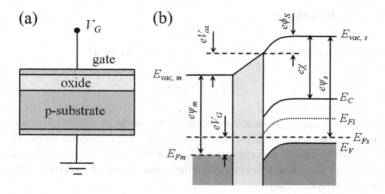

Fig. 10.5 (a) A metal–oxide–semiconductor (MOS) capacitor with a non-zero voltage between the substrate and the gate and (b) the energy band diagram of this structure

see the right part of Fig. 10.5b. The same energy difference can be expressed in terms of the metal properties, the potential drop across the oxide, and the band bending parameter, namely,

$$e\psi_s = e\psi_m - eV_G + eV_{ox} + e\phi_S,$$

see the left part of Fig. 10.5b. Equating the two expressions, we have

$$e\psi_m - eV_G + eV_{ox} + e\phi_S = E_C - E_F + e\chi,$$

or

$$V_G = \psi_m + V_{ox} + \phi_S - \chi - \frac{E_C - E_F}{e}.$$

This relation can be written in several equivalent forms. For instance, previously, we have obtained an expression for the intrinsic Fermi energy

$$E_{Fi} = \frac{E_C + E_V}{2} + \frac{kT}{2}\ln\frac{N_V}{N_C} \approx \frac{E_C + E_V}{2} = E_C - \frac{E_g}{2}.$$

Given that the effective densities of states N_V and N_C are of the same order of magnitude, the ratio N_V/N_C is of the order of unity, and thus the logarithmic term can be neglected. Then

$$V_G = \psi_m - \chi_S + V_{ox} + \phi_S' + \frac{E_F - E_{Fi}}{e} - \frac{E_g}{2e}.$$

10.4.3   Oxide Voltage

An oxide layer of thickness t_{ox} has the capacitance per unit area

$$C'_{ox} = \frac{\epsilon_0 \epsilon_{ox}}{t_{ox}},$$

where ϵ_{ox} is the relative dielectric constant of the oxide. A voltage V_{ox} across the oxide layer is associated with the surface charges on its opposite sides. The gate (metal) acquires a charge per unit area Q'_m, which has the same sign as the oxide voltage polarity. The charge developed in the semiconductor underneath the oxide layer has the surface density

$$Q'_s = -Q'_m.$$

It equals in magnitude to the surface-charge density on the metal side because the whole structure must be electrically neutral. Thus, by the definition of capacitance, the voltage across the oxide is given by

$$V_{ox} = \frac{Q'_m}{C'_{ox}} = -\frac{Q'_s}{C'_{ox}} = -\frac{t_{ox}}{\epsilon_0 \epsilon_{ox}} Q'_s.$$

The semiconductor surface-charge density consists of three contributions:

1. The negative charge of the ionized acceptors is in the depletion region underneath the oxide. The surface-charge density of the depletion region is

$$Q'_d = -eN_a x_d, \quad x_d = \sqrt{\frac{2\epsilon_0 \epsilon}{e N_a} \phi_s}.$$

Here, x_d is the width of the depletion region, as derived earlier (see Sect. 6.3.3).
2. The negative charge is due to the electrons in the inversion layer. We denote the respective surface-charge density as Q'_i.
3. During the semiconductor surface oxidation process, additional positive charge is formed at the Si/SiO_2 interface due to the charge carriers trapped by the surface states, trapped ions of various types, charged defects, etc. The magnitude of the respective surface-charge density Q'_{ss} (the subscript ss stands for "semiconductor surface") depends on the technological conditions and is fixed.
    Thus, in general, the voltage drop across the oxide layer is

$$V_{ox} = -\frac{Q'_s}{C'_{ox}} = \frac{e N_a x_d + Q'_i - Q'_{ss}}{C'_{ox}}.$$

Fig. 10.6 The energy band diagram of an enhancement-mode n-MOSFET when the flat-band voltage is applied to the gate

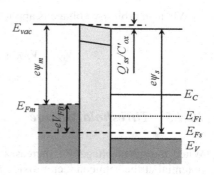

10.4.4 Flat-Band Voltage

The general relation between the gate obtained in the end of Sect. 10.4.2 involves the surface potential ϕ_S, which determines the magnitude of band bending and, accordingly, the width of the depletion region. The flat-band voltage is defined as that value of the gate voltage $V_G = V_{FB}$ for which the band bending is zero, i.e., $\phi_S = 0$, see Fig. 10.6. Then, there is no depletion region under the oxide layer, i.e., $x_d = 0$ under the flat-band conditions. Recalling that the inversion layer is formed after the depletion region reaches its maximal size, we immediately realize that also $Q'_i = 0$.

Then, the oxide voltage under the flat-band conditions is determined just by the trapped surface-charge density,

$$V_{oxFB} = -\frac{Q'_{ss}}{C'_{ox}},$$

and the flat-band value of the gate voltage is

$$V_{FB} = \psi_m - \chi - \frac{E_{Fi} - E_F}{e} - \frac{E_g}{2e} - \frac{Q'_{ss}}{C'_{ox}} = \psi_m - \psi_s - \frac{Q'_{ss}}{C'_{ox}}.$$

This is just the difference between the metal and semiconductor work functions minus the oxide voltage due to the trapped surface charges. For an n-channel MOSFET, the flat-band voltage is negative because the semiconductor work function,

$$e\psi_s = e\chi + E_{Fi} - E_F + \frac{E_g}{2},$$

exceeds the metal counterpart, and because the charge Q'_{ss} on the semiconductor–oxide interface is positive.

When an arbitrary voltage is applied to the gate, it can be expressed as

$$V_G = V_{FB} + \frac{eN_a x_d + Q_i'}{C_{ox}'} + \phi_S.$$

10.4.5   Threshold Voltage

When the gate voltage V_G increases to the threshold value V_T, the electrostatic potential at the semiconductor–oxide interface becomes $\phi_S = \phi_T = 2\frac{E_{Fi} - E_F}{e}$. At this value of the surface potential, the space-charge region underneath the oxide reaches its maximal width

$$x_{dT} = \sqrt{2\frac{\epsilon_0 \epsilon}{eN_a}\phi_T} = \frac{2}{e}\sqrt{\frac{\epsilon_0 \epsilon}{N_a}(E_{Fi} - E_F)}.$$

A further increase of V_G above V_T results in an increase of the inversion-layer surface-charge density, but not in an increase of the depletion region.

At the threshold voltage $V_G = V_T$, the inversion layer just starts to be formed but is not quite there yet. Therefore, we can set at the threshold $Q_i' = 0$. The oxide voltage is then

$$V_{oxT} = \frac{eN_a x_{dT} - Q_{ss}'}{C_{ox}'}.$$

We thus have an expression for the threshold gate voltage

$$V_T = V_{FB} + \frac{eN_a x_{dT}}{C_{ox}'} + 2\frac{E_{Fi} - E_F}{e}.$$

This expression can be written in other equivalent forms. For example, using the formula for the flat-band voltage, we have

$$V_T = \psi_m - \chi + \frac{eN_a x_{dT} - Q_{ss}'}{C_{ox}'} + \frac{E_{Fi} - E_F}{e} - \frac{E_g}{2e}.$$

We have considered the case of an n-MOSFET here. Derivation of these relations for a p-channel MOSFET, and also for the case when, instead of a metal, a heavily doped p- or n-type semiconductor is used as a gate electrode, can be performed in a similar manner.

10.5 Capacitance–Voltage Measurements

A great deal about the MOS structure can be learned from the capacitance–voltage measurements, in which the capacitance between the body and the gate electrodes is measured as a function of the gate voltage. Such measurements are performed by biasing the gate with a dc-voltage V_G and measuring the current produced by a weak ac-voltage $v(t) = v_0 \sin(\omega t)$; this current equals

$$i(t) = \frac{dQ}{dt} = C\frac{dv}{dt} = Cv_0\omega\cos(\omega t),$$

where

$$C = \frac{dQ/dt}{dv/dt} = \frac{dQ}{dv}$$

is the MOS capacitance and Q is the charge on the metal electrode. To exclude the influence of the MOS geometry, we will evaluate the capacitance per unit area, $C' = C/A$, where $A = WL$, see Fig. 10.3.

As before, we focus on the n-MOSFET, i.e., the body is of p-type. Let us assume for now that the frequency ω of the ac-voltage is sufficiently small for the charge carriers in the semiconductor to be able to react to the time-dependent electric field. Three regimes can be distinguished.

Accumulation regime, in which the gate voltage is below the flat-band value:

$$V_G < V_{FB}.$$

At the typical doping levels $N_a > 10^{14}$ cm^{-3}, the semiconductor body behaves as a poor conductor, and therefore, it remains under zero potential. Application of a gate voltage below the flat-band value results in an accumulation of holes in a very thin surface layer underneath the oxide, see Fig. 10.7a. The thickness of this layer is of the order of Debye screening length, see Sect. 6.3.2. Since the voltage drop occurs

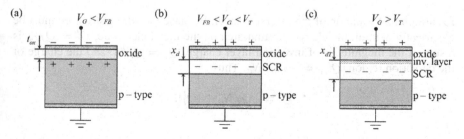

Fig. 10.7 The MOS structure in the (**a**) accumulation, (**b**) depletion, and (**c**) inversion regime

mostly across the oxide, the capacitance of the structure equals to that of the oxide layer:

$$C'(V) = C'_{ox} = \frac{\epsilon_0 \epsilon_{ox}}{t_{ox}}.$$

Depletion regime, in which the gate voltage is high enough for the depletion region to form near the surface, but insufficient for the formation of the inversion layer, see Fig. 10.7b:

$$V_{FB} < V_G < V_T.$$

The depletion region width and capacitance per unit area depend on V as

$$x_d = \sqrt{\frac{2\epsilon_0 \epsilon (V - V_{FB})}{e N_a}}, \quad C'_d = \frac{\epsilon_0 \epsilon}{x_d} = \sqrt{\frac{e N_a \epsilon_0 \epsilon}{2(V - V_{FB})}}.$$

The net capacitance now consists of the oxide and the depletion region capacitances connected in series:

$$\frac{1}{C'} = \frac{1}{C'_{ox}} + \frac{1}{C'_d} \quad \Rightarrow \quad C' = \frac{C'_{ox} C'_d}{C'_{ox} + C'_d}.$$

Since C'_d decreases with V, so does the total capacitance at $V_G > V_{FB}$. The depletion regime ends when the gate voltage reaches the threshold value, at which an inversion layer forms near the surface. The minimal capacitance per unit area is then

$$C'_{min} = \frac{C'_{ox} C'_{dT}}{C'_{ox} + C'_{dT}}, \quad C'_{dT} = \frac{\epsilon_0 \epsilon}{x_d} = \sqrt{\frac{e N_a \epsilon_0 \epsilon}{2\phi_T}}.$$

Inversion regime, in which the gate voltage exceeds the threshold value,

$$V_G > V_T,$$

resulting in a formation of the inversion layer. Because the whole structure must be electrically neutral, the charge per unit area on the metal electrode, $Q' = Q/A$, is a sum of the magnitudes of inversion-layer charge, $Q'_i = Q_i/A$, and the charge of the depletion region, $Q'_{dT} = -e x_{dT} N_a$. Thus,

$$Q' = e x_{dT} N_a + |Q'_i|.$$

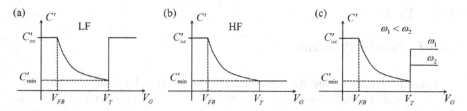

Fig. 10.8 The capacitance–voltage curve of a MOS structure at (**a**) low-frequency driving, (**b**) high-frequency driving, and (**c**) intermediate-frequency driving

The depletion region has the maximal width $x_{dT} = \sqrt{\frac{2\epsilon_0\epsilon\phi_T}{eN_a}}$ corresponding to the threshold value of the surface potential. Therefore, it does not increase with increasing V_G. On the other hand, the charge of the inversion layer grows exponentially strongly with the gate voltage, so that at V_G exceeding V_T by a small amount, the inversion-layer contribution starts to dominate over the depletion region contribution: $Q' \approx |Q_i'|$. Correspondingly, the voltage drop across the depletion region can be neglected, i.e., the main voltage drop occurs across the oxide layer, and the measured capacitance is $C'(V) = C_{ox}'$, as in the accumulation regime:

$$Q' \approx |Q_i'| \;\Rightarrow\; C' \approx C_{ox}'.$$

The capacitance–voltage curve for slow driving is sketched in Fig. 10.8a.

So far, we assumed that the frequency ω is sufficiently low for the semiconductor charge carriers to react to the time-dependent voltage $v(t)$. The slowest component of the semiconductor charge is the inversion layer, which is formed due to the generation–recombination processes. If the frequency ω is much higher than the inverse recombination time, then the charge density in the inversion layer is unable to follow the temporal variations of the gate voltage. As a result, the inversion layer becomes a thin static charged sheet. Even though it may have an appreciable surface-charge density Q_i', it practically does not change in time, making very little contribution to the capacitance

$$C' = \frac{dQ_i'/dt}{dv/dt} + \frac{dQ_d'/dt}{dv/dt}.$$

As a result, the measured capacitance saturates at the value C_{min}', see Fig. 10.8b at high frequencies.

At the intermediate frequencies, the capacitance at $V > V_T$ saturates at some value between C_{min}' and C_{ox}', which increases as the frequency is decreased, see Fig. 10.8c.

10.6 Problems

10.6.1 Solved Problems

Problem 1 Determine the flat-band and the threshold voltage of a Si-based n-MOSFET with the base doping level $N_a = 10^{15}\,\text{cm}^{-3}$, SiO$_2$ insulator of thickness 100 nm and the dielectric constant 3.9. The trapped surface-charge density is $10^{11}\,\text{cm}^{-2}$. The temperature is 300 K. As the gate electrode polycrystalline n$^+$-Si is used with the Fermi level higher than the conduction band energy by 0.05 eV, you may assume that the electron affinity of the gate electron is the same as of the MOSFET body.

Problem 2 Same as in Problem 1, but this time it is a p-MOSFET, i.e., the body is n-type with the same doping concentration $N_d = 10^{15}\,\text{cm}^{-3}$, and the gate is made of the heavily doped p$^+$-Si with the Fermi level 0.05 eV below the valence band edge.

Problem 3 The experimental capacitance–voltage curve of a Si-based MOSFET is sketched in Fig. 10.9a. Its oxide layer is SiO$_2$ with the dielectric constant $\epsilon_{ox} = 3.9$. The gate metal work function is $e\psi_m = 4\,\text{eV}$. The temperature is 300 K. Based on this information, determine the following:

(a) Is this a p-MOSFET or an n-MOSFET?
(b) What is the oxide layer thickness?
(c) What is the semiconductor doping level?
(d) What is the surface concentration of trapped charges in the semiconductor–oxide interface?

Problem 4 For an enhancement-mode n-MOSFET biasing shown in Fig. 10.9b, sketch the current–voltage curve, I_D vs. V_{DS}, for $V_{GD} = 0, 0.5\,V_T, V_T$, and $2\,V_T$.

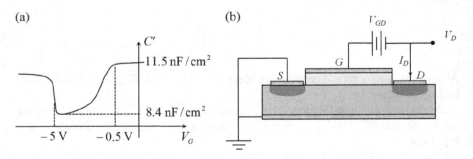

Fig. 10.9 (a) The capacitance–voltage curve of some MOS structure in Problem 3. (b) The biasing of an n-MOSFET in Problem 4

10.6.2 Practice Problems

Problem 1 A MOS capacitor is formed by an n^+-Si/SiO$_2$/n-Si structure at $T = 300$ K. The substrate is grounded. The n^+-Si, used as a gate electrode, is doped so heavily that its Fermi energy is above the conduction band edge by $\Delta E = 0.05$ eV. The n-type silicon body is doped at $N_d = 10^{17}$ cm^{-3}. The SiO$_2$ layer has thickness $t_{ox} = 200$ nm and a relative dielectric constant of $\epsilon_{ox} = 3.9$. On the interface with n-Si, it carries the trapped charges of surface density $Q'_{ss} = 6 \cdot 10^{-9}$ C/cm^2. The properties of Si are given in Appendix A.3. Draw the energy band diagram of the structure when a positive voltage V_G is applied to the gate. Use this diagram to express the applied gate voltage V_G in terms of the characteristic energies, E_g, ΔE, $E_F E_{Fi}$, $e\phi_S$, and eV_{ox}. Find the flat-band voltage and the threshold voltage of the device.

Problem 2 A p-MOSFET for which $W/L = 7$, where W is the device width, and L is the source-to-drain distance; the oxide thickness and dielectric constant are $t_{ox} = 50$ nm and $\epsilon_{ox} = 3.9$, and the hole mobility near the surface $\mu_h = 70$ cm^2/(V · s). The device is to be used as a controlled resistor.

(a) Calculate the surface-charge density in the inversion channel that is required for the MOSFET resistance between the source and the drain to be 10 kΩ at low values of $V_D \ll V_D^{(sat)}$. Note that at small drain voltage, the inversion-layer surface-charge density Q'_i is independent on the coordinate x.
(b) What gate voltage in excess of the threshold voltage, $\Delta V = V_G - V_T$, is necessary to maintain the inversion surface-charge density found in part (a)?

10.6.3 Solutions

Problem 1 We consider a MOS structure, whose energy band diagrams at zero and positive gate voltage V are shown in Figs. 10.4b and 10.5b, respectively. The work function of the "metal" n^+-Si

$$e\psi_m = e\chi - \Delta E, \quad \Delta E = 0.05 \text{ eV}.$$

We can directly apply the results obtained in Sects. 10.4.4 and 10.4.5. Namely, the flat-band voltage is given by

$$V_{FB} = \psi_m - \chi - \frac{E_{Fi} - E_F}{e} - \frac{E_g}{2e} - \frac{Q'_{ss}}{C'_{ox}}$$

$$= -\frac{\Delta E}{e} - \frac{E_{Fi} - E_F}{e} - \frac{E_g}{2e} - \frac{Q'_{ss}}{C'_{ox}}.$$

The second term in the right-hand side of this expression can be evaluated as

$$p_0 = N_a = n_i e^{(E_{Fi} - E_F)/kT}$$

$$\Rightarrow \frac{E_{Fi} - E_F}{e} = V_{th} \ln \frac{N_a}{n_i} = 0.02585 \ln \frac{10^{15}}{8.59 \cdot 10^9} = 0.3015 \, \text{V}.$$

The oxide layer capacitance per unit area is

$$C'_{ox} = \frac{\epsilon_0 \epsilon_{ox}}{t_{ox}} = \frac{8.854 \cdot 10^{-14} \cdot 3.9}{10^{-5}} = 3.453 \cdot 10^{-8} \, \text{F/cm}^2.$$

The voltage drop across the oxide is then

$$V_{oxFB} = -\frac{Q'_{ss}}{C'_{ox}} = -\frac{1.602 \cdot 10^{-19} \cdot 10^{11}}{3.453 \cdot 10^{-8}} = -0.4639 \, \text{V}.$$

The flat-band voltage is

$$V_{FB} = -0.05 - 0.3015 - \frac{1.12}{2} - 0.4639 = -1.375 \, \text{V}.$$

We now go on to the threshold voltage

$$V_T = V_{FB} + \frac{e N_a x_{dT}}{C'_{ox}} + 2 \frac{E_{Fi} - E_F}{e}.$$

At threshold, the potential on the semiconductor surface has the threshold value

$$\phi_T = 2 \frac{E_{Fi} - E_F}{e} = 2 \cdot 0.3015 = 0.603 \, \text{V}.$$

The voltage drop across the oxide layer is now due to not only trapped charges, but also due to the charges of ionized acceptors in the depletion region. The depletion region width has the maximal value

$$x_{dT} = \sqrt{2 \frac{\epsilon_0 \epsilon}{e N_a} \phi_T} = \sqrt{2 \cdot \frac{8.854 \cdot 10^{-14} \cdot 11.7}{1.602 \cdot 10^{-19} \cdot 10^{15}} \cdot 0.603} = 0.8831 \, \mu\text{m}.$$

The contribution of the SCR to the voltage drop across the oxide is

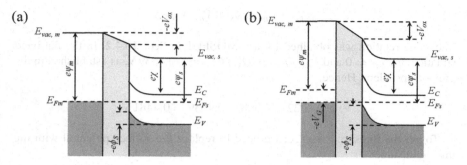

Fig. 10.10 The energy-band diagram of a p-MOSFET at (a) $V_G = 0$ and (b) $V_G < 0$

$$\frac{e N_a x_{dT}}{C'_{ox}} = \frac{1.602 \cdot 10^{-19} \cdot 10^{15} \cdot 0.8831 \cdot 10^{-4}}{3.453 \cdot 10^{-8}} = 0.4097 \, \text{V}.$$

Collecting everything together, we find the threshold voltage to be

$$V_T = -1.375 + 0.4097 + 0.603 = -0.362 \, \text{V}.$$

Problem 2 The energy band diagram of the structure at zero and at a negative gate voltage is shown in Fig. 10.10. Because the Fermi level of the gate p$^+$-Si electrode is below the valence band edge by the amount $\Delta E = 0.05 \, \text{eV}$, its work function,

$$e\psi_m = e\chi + E_g + \Delta E = 4.05 + 1.12 + 0.05 = 5.22 \, \text{eV},$$

exceeds the work function of the n-type body,

$$e\psi_s = e\chi + E_C - E_F = e\chi + E_C - E_{Fi} + E_{Fi} - E_F = e\chi + \frac{E_g}{2} - kT \ln \frac{N_d}{n_i}$$

$$= 4.05 + 0.56 - 0.02585 \ln \frac{10^{15}}{8.59 \cdot 10^9} = 4.308 \, \text{eV}.$$

Consider the energy difference between the metal vacuum and the semiconductor Fermi levels. On the one hand,

$$E_{vac,m} - E_{Fs} = e\psi_m - eV_G,$$

and on the other hand,

$$E_{vac,m} - E_{Fs} = -eV_{ox} - e\phi_s + e\psi_s.$$

Note that V_G, V_{ox}, and ϕ'_s are all negative. From these two equations, we express the applied voltage in terms of everything else:

$$V_G = \psi_m - \psi_s + V_{ox} + \phi_S.$$

The same relation was obtained for a n-MOSFET in Sect. 10.4.2. In the flat-band situation, $\phi_{S,FB} = 0$ and $V_{ox,FB} = -Q'_{ss}/C'_{ox} = -0.4639$ V, as established in the previous problem. Hence,

$$V_{FB} = 5.22 - 4.308 - 0.4639 = 0.4481 \text{ V}.$$

To get the threshold voltage, we need to replace the surface potential with the threshold value

$$\phi_T = -2V_{th} \ln \frac{N_d}{n_i} = -0.6031 \text{ V}.$$

The depletion region width at threshold is

$$x_{dT} = \sqrt{\frac{2\epsilon_0\epsilon|\phi'_T|}{eN_d}} = 0.8832 \,\mu\text{m},$$

and the voltage drop across the oxide layer is

$$V_{oxT} = \frac{-eN'_s - eN_dx_{dT}}{C'_{ox}}$$

$$= -1.602 \cdot 10^{-19} \frac{10^{11} + 10^{15} \cdot 0.8832 \cdot 10^{-4}}{3.453 \cdot 10^{-8}} = -0.8739 \text{ V},$$

where the numerical value of C'_{ox} was found in the previous problem.
We can find finally the threshold voltage:

$$V_T = \psi_m - \psi_s + V_{oxT} + \phi_T$$

$$= 5.22 - 4.308 - 0.8739 - 0.6031 = -0.565 \text{ V}.$$

Problem 3

(a) The measured capacitance–voltage curve shows that the capacitance at $V > -0.5$ V is higher than the capacitance at $V < -5$ V. This means that a large negative voltage $V < -5$ V must be applied to the gate to create an inversion layer. We conclude that the majority carriers in the MOSFET body are electrons, i.e., this is a p-MOSFET with the flat-band and the threshold voltage values of

$$V_{FB} = -0.5 \text{ V}, \quad V_T = -5 \text{ V}.$$

(b) The capacitance of the oxide layer is $C'_{ox} = 11.5$ nF/cm^2. From this value, we can find the oxide thickness as

$$C'_{ox} = \frac{\epsilon_0 \epsilon_{ox}}{t_{ox}} \quad \Rightarrow \quad t_{ox} = \frac{\epsilon_0 \epsilon_{ox}}{C'_{ox}} = \frac{8.854 \cdot 10^{-14} \cdot 3.9}{11.5 \cdot 10^{-9}} = 0.3\,\mu\text{m}.$$

(c) The minimal capacitance $C'_{min} = 8.4\,\text{nF/cm}^2$ is formed by the oxide layer, and the depletion region at its maximal width $x_{dT} = \sqrt{2\epsilon_0\epsilon|\phi_T|/(eN_d)}$, where $\phi_T = 2V_{th}\,\ln(N_d/n_i)$ is the threshold surface potential. We have

$$\frac{1}{C'_{min}} = \frac{1}{C'_{ox}} + \frac{1}{C'_{dT}} \quad \Rightarrow \quad C'_{dT} = \frac{C'_{ox}C'_{min}}{C'_{ox} - C'_{min}} = \frac{11.5 \cdot 8.4}{11.5 - 8.4} = 31.16\,\text{nF/cm}^2.$$

The SCR capacitance per unit area at the threshold surface potential is

$$C'_{dT} = \frac{\epsilon_0\epsilon}{x_{dT}} = \sqrt{\frac{\epsilon_0\epsilon e N_d}{4V_{th}\,\ln(N_d/n_i)}}.$$

To find the donor concentration from this equation, we rearrange the terms as

$$N_d = \frac{4V_{th}C'^2_{dT}}{\epsilon_0\epsilon e}\,\ln\frac{N_d}{n_i}.$$

An iterative solution of this equation starts with a reasonable initial value of N_d, say, $N_d = 10^{18}\,\text{cm}^{-3}$. Its substitution into the right-hand side gives a new value $N_d = 1.12 \cdot 10^{16}\,\text{cm}^{-3}$, which is then substituted again into the right-hand side to further improve the accuracy of the result. This procedure is repeated until convergence is reached, giving the value

$$N_d = 8.34 \cdot 10^{15}\,\text{cm}^{-3}.$$

(d) The flat-band voltage is given by

$$V_{FB} = \psi_m - \psi_s - \frac{Q'_{ss}}{C'_{ox}},$$

where the metal work function is known, and the semiconductor work function is

$$\psi_s = \chi + \frac{E_C - E_F}{e} = \chi + \frac{E_g}{2e} - \frac{E_F - E_{Fi}}{e}$$

$$= 4.05 + 0.56 - 0.02585 \cdot \ln\frac{8.34 \cdot 10^{15}}{8.59 \cdot 10^9} = 4.254\,\text{V}.$$

We obtain

$$Q'_{ss} = C'_{ox}\cdot(\psi_m - \psi_s - V_{FB}) = 11.5\cdot10^{-9}(4 - 4.254 + 0.5) = 2.83\cdot10^{-9}\,\text{C/cm}^2.$$

This corresponds to the concentration of charges of

$$Q'_{ss}/e = 1.76 \cdot 10^{10} \, \text{cm}^{-2}.$$

Problem 4 If V_G and V_D were independent parameters, the current–voltage curve of a MOSFET would depend on the relation between these voltages (see Lecture Notes for details):

$$I_D(V_D) = \begin{cases} 0 \text{ for } V_G < V_T, \\ K_n \left(2(V_G - V_T)V_D - V_D^2 \right) \text{ for } V_D < V_G - V_T, \\ K_n(V_G - V_T)^2 \text{ for } V_D \geq V_G - V_T. \end{cases}$$

That is, the current is zero in the sub-threshold regime; it increases with V_D in the sub-saturation region; and it is constant $I_D^{sat} = K_n(V_G - V_T)^2$ in the saturation regime.

For the connection in Fig. 10.9b, the gate and the drain voltages are not independent. Rather, they differ by a fixed amount set by the battery:

$$V_G = V_D + V_{DG}.$$

The minimal gate voltage required for the current to flow is V_T; otherwise, there is no inversion layer. The current is guaranteed to be zero for $V_G = V_D + V_{DG} < V_T$, or $V_D < V_T - V_{DG}$.

At $V_D \geq V_T - V_{DG}$, the current will be increasing with V_D according to whether the gain-to-drain voltage is below or above the threshold voltage.

If $V_{GD} = V_G - V_D < V_T$, we are in the saturation regime, in which

$$I_D(V_D) = K_n(V_G - V_T)^2 = K_n(V_D + V_{GD} - V_T)^2.$$

In particular,

$$\text{for } V_{GD} = 0 : \quad I_D(V_D) = \begin{cases} 0 & \text{for } V_D < V_T \\ K_n(V_D - V_T)^2 & \text{for } V_D > V_T \end{cases},$$

$$\text{for } V_{GD} = 0.5 \, V_T : \quad I_D(V_D) = \begin{cases} 0 & \text{for } V_D < 0.5 \, V_T \\ K_n(V_D - 0.5 \, V_T)^2 & \text{for } V_D > 0.5 \, V_T \end{cases},$$

and

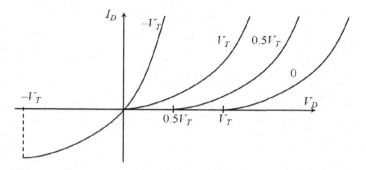

Fig. 10.11 The current–voltage curves of an n-MOSFET from Fig. 10.9b for different values of the gate-to-drain voltage, as indicated near each curve

$$\text{for } V_{GD} = V_T \; : \; I_D(V_D) = \begin{cases} 0 & \text{for } V_D < 0 \\ K_n V_D^2 & \text{for } V_D > 0. \end{cases}$$

If $V_{GD} = V_G - V_D > V_T$, we are in the sub-saturation regime, in which

$$I_D(V_D) = K_n \left(2(V_G - V_T)V_D - V_D^2 \right) = K_n \left(2(V_D + V_{GD} - V_T)V_D - V_D^2 \right)$$
$$= K_n(V_D^2 + 2V_D(V_{GD} - V_T)).$$

In particular,

$$\text{for } V_{GD} = 2\,V_T \; : \; I_D(V_D) = \begin{cases} 0 & \text{for } V_D < -V_T \\ K_n V_D(V_D + 2V_T) & \text{for } V_D > -V_T \end{cases} .$$

The $I_D(V_D)$ curves for all these cases are shown in Fig. 10.11 with the V_{GD} voltage indicated at each curve.

Chapter 11
PN Junction Diode

11.1 The Structure of a pn Junction

A pn junction diode is fabricated by letting the doping impurities, say, acceptors, diffuse into a, say, donor-doped substrate. After that, good ohmic contact electrodes must be deposited onto the resulting n- and p-regions, see Fig. 11.1a. For the ease of the discussion, a pn diode is usually schematically represented as an elongated structure, see Fig. 11.1b. Its electronic symbol is shown in Fig. 11.1c.

It is immediately obvious that the current–voltage curve of this device should be non-linear. Indeed, if a positive bias is applied to the p-region relative to the n-region, then the majority holes will be pushed from the p-side into the n-side, the majority electrons from the n-side will easily flow into the p-side, and we will have a large current through the structure. If the polarity of bias is reversed—minus to the p-side and plus to the n-side—then it will be the minority electrons in the p-side that will be pushed into the n-side, and minority holes in the n-side that will go to the p-side, meaning that the current will be small. The asymmetry of the current–voltage curve opens the possibility to use a pn junction, e.g., to convert an ac-voltage into a dc-voltage. Other uses of pn diodes are numerous; we will discuss their optoelectronic applications later on.

Apart from electrons and holes that are free to move, there are also stationary positively charged donor ions in the n-region and negatively charged acceptor ions in the n-region. They create the charged background that balances the respective charge densities of electrons and holes inside the quasi-neutral p- and n-regions (p-QNR and n-QNR, see Fig. 11.1b). At the interface between the n- and p-type semiconductors, electrons from the n-side and holes from the p-side recombine with each other, and therefore, the interface region contains only a negligibly small concentration of electrons and holes. But it has positively charged donor ions on the n-side and negatively charged acceptor ions on the p-side, forming a kind of capacitor. In other words, the interface region becomes the space-charge region

(a)   (b)   (c)

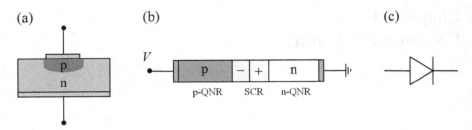

Fig. 11.1 (**a**) The structure of a "real-life" pn junction diode, (**b**) its schematic representation showing the two quasi-neutral regions (QNRs) separated by the space-charge region (SCR), and (**c**) the electronic symbol of a pn diode

(SCR) or depletion region. The total charges of the two sides of SCR are equal in magnitude but opposite in sign, so that its total charge is zero.

When analyzing the pn junction diode, we will make the following assumptions. We will consider a step homojunction, in which the host lattice in the p- and n-sides is the same and the doping profile is sharp, i.e., the concentration of donors is N_d on the n-side and zero on the p-side, and the concentration of acceptors is N_a on the p-side and zero on the n-side. We will assume complete ionization of the impurities with $n_i \ll N_d$ and N_a. Finally, we will use the depletion approximation to describe the SCR.

11.2   The Energy Band Diagram of a pn Junction at Zero Bias

Consider the case of zero bias. In thermal equilibrium, the Fermi energy must be uniform throughout the pn structure. But on the p-side, E_F is closer to the valence band edge, and on the n-side, it is closer to the conduction band edge. This implies that the energy bands must bend in the SCR to maintain the uniformity of Fermi energy, see figure showing the energy band structure of a pn junction at zero bias.

Physically, the bands are bent because the electrons from the n-part of the diode diffuse into the p-part, where they recombine with the holes. Vice versa, the holes from the p-side diffuse into the n-side and recombine with the electrons there. As a result, only the positively charged donor ions and negatively charged acceptor ions remain on the two sides of the junction in the SCR. Hence, the SCR represents a charged capacitor with a built-in voltage, V_{bi}, developed across it.

The magnitude of V_{bi} can be established from the following considerations. In the presence of a non-uniform electric potential in a semiconductor, the electron concentration is given by

$$n_0(x) = n_{0n} e^{\phi(x)/V_{th}}, \quad n_{0n} = N_d,$$

where $n_{0n} = N_d$ is the electron concentration in the n-part, assumed to be grounded. From this, we conclude that electron concentrations in the p-region are given by

$$n_{0p} = N_d e^{-V_{bi}/V_{th}}.$$

On the other hand, by the mass action law,

$$n_{0p} = \frac{n_i^2}{p_{0p}}, \quad p_{0p} = N_a,$$

where $p_{0p} = N_a$ is the majority hole concentration in the p-region far away from the SCR boundary. Combining these two expressions for n_{0p}, we find

$$N_d e^{-V_{bi}/V_{th}} = \frac{n_i^2}{N_a} \Rightarrow V_{bi} = V_{th} \ln \frac{N_a N_d}{n_i^2}.$$

If the two sides of the junction are connected by a conducting wire, the built-in voltage will not cause an electric current to flow through it. The existence of such a current would contradict the law of energy conservation. The physical reason why the built-in voltage cannot produce an current is that there would be additional semiconductor band bending in the metal–semiconductor contact regions on the p- and n-sides, see Fig. 11.2a. These two contact regions will be characterized by their own values of the built-in voltage, so that the three built-in voltages, which add up to zero with the built-in voltage in the pn junction region.

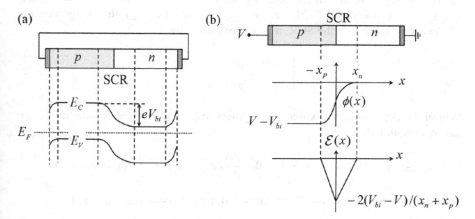

Fig. 11.2 (a) A pn diode at zero bias (top) and its energy band diagram (bottom). (b) A pn diode under a non-zero bias (top), together with the electric potential and electric field distribution within the structure (bottom)

For the same reason, touching the two ends of a pn diode with two terminals of a voltmeter will show zero voltage, not V_{bi}. The built-in voltage can be found by measuring the voltage-dependent capacitance, similar to the metal–semiconductor case.

11.3 PN Junction Under an External Bias

Consider what happens when a bias V is applied to the p-side of a pn diode, see Fig. 11.2b. A forward bias, $V > 0$, reduces the potential barrier that prevents the majority carriers from entering the opposite side of the junction, whereas a reverse bias, $V < 0$, increases this barrier. We will focus on the practically interesting case when the applied bias is smaller than the built-in voltage,

$$V < V_{bi},$$

meaning that the SCR exists in the pn structure. In the less interesting opposite case, $V > V_{bi}$, the diode will behave more or less according to Ohm's law.

Because the QNRs have a considerable number of mobile charge carriers, the voltage drop across them is negligible. But this is not so in the SCR, which is depleted of the free charge carriers. Hence, the voltage drop V occurs practically completely across the SCR.

To find out how the potential changes in the diode, we adopt the depletion approximation, see Sect. 6.3.3. Namely, we assume that there are no free charge carriers in the SCR, which has abrupt boundaries at $x = -x_p$ and $x = x_n$, see Fig. 11.2b. Outside of the SCR (i.e., at $x < -x_p$ and $x > x_n$), the charge density $\rho(x)$ is zero, and inside the SCR, it is approximated by

$$\rho(x) = \begin{cases} -eN_a & \text{at } -x_p < x < 0 \\ eN_d & \text{at } 0 < x < x_n \end{cases}.$$

Within the depletion approximation, we must impose the boundary conditions for the SCR boundaries:

$$\phi(-x_p) = V - V_{bi}, \quad \phi(x_n) = 0, \quad \frac{d\phi}{dx}(x = -x_p) = \frac{d\phi}{dx}(x = x_n) = 0.$$

With these boundary conditions, the solution of Poisson's equation reads

$$\phi(x) = \begin{cases} \frac{eN_a}{2\epsilon_0\epsilon}(x + x_p)^2 + V - V_{bi} & \text{at } -x_p < x < 0 \\ -\frac{eN_d}{2\epsilon_0\epsilon}(x - x_n)^2 & \text{at } 0 < x < x_n \end{cases}.$$

The first derivative of the electric potential is related to the electric field:

$$\mathcal{E}(x) = -\frac{d\phi(x)}{dx} = \begin{cases} -\frac{eN_a}{\epsilon_0\epsilon}(x + x_p) & \text{at } -x_p < x < 0 \\ \frac{eN_d}{\epsilon_0\epsilon}(x - x_n) & \text{at } 0 < x < x_n \end{cases}.$$

It remains to determine the SCR boundaries, x_n and x_p. First, we note that electric field must be continuous at $x = 0$, i.e., $\mathcal{E}(x = 0)$ must have the same value if one uses the upper and lower expressions for $\mathcal{E}(x)$. This immediately gives

$$N_a x_p = N_d x_n.$$

This equality expresses the fact that the net charge of the negatively ionized acceptors in the left part of the SCR must be compensated by the positive charge of ionized donors in its right part, so that the SCR is electrically neutral.

In addition, the potential must be continuous at $x = 0$, i.e., $\phi(x = 0)$ must have the same value if one uses either expression for $\phi(x)$. This gives

$$\frac{eN_a}{2\epsilon_0\epsilon}x_p^2 + V - V_{bi} = -\frac{eN_d}{2\epsilon_0\epsilon}x_n^2.$$

Solving the two equations for two x_p and x_n, we obtain

$$x_p = \sqrt{2\frac{\epsilon_0\epsilon(V_{bi} - V)}{e(N_a + N_d)}\frac{N_d}{N_a}}, \quad x_n = \sqrt{2\frac{\epsilon_0\epsilon(V_{bi} - V)}{e(N_a + N_d)}\frac{N_a}{N_d}}.$$

The total width of the SCR is, then,

$$w = x_p + x_n = \sqrt{2\frac{\epsilon_0\epsilon(V_{bi} - V)}{e}\frac{N_a + N_d}{N_d N_a}} = \sqrt{2\frac{\epsilon_0\epsilon(V_{bi} - V)}{eN}},$$

where

$$\frac{1}{N} = \frac{1}{N_d} + \frac{1}{N_a} \Rightarrow N = \frac{N_d N_a}{N_d + N_a}.$$

The maximal (in magnitude) built-in electric field is realized at $x = 0$. It is given by several equivalent expressions:

$$\mathcal{E}_{max} = \mathcal{E}(0) = -\frac{eN_a x_p}{\epsilon_0\epsilon} = -\frac{eN_d x_n}{\epsilon_0\epsilon} = -\sqrt{\frac{2eN}{\epsilon_0\epsilon}(V_{bi} - V)} = -2\frac{V_{bi} - V}{w}.$$

11.4   SCR Capacitance

The insulating SCR is sandwiched between the two conducting QNRs. It can be modeled as a parallel-plane capacitor of thickness w. Its capacitance per unit area is

$$C' = \frac{\epsilon_0 \epsilon}{w} = \sqrt{\frac{e \epsilon_0 \epsilon N}{2(V_{bi} - V)}},$$

where we used the expression for w obtained in the previous section.

Note that the inverse capacitance squared

$$\frac{1}{C'^2} = \frac{2}{e \epsilon_0 \epsilon N}(V_{bi} - V)$$

depends linearly on the applied voltage, see Fig. 11.3. It becomes zero at $V = V_{bi}$. The slope of this straight line is

$$-\frac{2}{e \epsilon_0 \epsilon N} = -2 \frac{N_d + N_a}{\epsilon_0 \epsilon e N_d N_a}.$$

Thus, the measurements of C' vs. V allow one to determine the product $N_d N_a$ (present in V_{bi}) and the sum $N_d + N_a$ (present in the slope of $1/C'^2$ vs. V). If one knows which side is doped more heavily, one can determine the doping concentrations N_a, N_d.

A one-sided junction is a highly non-symmetrically doped pn junction, where either $N_a \gg N_d$ or $N_d \gg N_a$. In the former case, it is denoted as p$^+$n junction, and in the latter case as pn$^+$ junction. It follows from the neutrality condition $x_p N_a = x_n N_d$ that the depletion layer penetrates primarily into that side that is doped more lightly. Indeed, if, e.g., $N_a > N_d$, then $x_n = x_p \frac{N_a}{N_d} > x_p$. If $N_a \gg N_d$, we can approximate N with N_d to a good accuracy. For a p$^+$n junction, the expressions for SCR width and capacitance simplify to

Fig. 11.3 The inverse square of the capacitance per unit area of the SCR in a pn junction diode as a function of the applied voltage

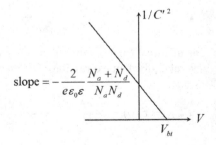

$$\text{slope} = -\frac{2}{e \varepsilon_0 \varepsilon} \frac{N_a + N_d}{N_a N_d}$$

$$w = x_n = \sqrt{2\frac{\epsilon_0 \epsilon (V_{bi} - V)}{e N_d}}, \quad C' = \sqrt{\frac{e \epsilon_0 \epsilon N_d}{2(V_{bi} - V)}}.$$

These expressions are identical to the ones obtained in the discussion of a junction formed by metal and an n-type semiconductor, see Sect. 9.3. In the opposite case of a pn^+ junction, the respective expressions are the same, but with N_d replaced by N_a.

11.5 Current–Voltage Relation of a pn Junction Diode

11.5.1 Charge Carrier Concentrations Near the Boundaries of the SCR

When an external voltage V is applied to a pn junction diode, an electric current will flow through it. Even though the system as a whole is out of equilibrium, its electron and hole subsystems find themselves in local thermodynamic equilibria separately. This is so because equilibration of electrons belonging to the same band proceeds much faster than equilibration between different bands. The former involves intraband scattering processes, where electrons exchange energy in small portions, and the latter requires the much slower generation–recombination processes, in which electrons exchange large amounts of energy of the order of the band gap energy E_g. With this in mind, we can use the language of the electron and hole quasi-Fermi energies (see Sect. 7.7), E_{Fe} and E_{Fh}, in the p- and n-sides of the pn junctions.

Consider the p-side. Far away from the SCR boundary, $x \ll -x_p$, the minority electrons exchange energy only with the majority holes, and therefore both electrons and holes have a common Fermi energy E_{Fp}, which is different from the n-side counterpart, E_{Fn}. But near the SCR boundary, the minority electrons on the p-side "communicate" with the majority electrons from the n-side much more efficiently than with the majority holes. This is so because the SCR is so thin that electron transport through it occurs very swiftly. From this, we can conclude that the minority electrons near the SCR boundary at $x = -x_p$ must have the same quasi-Fermi energy E_{Fn} as the electrons on the n-side of the junction. As we move deeper into the p-QNR, the electron quasi-Fermi energy aligns itself with the hole counterpart, see Fig. 11.4. The same reasoning applies to the minority holes on the n-side.

Hence, far away from the SCR boundaries, the minority carrier concentrations are

$$n_p = \frac{n_i^2}{N_a}, \quad p_n = \frac{n_i^2}{N_d}.$$

Note that we dropped the subscript 0 because the system is not in equilibrium. Near the SCR edges, the minority concentrations differ from these values. Namely,

Fig. 11.4 The energy band diagram of a biased pn junction diode, showing the electron and hole quasi-Fermi energies

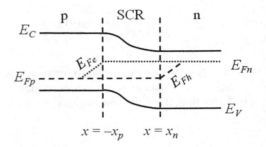

because the potential difference across the SCR is $V_{bi} - V$, and because the carriers of the same type on the two sides of the SCR have the same quasi-Fermi energy, we can relate the electron and hole concentrations on the two sides of the SCR to each other by

$$n(-x_p) = n(x_n)e^{-(V_{bi}-V)/V_{th}} = N_d e^{-(V_{bi}-V)/V_{th}} = \frac{n_i^2}{N_a}e^{V/V_{th}},$$

$$p(x_n) = p(-x_p)e^{-(V_{bi}-V)/V_{th}} = N_a e^{-(V_{bi}-V)/V_{th}} = \frac{n_i^2}{N_d}e^{V/V_{th}}.$$

In the last step of both lines, we used the expression for V_{bi} obtained in Sect. 11.2. These formulas are called Shockley boundary conditions. We will use them to derive the current–voltage relation of a pn junction diode.

It follows from the Shockley boundary conditions that the product

$$n(-x_p)p(-x_p) = n(-x_p)N_a = n_i^2 e^{V/V_{th}}.$$

On the other hand, it is given by

$$n(-x_p)p(-x_p) = n(-x_p)N_a = n_i^2 e^{(E_{Fe}-E_{Fh})/kT}.$$

It then follows that the difference of quasi-Fermi energies is just the applied voltage multiplied by the elementary charge:

$$E_{Fe} - E_{Fh} = eV.$$

11.5.2 Current–Voltage Relation of an Ideal pn Junction Diode

To figure out the $J - V$ curve of a pn junction diode, we will need a relation between the charge carriers' concentration and the current density. It is convenient to introduce the excess concentrations of electrons and holes as

$$\Delta n = n - n_0, \quad \Delta p = p - p_0.$$

In terms of the excess concentrations, Shockley boundary conditions read

$$\Delta n(-x_p) = \frac{n_i^2}{N_a}(e^{V/V_{th}} - 1), \quad \Delta p(x_n) = \frac{n_i^2}{N_d}(e^{V/V_{th}} - 1).$$

Let us focus on the QNR on the n-side of the junction, where the excess minority hole concentration is described by the continuity equation (see Sect. 8.8):

$$\frac{\partial \Delta p}{\partial t} = D_{hn}\frac{\partial^2 \Delta p}{\partial x^2} - \frac{\Delta p}{\tau_{hn}} \text{ at } x > x_n.$$

There is no drift current because the electric field in the QNRs is zero. D_{hn} is the hole diffusion coefficient, and τ_{hn} is the hole recombination time in the quasi-neutral n-region. The derivative $\partial \Delta p/\partial t$ equals zero because we are interested in the steady-state situation when the excess concentrations and current densities are time-independent. Thus, we have

$$D_{hn}\frac{\partial^2 \Delta p}{\partial x^2} - \frac{\Delta p}{\tau_{hn}} = 0 \text{ at } x > x_n.$$

Since the steady-state continuity equation is a second-order differential equation, it must be supplemented by two boundary conditions. One is provided by the Shockley boundary condition for $\Delta p(x_n)$ written above. The other one says that far away from the junction boundary in the quasi-neutral n-region, the hole concentration must approach its equilibrium value, meaning that the excess hole concentration approaches zero:

$$\Delta p(x \to \infty) = 0.$$

The solution of hole continuity equation that respects both boundary conditions is

$$\Delta p(x) = \Delta p(x_n)e^{-(x-x_n)/L_{hn}} = \frac{n_i^2}{N_d}(e^{V/V_{th}} - 1)e^{-(x-x_n)/L_{hn}} \text{ at } x > x_n.$$

Here,

$$L_{hn} = \sqrt{D_{hn}\tau_{hn}}$$

is the so-called hole diffusion length in the n-region.

We treat excess minority electrons in the p-region in the same manner. The respective diffusion equation in the steady state reads

$$D_{ep} \frac{\partial^2 \Delta n}{\partial x^2} - \frac{\Delta n}{\tau_{ep}} = 0 \text{ at } x < -x_p,$$

where D_{ep} and τ_{ep} are the diffusion coefficient and lifetime of electrons in the p-region. In addition to the Shockley boundary condition for $\Delta n(-x_p)$, we also require that

$$\Delta n(x \to -\infty) = 0.$$

With these boundary conditions, the solution of the continuity equation is

$$\Delta n(x) = \Delta n(-x_p) e^{(x+x_p)/L_{ep}} = \frac{n_i^2}{N_a} (e^{V/V_{th}} - 1) e^{(x+x_p)/L_{ep}} \text{ at } x < -x_p,$$

where the electron diffusion length in the p-region is defined as

$$L_{ep} = \sqrt{D_{ep}\tau_{ep}}.$$

Once we know the excess concentrations, we can determine the respective diffusive flux densities:

$$j_{hn}(x) = -D_{hn} \frac{\partial \Delta p}{\partial x} = \frac{n_i^2}{N_d} (e^{V/V_{th}} - 1) \frac{D_{hn}}{L_{hn}} e^{-(x-x_n)/L_{hn}} \text{ for } x > x_n,$$

$$j_{ep}(x) = -D_{ep} \frac{\partial \Delta n}{\partial x} = -\frac{n_i^2}{N_a} (e^{V/V_{th}} - 1) \frac{D_{ep}}{L_{ep}} e^{(x+x_p)/L_{ep}} \text{ for } x < -x_p.$$

Setting x to x_n in the first expression gives the number of holes that pass through the SCR per unit area per unit time. Likewise, setting x to $-x_p$ in the expression for $j_{ep}(x)$, we obtain the number of electrons that pass through the SCR per unit area per unit time. Multiplication of the hole flux density expression by e and the respective electron flux density by $-e$ gives the hole and electron contributions to the current density through the SCR:

$$J_h = e j_{hn}(x_n) = e \frac{n_i^2}{N_d} \frac{D_{hn}}{L_{hn}} (e^{V/V_{th}} - 1),$$

$$J_e = -e j_{ep}(-x_p) = e \frac{n_i^2}{N_a} \frac{D_{ep}}{L_{ep}} (e^{V/V_{th}} - 1).$$

The total electric current through the pn diode structure must be uniform, i.e., independent of x. Hence, the total electric current density through any part of the structure is the same as through the SCR,

$$J = J_h + J_e,$$

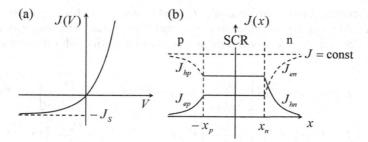

Fig. 11.5 (a) The $J - V$ curve of an ideal pn junction diode. (b) The electron and hole contributions to the net electric current density as functions of the position x in the diode

giving the ideal pn junction $J - V$ curve:

$$J(V) = J_S(e^{V/V_{th}} - 1),$$

see Fig. 11.5a. The reverse saturation current is

$$J_S = e n_i^2 \left(\frac{D_{hn}}{N_d L_{hn}} + \frac{D_{ep}}{N_a L_{ep}} \right) = e n_i^2 \left(\frac{1}{N_d} \sqrt{\frac{D_{hn}}{\tau_{hn}}} + \frac{1}{N_a} \sqrt{\frac{D_{ep}}{\tau_{ep}}} \right).$$

At large reverse bias, the current density attains the value $-J_S$, and at forward bias $0 < V < V_{bi}$, it increases exponentially with V, see Fig. 11.5a.

The current–voltage relation of an ideal pn junction diode is the same as that of an ideal Schottky diode, see Sect. 9.5. However, the saturation current density of the pn diode is several orders of magnitude lower than that of a Schottky diode. The reason is that the current is formed by the minority carriers in the pn diode and by the majority carriers in the Schottky diode.

11.5.3 Current Densities in a pn Diode

Coming back to our general expressions for the minority particle current densities and multiplying them by the respective charges, we obtain the x-dependent electric current densities of minority electrons in the p-region and minority holes in the n-region as exponentially decaying functions of distance from the SCR edges:

$$J_{hn}(x) = e \frac{n_i^2}{N_d} (e^{V/V_{th}} - 1) \frac{D_{hn}}{L_{hn}} e^{-(x-x_n)/L_{hn}} \quad \text{for } x > x_n,$$

$$J_{ep}(x) = e \frac{n_i^2}{N_a} (e^{V/V_{th}} - 1) \frac{D_{ep}}{L_{ep}} e^{(x+x_p)/L_{ep}} \quad \text{for } x < -x_p.$$

The total current through the structure, J, must be constant. This is only possible if there are additional contributions to the current density due to the majority charge carriers. For majority electrons in the n-region and majority holes in the p-region, we have

$$J_{en}(x) = J - J_{hn}(x) = en_i^2 \left[\frac{D_{hn}}{N_d L_{hn}} \left(1 - e^{-(x-x_n)/L_{hn}} \right) + \frac{D_{ep}}{N_a L_{ep}} \right] (e^{V/V_{th}} - 1)$$

$$J_{hp}(x) = J - J_{ep}(x) = en_i^2 \left[\frac{D_{ep}}{N_a L_{ep}} \left(1 - e^{(x+x_p)/L_{ep}} \right) + \frac{D_{hn}}{N_d L_{hn}} \right] (e^{V/V_{th}} - 1).$$

Physically, the onset of the majority currents can be understood as follows. Consider the n-side of the junction, $x = x_n$. At the edge of the SCR, the excess minority holes are injected from the p-region, where they are the majority carriers. Starting from $x = x_n$, the minority holes diffuse into the bulk of the n-region, recombining with the majority electrons on the way. To compensate for this loss of the electrons, electrons have to diffuse from the bulk of the n-region toward the SCR boundary, creating an additional current density J_{en}. Similar reasoning applies, of course, to the majority hole current density on the p-side, see Fig. 11.5b.

11.5.4   SCR Recombination Current

So far, we considered an ideal pn junction diode. Let us consider the deviation from the ideal behavior due to the generation–recombination processes in the SCR.

The average electron and hole concentrations in the SCR are related by the mass action law,

$$n_{SCR} p_{SCR} = n_i^2 e^{(E_{Fe} - E_{Fh})/kT} = n_i^2 e^{V/V_{th}},$$

see the end of Sect. 11.5.1. Because recombination requires both electrons and holes, it is reasonable to assume that the largest recombination rate in the SCR occurs at that position where their concentrations are equal:

$$n_{SCR} = p_{SCR} = n_i e^{V/(2V_{th})}.$$

With the charge carrier lifetime in the SCR, τ_{SCR}, we can write the total recombination rate as

$$R_{SCR}(V) = \frac{n_i e^{V/(2V_{th})}}{\tau_{SCR}}.$$

Generation of carriers is a voltage-independent process that proceeds at a constant rate, G_{SCR}. When $V = 0$, it must exactly balance the recombination rate. Thus, the generation rate is

$$G_{SCR} = R_{SCR}(V = 0) = \frac{n_i}{\tau_{SCR}}.$$

The net SCR recombination rate is

$$U_{SCR} = G_{SCR} - R_{SCR} = \frac{n_i}{\tau_{SCR}}(e^{V/(2V_{th})} - 1).$$

The rate of change of electron and hole concentrations due to the generation–recombination in the SCR is

$$Aw\left(\frac{dn}{dt}\right)_{SCR} = Aw\left(\frac{dp}{dt}\right)_{SCR} = AwU_{SCR} = Aw\frac{n_i}{\tau_{SCR}}\left(e^{V/(2V_{th})} - 1\right).$$

In the steady state, $(\partial n/\partial t)_{SCR} = (\partial p/\partial t)_{SCR} = 0$; this must be compensated by additional electron and hole currents that flow from the n- and p-sides into the SCR. These additional currents produce the recombination current density

$$J_{SCR}(V) = J_S'\left(e^{V/(2V_{th})} - 1\right), \quad J_S' = ew\frac{n_i}{\tau_{SCR}}.$$

The total current density is the sum of the two contributions,

$$J(V) = J_S(e^{V/V_{th}} - 1) + J_S'(e^{V/(2V_{th})} - 1).$$

At large bias, V, the second term can be neglected, and the $J - V$ curve resembles the one of an ideal pn junction. But at small voltage, the SCR-related second term becomes important.

The current–voltage curve of a real diode is often described phenomenologically by introducing the ideality factor, n, as

$$J(V) = J_S(e^{V/(nV_{th})} - 1).$$

The value $n = 1$ corresponds to an ideal diode; the typical value of the ideality factor is somewhere between 1 and 2.

11.6 Problems

11.6.1 Solved Problems

Problem 1 The capacitance per unit area of a silicon pn junction at $V = 0$ is $C' = 0.394\,\text{nF/cm}^2$; at $V = -0.5\,\text{V}$, it is $C' = 0.283\,\text{nF/cm}^2$. The n-side is doped more heavily than the p-side. Find the doping levels N_a and N_d.

Problem 2 A silicon pn junction has the donor and acceptor concentrations of $N_d = 2 \cdot 10^{15} \, \text{cm}^{-3}$ and $N_a = 8 \cdot 10^{15} \, \text{cm}^{-3}$ on the two opposite sides of the junction. Determine the minority carrier concentrations at the edges of the space-charge region for $V = 0.5 \, \text{V}$ and $-0.5 \, \text{V}$.

Problem 3 Consider an ideal Si-based pn junction diode at 300 K. The doping levels on the two sides are $N_d = 10^{12} \, \text{cm}^{-3}$, $N_a = 10^{17} \, \text{cm}^{-3}$, and the minority carriers mobilities in the respective regions are $\mu_{ep} = 760 \, \text{cm}^2/(\text{V} \cdot \text{s})$, $\mu_{hn} = 500 \, \text{cm}^2/(\text{V} \cdot \text{s})$. The current density for a forward bias $V = 0.2 \, \text{V}$ is $2.3 \cdot 10^{-4} \, \text{A}/\text{cm}^2$. Determine the minority hole lifetime in the quasi-neutral n-region.

11.6.2   Practice Problems

Problem 1 A pn junction diode is made of an unknown semiconductor. Its capacitance–voltage measurements performed at two different temperatures, $T_1 = 290 \, \text{K}$ and $T_2 = 310 \, \text{K}$, revealed the values of the built-in voltages $V_{bi}(T_1) = 2.38 \, \text{V}$ and $V_{bi}(T_2) = 2.44 \, \text{V}$, respectively. Find the band gap of this semiconductor material.

Problem 2 In the discussion of an ideal pn junction diode, we assumed the resistance of the QNRs to be zero. In reality, it is small, but finite. It can be modeled as a series resistance connected to an ideal diode as shown in Fig. 11.6. Express the voltage V applied to the diode as a function of the current I through it. What is the slope dV/dI of the $V(I)$ curve at zero current? What is the value of the current when a large voltage V is applied in the reverse direction? What is the slope of the $V(I)$ curve when the voltage V in the forward direction is large?

Problem 3 The doping concentrations in a silicon pn junction diode are $N_d = 10^{16} \, \text{cm}^{-3}$ and $N_a = 10^{17} \, \text{cm}^{-3}$. At a forward bias V, the minority carrier concentration on one side of the SCR is one-tenth of the majority concentration and exceeds the minority concentration on the opposite side. Find V.

Problem 4 Consider a non-ideal Si pn junction diode, as described in Sect. 11.5.4. Assume the following parameter values $N_d = N_a = 10^{15} \, \text{cm}^{-3}$, $\tau_{ep} = \tau_{hn} = 10^{-5} \, \text{s}$, $\tau_{SCR} = 10^{-7} \, \text{s}$, $D_{ep} = D_{hn} = 10 \, \text{cm}^2/\text{s}$, $T = 300 \, \text{K}$. The total $J - V$ curve contains the ideal-diode part and the recombination current in the SCR,

$$J(V) = J_{id}(V) + J_{SCR}(V),$$

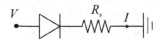

Fig. 11.6 The effect of a finite resistance of the QNRs modeled as a resistance R_s connected in series to the diode

as explained in Sects. 11.5.2 and 11.5.4, respectively. How are the conductances due to these two mechanisms related? In other words, find the ratio $\frac{dJ_{id}(V)/dV}{dJ_{SCR}(V)/dV}$ at $V = 0$.

11.6.3 Solutions

Problem 1 The inverse capacitance squared as a function of voltage is

$$\frac{1}{C'^2} = \frac{2}{e\epsilon_0\epsilon}\frac{N_a + N_d}{N_a N_d}(V_{bi} - V), \quad V_{bi} = V_{th}\ln\frac{N_a N_d}{n_i^2}.$$

It is convenient to denote

$$S = N_a + N_d, \quad P = N_a N_d.$$

The graph of $1/C'^2$ vs. V is a straight line with the slope given by $(2/e\epsilon_0\epsilon)S/P$. Its intercept with the x-axis is $V_{bi} = V_{th}\ln(P/n_i^2)$, from which P can be found as $P = n_i^2\,e^{V_{bi}/V_{th}}$.

Since we have just two data points, the straight line fit is particularly easy. Namely, at $V = 0$, we have $1/C'^2 = 6.442 \cdot 10^{18}\,\text{cm}^4/\text{F}^2$, and at $V = -0.5\,\text{V}$, we have $1/C'^2 = 12.49 \cdot 10^{18}\,\text{cm}^4/\text{F}^2$. The straight line fit can be found based on this information:

$$\frac{1}{C'^2} = -kV + b, \quad k = 12.1 \cdot 10^{18}\,\frac{\text{cm}^4}{\text{F}^2 \cdot \text{V}}, \quad b = 6.442 \cdot 10^{18}\,\frac{\text{cm}^4}{\text{F}^2}.$$

From this, we obtain the built-in voltage

$$V_{bi} = \frac{b}{k} = 0.5326\,\text{V}$$

and the product of the doping concentrations:

$$P = n_i^2\,e^{V_{bi}/V_{th}} = (8.59 \cdot 10^9)^2\,e^{0.5326/0.02585} = 6.546 \cdot 10^{28}\,\text{cm}^{-6}.$$

From the slope k of the $1/C'^2$ vs. V dependence, we find the sum of the doping concentrations as

$$S = \frac{1}{2}e\epsilon_0\epsilon\,Pk = 6.572 \cdot 10^{16}\,\text{cm}^{-3}.$$

It remains to solve two equations:

$$N_a N_d = P, \quad N_a + N_d = S$$

with $N_d > N_a$. We express N_a in terms of N_d using the first equation and plug in the result into the second equation:

$$N_a = \frac{P}{N_d},$$

$$\frac{P}{N_d} + N_d = S \;\Rightarrow\; N_d^2 - S N_d + P = 0 \;\Rightarrow\; N_d = -\frac{S}{2} + \sqrt{\frac{S^2}{4} - P}.$$

Numerically,

$$N_d = \frac{6.572 \cdot 10^{16}}{2} + \sqrt{(3.286 \cdot 10^{16})^2 - 6.546 \cdot 10^{28}} = 6.57 \cdot 10^{16}\,\mathrm{cm}^{-3},$$

$$N_a = \frac{6.546 \cdot 10^{28}}{6.572 \cdot 10^{16}} = 9.96 \cdot 10^{11}\,\mathrm{cm}^{-3}.$$

Problem 2 The Shockley boundary conditions (see Sect. 11.5.1) give

$$n_0(-x_p) = \frac{n_i^2}{N_a} e^{V/V_{th}}, \quad p_0(x_n) = \frac{n_i^2}{N_d} e^{V/V_{th}}.$$

A positive applied voltage exponentially enhances the minority carriers' concentration relative to their equilibrium values. Plugging in the numbers, we obtain

$$\frac{n_i^2}{N_a} = \frac{(8.59 \cdot 10^9)^2}{8 \cdot 10^{15}} = 9\,223\,\mathrm{cm}^{-3}, \quad \frac{n_i^2}{N_d} = \frac{(8.59 \cdot 10^9)^2}{2 \cdot 10^{15}} = 36\,894\,\mathrm{cm}^{-3},$$

$$e^{V/V_{th}} = e^{0.5/0.02585} = 2.514 \cdot 10^8, \quad e^{-V/V_{th}} = e^{-0.5/0.02585} = 3.979 \cdot 10^{-9},$$

and thus at $V = 0.5\,\mathrm{V}$

$$n_0(-x_p) = 9\,223 \cdot 2.514 \cdot 10^8 = 2.32 \cdot 10^{12}\,\mathrm{cm}^{-3},$$

$$p_0(x_n) = 36\,894 \cdot 2.514 \cdot 10^8 = 9.28 \cdot 10^{12}\,\mathrm{cm}^{-3}.$$

At $V = -0.5\,\mathrm{V}$, $n_0(-x_p) = p_0(x_n) = 0$ for all practical purposes.

Problem 3 Assuming the ideal current–voltage relation, $J(V) = J_S(e^{V/V_{th}} - 1)$, we find the reverse saturation current density (with $V = 0.2\,\mathrm{V}$):

$$J_S = \frac{J(V)}{e^{V/V_{th}} - 1} = \frac{2.3 \cdot 10^{-4}}{e^{0.2/0.02585} - 1} = 1.004 \cdot 10^{-7} \, \text{A/cm}^2.$$

From the theory of the pn junction diodes, the reverse saturation current density is given by the formula

$$J_S = e \left(\frac{n_i^2}{N_a} \sqrt{\frac{D_{ep}}{\tau_{ep}}} + \frac{n_i^2}{N_d} \sqrt{\frac{D_{hn}}{\tau_{hn}}} \right),$$

and because $N_d \ll N_a$ (by 100 000 times), i.e., $1/N_d \gg 1/N_a$, we expect that the second term in the brackets produces a much higher contribution to J_S than the first term. We thus can approximate

$$J_S = e \frac{n_i^2}{N_d} \sqrt{\frac{D_{hn}}{\tau_{hn}}},$$

from which the minority hole lifetime is found as

$$\tau_{hn} = \left(e \frac{n_i^2}{N_d} \right)^2 \frac{D_{hn}}{J_S^2} = \left(e \frac{n_i^2}{N_d} \right)^2 \frac{V_{th} \mu_{hn}}{J_S^2},$$

where we used Einstein's relation $D = \mu V_{th}$ in the second equality. Numerically,

$$\tau_{hn} = \left(1.602 \cdot 10^{-19} \cdot \frac{(8.59 \cdot 10^9)^2}{10^{12}} \right)^2 \frac{0.02585 \cdot 500}{1.008 \cdot 10^{-14}} = 3.29 \cdot 10^{-7} \, \text{s} = 0.179 \, \mu\text{s}.$$

Chapter 12
Optoelectronic Devices

12.1 Solar Cells (SCs)

12.1.1 SC Operation

The simplest solar cell (SC) is based on a pn junction diode, see Fig. 12.1. The principle of its operation can be readily understood if we recall a few basic facts about the diode. It consists of two quasi-neutral regions (QNRs) separated by a space-charge region (SCR) with the built-in electric field directed from the n- to the p-part of the diode. The top QNR is referred to as emitter, and the bottom one as base. The incident light generates excess electron–hole pairs in the diode. The excess electrons and holes diffuse toward the SCR, where they get separated by the built-in electric field. The excess electrons produced in the p-QNR are transferred into the n-QNR, and the excess holes generated in the n-QNR are transferred into the p-QNR. As a result, the two parts of the diode become oppositely charged, allowing one to produce electric power.

If we connect the two sides of the SC with a wire with zero electric resistance, the so-called short-circuit current, I_{SC}, will flow from the p-side to the n-side as the system will try to restore the thermal equilibrium perturbed by the irradiation.

If the emitter and the base are disconnected, the two sides of the diode will be charging up. This process cannot continue indefinitely for two reasons. First, the transfer of the photogenerated holes into the p-side and electrons into the n-side of the SC results in the generation of an additional electric field from the p- to the n-side of the diode; this additional electric field counteracts the built-in electric field. Second, the photogenerated electron–hole pairs recombine with each other at a finite rate. As a result, the charging-op of the two sides of the SC eventually stops, and a voltage develops across the SC. This voltage is called the open-circuit voltage, V_{OC}, with the p-side being positive and the n-side negative.

M. Evstigneev, *Introduction to Semiconductor Physics and Devices*,
https://doi.org/10.1007/978-3-031-08458-4_12

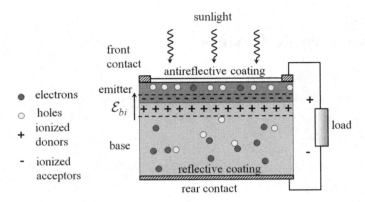

Fig. 12.1 The structure of a solar cell

12.1.2   *Spectral Irradiance (Spectral Intensity)*

A photon of frequency f and wavelength $\lambda = c/f$ carries the energy

$$hf = \frac{hc}{\lambda} \approx \frac{1.24\,\text{eV} \cdot \mu\text{m}}{\lambda} .$$

The short-wavelength photons interact primarily with the valence electrons by promoting them to the conduction band, provided that the photon energy exceeds the band gap energy. The visible light spectrum covers the range of λ from about 450 nm (violet) to about 700 nm (red), which corresponds to the energies from about 1.5 eV to about 3 eV, which is similar to the typical semiconductor band gap energies. The longer-wavelength (and hence smaller-energy) photons interact primarily with the lattice vibration quanta, the phonons. Because this interaction is relatively weak, semiconductors are for the most part transparent to the long-wavelength photons with $hf < E_g$.

Light is quantified by a parameter called spectral irradiance, also called spectral intensity, $F(f)$. It is related to the power $dP(f)$ carried by photons of frequency between f and $f + df$ through a unit surface area:

$$F(f) = \frac{dP(f)}{df} .$$

In other words, it is the energy flux density per unit frequency interval. It is measured in

$$[F(f)] = \text{W}/(\text{m}^2 \cdot \text{s}^{-1}) = \text{J}/\text{m}^2$$

in the SI units. The total intensity, i.e., the power incident per unit surface area, is obtained by integrating the spectral intensity with respect to frequency:

$$F = \int_0^\infty df\, F(f)\,, \quad [F] = \text{W/m}^2\,.$$

The photon flux density $\Phi(f)$ is defined though the number $dN(f)$ of photons of frequency between f and $f + df$ passing per unit time through a unit surface area:

$$\Phi(f) = \frac{dN(f)}{df} = \frac{F(f)}{hf}\,.$$

The light spectral intensity can also be defined as a function of the wavelength: namely, $F(\lambda)$ is the energy flux (i.e., the energy transferred per unit time through a unit surface area) per unit wavelength interval. That is, if $dP(\lambda)$ is the energy flux carried by the photons with the wavelength between λ and $\lambda + d\lambda$, then

$$F(\lambda) = \frac{dP(\lambda)}{d\lambda}\,.$$

It is measured in

$$[F(\lambda)] = \text{W/m}^3$$

in the SI units. However, since it is more convenient to measure the wavelength in μm or nm, $F(\lambda)$ is often measured in $\text{W/(m} \cdot \mu\text{m)}$ or $\text{W/(m} \cdot \text{nm)}$.

Note that the two quantities, $F(f)$ and $F(\lambda)$, are not related to each other by a simple replacement of f with c/λ. Their relation can be obtained by noting that the energy contained in a frequency interval df is

$$dP = F(f)|df| = F(\lambda)|d\lambda|\,.$$

The reason why the magnitude of the frequency and wavelength intervals is used rather than df and $d\lambda$ themselves is that it follows from the relation $f = c/\lambda$ that $df = -\frac{c}{\lambda^2}d\lambda$. The minus sign indicates that an increase of frequency by df means a decrease of the wavelength by $d\lambda$. In order to have both $F(f)$ and $F(\lambda)$ positive, the minus sign is omitted. Then, the relation between the two kinds of the spectral irradiance is found from

$$F(f)|df| = F(f)\frac{c}{\lambda^2}|d\lambda| = F(\lambda)|d\lambda| \quad \Rightarrow \quad F(f) = F(\lambda)\frac{\lambda^2}{c}\,.$$

Likewise,

$$F(\lambda) = F(f)\frac{f^2}{c}\,.$$

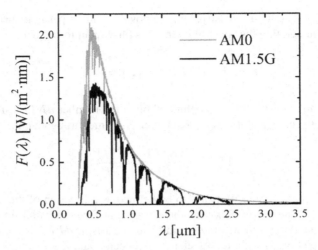

Fig. 12.2 The solar spectrum just outside of the Earth's atmosphere (AM0 spectrum, gray curve) and on the Earth's surface when the Sun is 48° above the horizon (AM1.5G spectrum, black curve)

The power transmitted by the photons of frequencies between f_{min} and f_{max} through a unit surface area per unit time is

$$\Delta P = \int_{f_{min}}^{f_{max}} df\, F(f) = \int_{\lambda_{min}}^{\lambda_{max}} d\lambda\, F(\lambda)\,,$$

where $\lambda_{min} = c/f_{max}$ and $\lambda_{max} = c/f_{min}$.

The solar spectral intensity in the outer space just above the Earth's atmosphere is shown as a gray curve in Fig. 12.2. This spectrum is denoted as AM0 spectrum, where AM stands for "air mass." The solar spectrum on the Earth depends on the time of the day and the geographic location. The current standard spectrum adopted to test the solar cells is the AM1.5G spectrum, which is obtained when the path length traveled by light through the atmosphere is 1.5 times as long as at the normal incidence; the G stands for "global." Note that the AM1.5G peak intensity is somewhat lower than the AM0 peak intensity, and the AM1.5G spectrum contains a number of minima that are not found in the AM0 counterpart. This is due to the light scattering and absorption by the atmosphere. The total irradiance received by the Earth under the AM1.5G conditions is 1000.4 W/m².

12.1.3 Light Absorption

When monochromatic light is passing through a semiconductor, it gets absorbed by it. Hence, light intensity should decrease with the distance x from the front surface of the material. This is quantified by the light absorption coefficient $\alpha(\lambda)$, defined as the fraction of photons of a given wavelength λ that are absorbed per unit length. It is measured in the units of inverse length,

$$[\alpha] = \text{cm}^{-1} \, .$$

If the light spectral irradiance at a particular wavelength λ at the distance x from the surface was $F(x)$, then at a slightly larger distance $x + dx$ is smaller by the amount

$$dF(x) = F(x + dx) - F(x) = -\alpha(\lambda)F(x)\,dx$$

$$\Rightarrow \quad \frac{dF(x)}{dx} = -\alpha(\lambda)F(x) \, .$$

The solution of this differential equation is the so-called Bouguer–Lambert–Beer law (Bouguer, early 1700s; Lambert, 1730; Beer, 1852),

$$F(x) = F_0 e^{-\alpha(\lambda)x} \, ,$$

where $F_0 = F(x = 0)$ is the intensity of the incident light. After passing through a slab of thickness L, the transmitted light intensity will be $F_0 e^{-\alpha L}$, and the energy absorbed per unit time by a unit surface area of the material will be $F_0(1 - e^{-\alpha L})$.

Absorption coefficient depends strongly on the wavelength and the material type, see Fig. 12.3 showing the $\alpha(\lambda)$ curve of silicon. Note that it goes sharply to zero at the photon energy near the band gap energy.

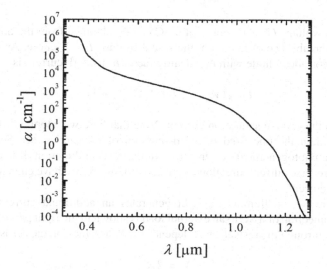

Fig. 12.3 Absorption coefficient of Si at $T = 300\,\text{K}$

The dominant absorption mechanism of photons whose energy exceeds the band gap energy is electron–hole pair generation. Assuming that one such photon creates just one electron–hole pair, the respective generation function can be estimated as

$$G(x) = \int_{E_g/h}^{\infty} df \, \frac{\alpha(f)F(f,x)}{hf} \, ,$$

where $F(f,x)$ is the spectral intensity of light at the frequency f and distance x from the front surface. Strictly speaking, the integral has to be multiplied by the so-called efficiency factor (less than 1) to take care of the fact that a photon can also be absorbed without the electron–hole pair production. For example, a photon can be absorbed by a free charge carrier. We will neglect the effect of such forms of photon absorption.

The light spectral intensity $F(f,x)$ at a distance x from the front surface is related to the spectral intensity $F_0(f)$ of the incident light. In order to maximize the generation function $G(x)$, both the front and rear surfaces of the SC are often made corrugated. This means that a photon that reaches the rear surface gets reflected at some angle to the normal. If it then reaches the front surface without absorption, it again has a good chance of being reflected at a different angle. Photons whose energy is only slightly higher than the band gap energy are absorbed very weakly by the material, see Fig. 12.3. They perform many back-and-forth trips between the front and rear surfaces before they produce an electron–hole pair. The result is that the light intensity $F(f,x)$ differs from the simple exponential function.

12.1.4 SC Current–Voltage Relation

The current–voltage $(I - V)$ curve of a SC can be obtained from the diode $I - V$ relation in the absence of radiation, the so-called dark $I - V$ curve. Assuming the SC to be a non-ideal diode with the ideality factor n, its dark current is

$$I_{dark}(V) = -I_S(e^{V/(nV_{th})} - 1) \, ,$$

where I_S is the reverse saturation current. Note that here we adopt a different sign convention than the one used in the discussion of the diodes, see Sect. 11.5.2. Because the photogenerated current flows in the reverse direction, it is convenient to treat the reverse current direction as positive; therefore, we have changed the sign of the dark current.

When the SC is illuminated, light generates an additional current, I_L. The simplest connection of a SC to an external load is shown in the right part of Fig. 12.4. The current through the resistance R depends on R, and the voltage across the SC is

$$V = IR \, .$$

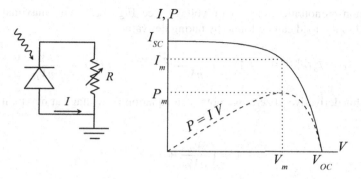

Fig. 12.4 Connection of a SC to a load R (left) and the current–voltage and power–voltage curves (right)

The current–voltage relation in the presence of illumination becomes

$$I(V) = I_L - I_S(e^{V/(nV_{th})} - 1),$$

see the right part of Fig. 12.4.

As the load R varies from 0 to ∞, the current changes from the short-circuit current

$$I_{SC} = I_L \text{ at } R = 0$$

to

$$I = 0 \text{ at } R = \infty.$$

At the same time, the voltage will change from

$$V = 0 \text{ at } R = 0$$

to the open-circuit voltage

$$V = V_{OC} = nV_{th} \ln\left(\frac{I_L}{I_S} + 1\right) \text{ at } R = \infty.$$

The open-circuit voltage is found from the condition $I(V_{OC}) = 0$, as no current is flowing if $R = \infty$.

The power delivered to the load is given by the product

$$P = VI(V) = V(I_L - I_S(e^{V/(nV_{th})} - 1)).$$

It is a non-monotonic function of voltage, see Fig. 12.4. The maximal power delivered to the load can be found by taking derivative:

$$\frac{dP}{dV} = I_L - I_S(e^{V/(nV_{th})} - 1) - I_S\frac{V}{nV_{th}}e^{V/(nV_{th})} = I(V) - I_S\frac{V}{nV_{th}}e^{V/(nV_{th})} .$$

Setting this derivative to zero, we obtain an equation for voltage at maximal power, V_m:

$$\frac{I_L}{I_S} + 1 = \left(\frac{V_m}{nV_{th}} + 1\right) e^{V/(nV_{th})} ,$$

which can be solved iteratively. Once V_m is found, the current in the maximal-power regime can be determined by a substitution of V_m into the expression for current, $I_m = I(V_m)$.

The fill factor of the $I - V$ curve is defined as the ratio

$$FF = \frac{I_m V_m}{I_{SC} V_{OC}} \cdot 100\% .$$

It measures the deviation of the $I - V$ curve from a rectangular shape. Typically, FF has the value between 70% and 80%.

Finally, the photoconversion efficiency is defined as the ratio of the maximal power produced by the SC to the power incident onto the cell in the form of light:

$$\eta = \frac{I_m V_m}{P_{in}} \cdot 100\% .$$

Depending on the SC type, photoconversion efficiency ranges from a few percent to about 25% or even 40%.

Because solar intensity on the Earth surface is about $1000\,\text{W/m}^2$, a solar panel of $1\,\text{m}^2$ area and 10% efficiency can produce the power of $100\,\text{W}$, which is comparable to the power consumption by a TV set. On average, a typical North American household consumes about $900\,\text{kWh}$ per month, i.e., the average power consumed is about $1250\,\text{W}$. In order to provide this power, about $12\,\text{m}^2$ area must be covered with SCs. This is two or three times smaller than the area of a typical roof of a private household. The price to install solar panels on a roof of a building is about $15K, but it pays off in saving on the electricity bills after about 5 years. Of course, these estimates may turn out too optimistic in the areas that get little direct sunlight.

12.2 Light-Emitting Diodes (LEDs)

12.2.1 LED Operation

The light-emitting diodes (LEDs) have numerous applications in everyday life, such as displays of various sorts (calculators, watches, street signs, etc.), lamps, and light sources in the optical-fiber communication systems. The structure of a LED is shown in Fig. 12.5. It is a forward-biased diode, whose front surface is used for the collection of light, and the rear surface is covered with a highly reflective coating.

When a p-n junction is forward-biased, the majority carriers from its two QNRs are injected into the opposite QNRs, where they become excess minority carriers. In the direct-band gap semiconductors, the injected excess carriers recombine with the majority carriers primarily radiatively, i.e., with the emission of light. The region where most of the recombination events occur is called the active region. The rear surface is covered with a reflective coating. The photons that reach the rear surface get reflected from it and have a chance to contribute to the overall light intensity emitted by the device. The effect of light emission by a forward-biased diode bears the name of electroluminescence.

The intensity of the emitted light is proportional to the current, which depends on the applied voltage as $e^{V/(nV_{th})} - 1 \approx e^{V/(nV_{th})}$, where n is the ideality factor of the diode.

The radiative recombination rates of the minority electrons in the p-QNR and of the minority holes in the n-QNR are, in general, different. It makes sense to collect light from that side of the diode whose radiative recombination rate is the higher; in Fig. 12.5, it is assumed that the radiative recombination rate is the larger in the p-side.

12.2.2 LED Spectrum

LEDs employ a radiative recombination of charge carriers, which is most efficient in the direct-band gap semiconductors. Because conduction band energy levels are populated with the probability proportional to $e^{-E/kT}$, and because the density of states in the conduction band depends on the energy as $\sqrt{E - E_C}$, the most probable electron energy value can be found by setting the derivative of the product

Fig. 12.5 The principle of operation of an LED

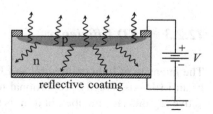

reflective coating

$d[\sqrt{E - E_C}e^{-E/kT}]/dE$ to zero. The result turns out to be $E_C + kT/2$. Similarly, the most probable energy of a hole in the valence band is $E_V - kT/2$. Then, the most probable energy of a photon emitted as a result of a radiative recombination event is

$$hf = E_g + kT \approx E_g \,,$$

and the respective wavelength

$$\lambda = \frac{c}{f} \approx \frac{hc}{E_g} \,.$$

The width of the LED spectrum can be estimated to be

$$\Delta f = \frac{kT}{h} \,,$$

or, in terms of the wavelength,

$$\Delta \lambda = |\Delta(c/f)| = \frac{c}{f^2} \Delta f = hc\frac{kT}{E_g^2} \,.$$

It is seen that the ratio $\frac{\Delta \lambda}{\lambda} = \frac{kT}{E_g}$ is quite small (about 0.02–0.03), i.e., a single LED emits a rather narrow spectral line.

In order to emit in the visible range of λ from about 0.4 to 0.75 μm, the band gap must be between 1.65 and 3.1 eV. GaAs with $E_g = 1.42$ eV cannot be used for illumination but can be employed in other application, such as telecommunication. The materials that are used in the visible LEDs are ternary and quaternary compounds, such as $Al_xGa_{1-x}As$ (red part of the spectrum), $GaAs_xP_{1-x}$ (red), $In_{1-y}(Al_xGa_{1-x})P$ (yellow/orange light), $In_xGa_{1-x}N$ (blue/green light), etc. The band gap of those compounds can be adjusted by changing the relative concentrations x, y of the components.

To produce white light, one can combine several LEDs of different colors, typically red, green, and blue. A less expensive method that produces the output spectrum comfortable to the eye is to cover a single LED with a photoluminescent material that absorbs the original LED photons and emits light of other, usually lower, frequencies in a broad spectrum similar to the day light.

12.2.3   LED Efficiency

The intensity of light emitted by an LED—that is, the power emitted by a unit area of an LED—is directly proportional to the number of recombination events in the active region. This number, in turn, is directly proportional to the number of charge carriers injected into the active region per unit area per unit time; it is just the electric current density divided by the elementary charge. Since the energy of a photon produced in a radiative recombination event is E_g, the intensity is proportional to

$$I_{EL} = E_g \frac{J}{e} \eta \, ,$$

where the proportionality constant $\eta < 1$ is the LED efficiency.

The LED efficiency depends on the efficiency of several processes involved in photoemission:

Injection Efficiency The net current in the diode consists of three contributions: the minority electron current in the p-QNR, the minority hole current in the n-QNR, and the recombination current in the SCR:

$$J = J_e + J_h + J_{SCR}$$

with

$$J_e = e \frac{n_i^2}{N_a} \frac{D_{ep}}{L_{ep}} (e^{V/V_{th}} - 1) \, , \quad J_e = e \frac{n_i^2}{N_d} \frac{D_{hn}}{L_{hn}} (e^{V/V_{th}} - 1) \, ,$$

$$J_{SCR} = ew \frac{n_i}{\tau_{SCR}} (e^{V/(2V_{th})} - 1) \, ,$$

as derived in Sect. 11.5.2. Here, D_{ep} and D_{hn} are the diffusion coefficients of the minority electrons in the p-region and minority holes in the n-region, L_{ep} and L_{hn} are the respective diffusion lengths, w is the SCR width, and τ_{SCR} is the carrier lifetime in the SCR. Assuming that the light-emitting part of the diode is the p-QNR, it is the current J_e that corresponds to electron injection from the n-QNR into the p-QNR and is therefore responsible for the light emission. The SCR recombination, corresponding to the contribution J_{SCR}, usually proceeds not radiatively, but via the trap-assisted Shockley–Read–Hall mechanism. On the other hand, the photons emitted due to hole recombination in the n-region get partly reabsorbed by the material as they travel toward the front surface, and therefore they produce relatively small contribution to the output light intensity. Correspondingly, the emitted light intensity should be proportional to the relative contribution of the electron injection current to the total current. This is parametrized by the so-called injection efficiency

$$\eta_{inj} = \frac{J_e}{J_e + J_h + J_{SCR}} \, .$$

In order to make this number close to unity, the n-side of the junction should be doped more heavily than the p-side, so that $J_e \gg J_h$. Also, the junction must be under a sufficiently large forward bias V in order to ensure that $J_e \gg J_{SCR}$.

Recombination Efficiency Not all electrons injected into the p-region recombine radiatively. The net recombination rate consists of the radiative and non-radiative contributions, $U = U_r + U_{nr}$, proportional to the respective inverse radiative and non-radiative recombination times τ_r and τ_{nr}. The recombination efficiency is defined as the relative number of radiative recombination events:

$$\eta_{rec} = \frac{U_r}{U_r + U_{nr}} \approx \frac{\tau_r^{-1}}{\tau_r^{-1} + \tau_{nr}^{-1}} .$$

In order to increase η_{rec}, one needs a high-purity sample with the well-passivated surfaces, so as to reduce the Shockley–Read–Hall and surface recombination rates. At the same time, in order to increase the role of radiative recombination, the p-side of the diode must be heavily doped.

Internal Quantum Efficiency Internal quantum efficiency is the product of injection and recombination efficiencies,

$$\eta_i = \eta_{inj} \eta_{rec} .$$

Note that η_{inj} decreases with the p-side doping concentration N_a, whereas η_{rec} increases with N_a. Thus, there must be an optimal doping concentration N_a that maximizes the internal efficiency.

External Quantum Efficiency External quantum efficiency is the fraction of photons produced inside the diode that actually leave the diode. In reality, over 95% of the photons produced in radiative recombination events are lost due to the following reasons:

(i) The typical photon energy actually exceeds E_g by about kT, as explained above. The majority of those photons will be reabsorbed by the material.

(ii) The photons that reach the front surface of the diode may be reflected back into the semiconductor and be reabsorbed. The probability of reflection depends on the angle of incidence and photon polarization. For normal incidence, reflection probability is polarization-independent and is given by the Fresnel formula

$$\Gamma_R = \left| \frac{n_{SC} - 1}{n_{SC} + 1} \right|^2 ,$$

where n_{SC} is the semiconductor refractive index. Typically, $n_{SC} \approx 3 - 4$, and hence, $\Gamma_R \approx$ 25–35%.

(iii) Depending on the angle of incidence, θ_i, the angle of refraction at which a photon exits the material is given by Snell's law,

$$n_{SC} \sin \theta_i = \sin \theta_r \quad \Rightarrow \quad \theta_r = \sin^{-1}(n_{SC} \sin \theta_i) .$$

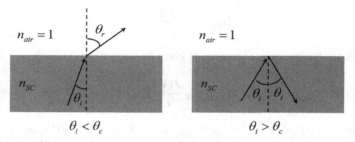

Fig. 12.6 Left: Photon refraction as it exits the semiconductor at the incident angle θ_i below the critical value θ_c. Right: Photon total internal reflection at the angle of incidence greater than the critical angle

This expression only makes sense when the argument of the inverse sine is below 1. If the argument exceeds 1, light will not exit the material but will undergo total internal reflection back into the semiconductor. This happens when the angle of incidence exceeds the critical angle given by

$$\theta_c = \sin^{-1} \frac{1}{n_{SC}} .$$

Assuming the typical value $n_{SC} = 3.5$, the critical angle turns out to be about 17°. One can conclude that the light is emitted practically perpendicular to the LED surface (Fig. 12.6).

Now, a solid angle corresponding to the angle of incidence θ_i is given by $\Omega(\theta_i) = 2\pi(1 - \cos\theta_i)$. The directions of the photons produced in radiative recombination are distributed isotropically, i.e., all directions are equally probable; this corresponds to the solid angle of 4π. Then, the fraction of photons that hit the surface at an angle below the critical angle is

$$\frac{\Omega(\theta_i)}{4\pi} = \frac{1 - \cos\theta_c}{2} \approx \frac{\theta_c^2}{4}$$

for small $\theta_c \ll 1$. Assuming $\theta_c = 17° \approx \pi/10$, this ratio is about 0.1, i.e., only 10% of the photons have a chance of leaving the material. This has to be multiplied by the transmission coefficient, $1 - \Gamma_R$, which is about 0.65 for normal incidence. Thus, we get the external quantum efficiency to be $\eta_e \approx 6.5\%$.

This estimate is actually too optimistic because we did not take into account the decrease of $(1 - \Gamma_R)$ with the angle of incidence and optical losses due to photon re-absorption. The actual external efficiency is about 1% to 2%, i.e., about 98% of the photogenerated photons do not leave the diode.

All in all, the number of photons that are emitted by the diode per unit area per unit time is

$$\Phi_{EL} = \frac{J(V)}{e} \eta_i \eta_e \, ,$$

and the electroluminescence intensity is given by

$$I_{EL} = E_g \frac{J(V)}{e} \eta_i \eta_e \, .$$

Assuming $E_g = 2\,\text{eV}$, $J = 1\,\text{A/cm}^2$, $\eta_{inj} = 50\,\%$, $\eta_{rec} = 100\,\%$, and $\eta_e = 1\,\%$, the intensity of the emitted light according to this expression is 0.01 W/cm^2, which is ca. 100 times smaller than the intensity of the light coming from the Sun. This is way too small for practical applications.

12.2.4   Increasing the LED Efficiency

The main problem of the homojunction-based LEDs discussed above is the re-absorption of the photons generated in radiative recombination. Because the energy of those photons exceeds the band gap energy by ca. kT, some of them may cause an electron transition from the valence to the conduction band and not contribute to the output light.

In a homojunction-based LED, the light-emitting part of the junction must be relatively thin in order to let the photons escape without being reabsorbed by the material. Unfortunately, this implies high probability of the injected minority carriers' diffusion to the surface and their non-radiative recombination there.

Hence, in order to increase the intensity of the emitted light, we need to (i) increase the rate of the radiative recombination in the active region, and (ii) reduce re-absorption of the photons outside of the active region.

Both problems are solved with the use of double heterojunction structures. The idea is to produce the diode's p- and n-sides from broad-band gap semiconductors, and to sandwich the active region in between, see Fig. 12.7a. Usually, the narrow-band gap light-emitting region is made of an intrinsic semiconductor. The photons produced in the active region have energy less than the band gap in the adjacent

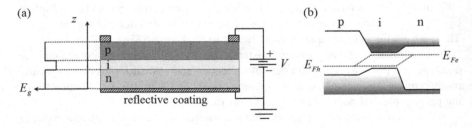

Fig. 12.7 (a) A heterojunction LED and (b) its energy band diagram

p- and n-regions. Therefore, those photons are not absorbed by the structure and eventually contribute to the output light intensity.

The second advantage of the double heterojunction arrangement is the fact that the conduction band edge in the active i-region is below the conduction band edge in the n-region, see the energy band diagram in Fig. 12.7b. This implies that the concentration of electrons in the active region is higher than that in the n-region. For the same reason, the valence band edge in the active i-region is higher than that in the p-region, implying higher concentration of holes. In turn, this means high recombination rate and high-output light intensity of the LED.

12.3 Semiconductor Lasers

12.3.1 Stimulated Emission and Einstein's Coefficients

Semiconductor LEDs generate light by spontaneous emission of photons in radiative recombination of electron–hole pairs. Semiconductor laser diodes also generate light by transitions of electrons from the conduction to the valence band. But, in contrast to LEDs, those transitions are stimulated by light quanta (photons) already present in the device. The stimulated emission effect was predicted by Albert Einstein in 1916, more than half a century before the first laser was made. The very name "laser" is an abbreviation for "Light Amplification by Stimulated Emission of Radiation."

Before discussing semiconductor lasers, let us focus on a more intuitive situation of lasing in a gas of atoms that do not form a solid. The atoms find themselves in a cavity of volume V. For simplicity, let us assume that each atom can exist in only two states: the ground state of energy E_1 and the excited state of energy E_2. Denoting the concentrations of atoms in the ground and in the excited state as n_1 and n_2, the total concentration is given by

$$n = n_1 + n_2 .$$

Because the concentration of atoms is constant, the rates of change of n_1 and n_2 are equal in magnitude and opposite in sign:

$$\frac{dn}{dt} = \frac{dn_1}{dt} + \frac{dn_2}{dt} = 0 \implies \frac{dn_1}{dt} = -\frac{dn_2}{dt} .$$

Apart from the atoms, the cavity is filled with photons. The photons interact with the atoms via the following elementary processes, see Fig. 12.8:

(i) *Absorption*, in which a photon of frequency f gives its energy hf to promote an atom from the ground to the excited state. In order for this to happen, the photon frequency must exactly match the excitation energy divided by Planck's constant:

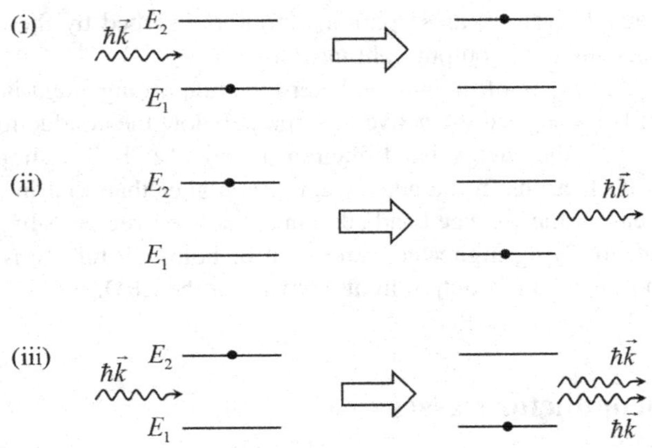

Fig. 12.8 Elementary interactions of a photon with an atom: (i) absorption, (ii) spontaneous emission, and (iii) stimulated emission

$$f = \Delta E / h = (E_2 - E_1)/h .$$

A photon of frequency below this value cannot excite an atom due to the energy conservation. It is not as obvious why a photon of frequency higher than $\Delta E / h$ cannot give part of its energy to an atom. This has to do with the particle nature of photons. As a particle, a photon cannot be partly absorbed; it can only be either completely absorbed or not absorbed at all.

(ii) *Spontaneous emission*, in which an excited atom returns to the ground state with the emission of a photon of energy $hf = \Delta E$. The emitted photon propagates in a random direction.

At a first glance, absorption and spontaneous emission exhaust all possible kinds of matter–light interaction. Let us now describe them mathematically.

We denote the concentration of photons of frequency f as $n_{ph}(f)$. The rate of change of the atomic concentrations n_1 and n_2 due to absorption must be proportional to the concentration of atoms in the ground state and to the total number of photons of frequency f in the volume V:

$$\left(\frac{dn_1}{dt}\right)_{abs} = -b_{abs} n_1 V n_{ph}(f) .$$

The parameter $b_{abs} > 0$ is the probability of photon absorption by an atom per unit time. It has the dimension of $[b_{abs}] = \mathrm{s}^{-1}$. It depends on the atom–photon interaction properties only and does not depend on the concentration of atoms n_1, n_2 and on the concentration of photons $n_{ph}(f)$. The minus sign indicates that photon absorption results in a decrease of n_1.

The rate at which the concentrations n_1 and n_2 change due to the spontaneous emission is proportional to the concentration of the excited atoms:

$$\left(\frac{dn_1}{dt}\right)_{spon} = b_{spon} n_2 \; .$$

The coefficient b_{spon} is the probability per unit time for an excited atom to return to its ground state with the emission of a photon. In other words, it is the inverse of the average time that an atom spends in the excited state.

If only absorption and spontaneous emission were operative, then the total rate of change of the concentration of the atoms in the ground state would be

$$\left(\frac{dn_1}{dt}\right)_{abs} + \left(\frac{dn_1}{dt}\right)_{spon} = -b_{abs} n_1 V n_{ph}(f) + b_{spon} n_2 \; .$$

Now, let us consider our system in the state of thermal equilibrium. Three statements can be made about the equilibrium state:

First, the concentrations of atoms in the ground state and in the excited state are stationary:

$$\left(\frac{dn_1}{dt}\right)_{eq} = \left(\frac{dn_2}{dt}\right)_{eq} = 0 \; .$$

Second, they are related to each other by a Boltzmann factor $e^{-\Delta E/kT}$:

$$n_2 = n_1 e^{-\Delta E/kT} \equiv n_1 e^{-hf/kT} \; .$$

Third, the concentration of photons is given by the so-called Planck distribution

$$n_{ph}(f) = \frac{g(f)}{e^{hf/kT} - 1} \; ,$$

where $g(f)$ is the density of photon states; the derivation of this result is performed in Appendix A.5.

Hence, in equilibrium,

$$\left(\frac{dn_1}{dt}\right)_{abs,eq} + \left(\frac{dn_1}{dt}\right)_{spon,eq} = -b_{abs} n_1 V \frac{g(f)}{e^{hf/kT} - 1} + b_{spon} n_1 e^{-hf/kT} \; .$$

This should be zero at any temperature. But because the two terms in the right-hand side depend on temperature differently, they cannot cancel each other. This is the problem because it implies that thermal equilibrium state is impossible.

To fix this problem, we must, following Einstein, postulate the existence of a third elementary process, namely:

(iii) *Stimulated emission* This is a process, in which a photon incident on an atom in its excited state induces a de-excitation of this atom into a ground state accompanied by an emission of another photon of the same wave vector as the incident photon, see the lowest diagram in Fig. 12.8.

The rate at which the concentration of the atoms in the ground state changes due to this process is proportional to the concentration of atoms in the excited state and the total number of photons in the volume V:

$$\left(\frac{dn_1}{dt}\right)_{stim} = b_{stim} V n_{ph}(f) n_2 .$$

The physical meaning of the parameter b_{stim} is the probability for the emission to be stimulated by a single photon per unit time. In thermal equilibrium,

$$\left(\frac{dn_1}{dt}\right)_{stim,eq} = b_{stim} V \frac{g(f)}{e^{hf/kT} - 1} n_1 e^{-hf/kT} .$$

Combining all three contributions to the time derivative dn_1/dt in equilibrium together, we obtain in the equilibrium state:

$$\left(\frac{dn_1}{dt}\right)_{eq} = \left(\frac{dn_1}{dt}\right)_{abs,eq} + \left(\frac{dn_1}{dt}\right)_{spon,eq} + \left(\frac{dn_1}{dt}\right)_{stim,eq}$$

$$= -b_{abs} n_1 V \frac{g(f)}{e^{hf/kT} - 1} + b_{spon} n_1 e^{-hf/kT} + b_{stim} V \frac{g(f)}{e^{hf/kT} - 1} n_1 e^{-hf/kT} .$$

This expression can be equal to zero provided that the coefficients b_{abs}, b_{spon}, and b_{stim} are related to one another in a special way. Indeed, dividing the second line by $n_1 e^{-hf/kT}$ and setting the result to zero, we obtain

$$-b_{abs} V \frac{g(f) e^{hf/kT}}{e^{hf/kT} - 1} + b_{spon} + b_{stim} V \frac{g(f)}{e^{hf/kT} - 1} = 0 .$$

The sum of the first and third terms is temperature-independent only if the coefficients of absorption and stimulated emission are the same:

$$b_{abs} = b_{stim} = b ,$$

and that the spontaneous emission coefficient is

$$b_{spon} = b g(f) V .$$

These relations are remarkable. In principle, there is no a priori reason to expect that the coefficients that describe the completely different physical processes can be expressible in terms of each other in such a simple manner. In particular, the last formula tells us that the probability of spontaneous emission per unit time equals the probability of photon absorption b multiplied by the total number of states $g(f)V$ available to an emitted photon of frequency f. The relations between the emission and absorption coefficients follow from purely statistical considerations and are indeed confirmed by the rigorous and rather complicated quantum-mechanical calculations.

In the literature, it is customary to work not with the photon concentration $n_{ph}(f)$, but with the photon spectral energy density

$$u(f) = hf n_{ph}(f)$$

and to express the rates of absorption, spontaneous emission, and stimulated emission as

$$\left(\frac{dn_1}{dt}\right)_{abs} = -B_{12}n_1 u(f) , \quad \left(\frac{dn_1}{dt}\right)_{spon} = An_2 , \quad \left(\frac{dn_1}{dt}\right)_{stim} = B_{21}n_2 u(f) .$$

The parameters A and B are called Einstein's coefficients; they are related to the respective probabilities of the elementary processes to occur per unit time by

$$B_{12} = b_{abs}\frac{V}{hf} , \quad A = b_{spon} , \quad B_{21} = b_{stim}\frac{V}{hf}$$

and to each other by

$$B_{12} = B_{21} = B , \quad A = Bhfg(f) .$$

12.3.2 Generation of Light

Consider a beam of monochromatic light of frequency $f = \Delta E/h$ propagating in a gas of two-state atoms. As this beam propagates in the x-direction, it interacts with the medium, as described in the previous section. As a result, the concentration of photons in the beam, $n_{ph}(f, x)$, depends not only on frequency, but also on the position x. For the sake of brevity, the frequency dependence will be implied, but not written explicitly.

We assume the photon concentration to be so high that stimulated emission dominates over spontaneous emission. The photon concentration changes with the coordinate x due to absorption and stimulated emission according to

$$\frac{dn_{ph}(x)}{dx} = \frac{dn_{ph}(x)}{dt}\frac{\partial t}{\partial x} = \frac{1}{c}\left(\left(\frac{dn_{ph}(x)}{dt}\right)_{stim} - \left(\frac{dn_{ph}(x)}{dt}\right)_{abs}\right)$$

$$= \frac{bVn_{ph}(x)}{c}(n_2 - n_1) = \gamma n_{ph}(x) \,.$$

Due to the interaction with light, the level populations n_1 and n_2 will also change. Let us assume that some external agent is operative that maintains them constant; this process is called pumping. In that case, the parameter

$$\gamma = \frac{bV}{c}(n_2 - n_1) \,,$$

called the gain factor, will be constant. Then, the concentration of photons in the beam, $n_{ph}(x) \propto e^{\gamma x}$, will either decrease or increase exponentially, depending on the sign of the gain factor, so will the light intensity, i.e., the energy transmitted by the photons per unit time through a unit surface area, $F(x) = chf n_{ph}(x)$.

The lasing effect means that the intensity of the beam increases. In order for it to be possible, i.e., for the gain factor γ to be positive, population inversion

$$n_2 > n_1$$

is a necessary condition. The concentration of atoms in the excited state must exceed the concentration of atoms in the ground state. The physical reason of lasing is quite simple: the more photons are present in the beam, the more stimulated emission events are produced, and the higher the rate of increase of the photon number.

12.3.3 Semiconductor Laser Operation

The Structure of a Semiconductor Laser

In a semiconductor, one deals not with two discrete energy levels, but with two bands. In order for the diode to lase, it should be in the population inversion state. That is, the electron concentration near the bottom of the conduction band must be higher than the concentration of electrons near the top of the valence band.

Clearly, this is impossible in thermal equilibrium, as the following calculation shows. Consider two electron states: state 1 in the valence band and state 2 in the conduction band, with

$$E_2 > E_C \,, \quad E_1 < E_V \,.$$

The probability for each of these states to be occupied by an electron is given by the Fermi–Dirac formula

$$f_{1,2} = \frac{1}{1 + e^{(E_{1,2}-E_F)/kT}} \,.$$

Population inversion means higher probability for an electron to be in the state 2 than in the state 1, which is only possible if

$$f_2 > f_1 \ .$$

But this implies that

$$e^{(E_2-E_F)/kT} < E^{(E_1-E_F)/kT} \quad \Rightarrow \quad E_2 - E_F < E_1 - E_F \ ,$$

which contradicts the obvious inequality $E_2 > E_C > E_V > E_1$.

Now consider what happens when the semiconductor consists of two differently doped regions forming a diode. In thermal equilibrium, lasing is not possible also in a diode. Hence, we must bring it out of equilibrium by applying a bias V. In that case, electrons in the n-region and holes in the p-region are characterized by their own quasi-Fermi energies, E_{Fe} and E_{Fh}, respectively. Inside the space-charge region of the diode, these two quasi-Fermi energies coexist, whereas outside the SCR, the quasi-Fermi energies belonging to the minority carriers quickly align themselves with the majority carriers' quasi-Fermi energies.

Inside the SCR, the probabilities for the states 1 and 2 to be occupied are modified to

$$f_2 = \frac{1}{1 + e^{(E_2-E_{Fe})/kT}} \ , \quad f_1 = \frac{1}{1 + e^{(E_1-E_{Fh})/kT}} \ ,$$

and the population inversion condition, $f_2 > f_1$, implies that we should have $E_2 - E_{Fe} > E_1 - E_{Fh}$ or, equivalently,

$$E_{Fe} - E_{Fh} > E_2 - E_1 = hf > E_g \ ,$$

where hf is the energy of the emitted photons, and the last inequality follows from the fact that $E_2 > E_C$, $E_1 < E_V$, and $E_C - E_V = E_g$. We conclude that in order for lasing to be possible, one or both sides of the p-n junction must be degenerately doped.

A semiconductor laser structure is based on a double heterojunction, where the active region (usually, intrinsic) is sandwiched between the degenerately doped n- and p-layers, called the cladding layers, see Fig. 12.9a. Note that a similar structure is used in the LEDs. To increase the optical path length of laser light in the active region, one side of the diode is covered with a highly reflecting coating and the opposite side with a partially reflective coating. This means that lasing effect is possible only for the photons that travel between the reflective sides, and a small fraction of those photons can exit the device through the partially reflective side.

As in the LEDs, it is advantageous to make the cladding n- and p-layers of a semiconductor with a slightly larger band gap than in the active region,

$$E_{gc} > E_{ga} \ ,$$

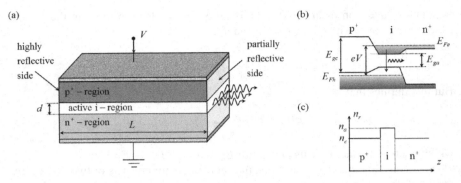

Fig. 12.9 (**a**) Schematic diagram of a semiconductor laser; (**b**) its energy band diagram; (**c**) variation of the refractive index along the coordinate z perpendicular to the diode plane

where the subscripts c and a stand for "cladding" and "active," respectively; see Fig. 12.9b. This band gap difference guarantees strong population inversion. A further improvement is to use the cladding material with the refractive index slightly lower than the refractive index of the active region,

$$n_c < n_a \ ,$$

see Fig. 12.9c. Then, the generated photons will remain trapped inside the active region as a result of the total internal reflection from the boundaries with the cladding layers.

Threshold Current

The salient feature of the double heterojunction laser is that its active i-region is very thin. The small size of the central active region implies that the excess carrier concentration there is practically constant. Hence, the physical picture behind the current–voltage curve is quite simple. Namely, the excess electrons are injected into the i-region from the n^+-region, with AJ/e electrons coming in per unit time, where A is the cross-sectional area of the laser. On the other side of the i-region, the excess holes are injected from the p^+-region, with AJ/e holes coming in per unit time. Inside the active region, electrons and holes recombine at a rate $U(\Delta n)Ad$, where d is the width of the active region. Because the number of excess carriers that come in per unit time must be the same as the number of charge carriers that recombine per unit time, the current density is given by

$$J = eU(\Delta n)d \ .$$

To find the current–voltage curve, we note that the excess carrier concentration Δn depends on the difference between the quasi-Fermi energies of electrons and holes. This difference is nothing else by the applied voltage multiplied by the elementary charge:

$$E_{Fe} - E_{Fh} = eV \,,$$

and hence, the excess concentration depends on the applied voltage:

$$\Delta n = \Delta n(V) \,.$$

In a non-degenerate semiconductor, the relationship between Δn and V would follow immediately from the generalization of the mass–action law, $(n_0 + \Delta n)(p_0 + \Delta n) = n_i^2 e^{(E_{Fe} - E_{Fh})/kT} = n_i^2 e^{V/V_{th}}$. However, since we are considering the case when the quasi-Fermi energies are located inside the bands, we cannot use this relationship and must resort to the qualitative rather than quantitative arguments.

When we discussed lasing in a gas of two-level atoms, the gain factor γ turned out to be proportional to the difference between the populations of the two levels. Hence, the light intensity along the beam was increasing according to

$$\frac{dF}{dx} = \gamma F(x) \,, \quad \gamma \propto n_2 - n_1 \,.$$

In a semiconductor laser, the difference $n_2 - n_1$ must be replaced by the excess carrier concentration, and thus the gain factor is proportional to $\gamma \propto \Delta n$. Furthermore, when light propagates in a semiconductor, it gets absorbed with the absorption coefficient α. Hence, the expression for the rate of change of light intensity with distance in a laser becomes

$$\frac{dF}{dx} = (\gamma - \alpha)F(x) \,, \quad \gamma \propto \Delta n \,.$$

After traveling the distance $2L$ between the two mirrors, light intensity changes by a factor $e^{2L(\gamma - \alpha)}$. Light also gets reflected from the two mirrors. Each reflection results in a change of light intensity by a factor equal to the mirror reflection coefficient, Γ_1 and Γ_2 for mirrors 1 and 2, respectively. The net effect of absorption and reflection is a change of light intensity by the product $\Gamma_1 \Gamma_2 e^{2L(\gamma - \alpha)}$. In order for lasing to be possible, this product must be bigger than 1.

In turn, it implies that the gain factor must exceed a threshold value, γ_T, such that

$$\Gamma_1 \Gamma_2 e^{2L(\gamma_T - \alpha)} = 1 \quad \Rightarrow \quad \gamma_T = \alpha + \frac{1}{2L} \ln \frac{1}{\Gamma_1 \Gamma_2} \,.$$

The gain factor is proportional to the excess carrier concentration. Hence, for lasing to be possible, excess concentration must exceed some threshold value, $\Delta n_T \propto \gamma_T$. On the other hand, according to the theory of the current–voltage curve in a p^+-i-n^+ structure outlined earlier in this section, the excess concentration enters into the current–voltage relation as an argument of the net recombination rate. Hence, in order to maintain the excess concentration at the level above threshold, the injection current must exceed the threshold value

$$J_T = eU(\Delta n_T)d \ .$$

We can make a rough approximation by assuming that the net recombination rate is proportional to the excess carrier concentration, $U(\Delta n) \propto \Delta n$. With this approximation, we conclude that the threshold current must be proportional to $J_T \propto \Delta n_T \propto \gamma_T$. Denoting the proportionality constant as β, we can write

$$J_T = \beta \left(\alpha + \frac{1}{2L} \ln \frac{1}{\Gamma_1 \Gamma_2} \right) \ .$$

At $J < J_T$, very weak light intensity is generated, whereas at $J = J_T$, lasing begins, see Fig. 12.10. The threshold current is thus quite easy to measure. A linear relation between J_T and $\alpha + \frac{1}{2L} \ln \frac{1}{\Gamma_1 \Gamma_2}$ is indeed experimentally confirmed with β being of the order of 100 A/cm. This implies that the threshold current may be of the order of thousands (or even tens of thousands) A/cm^2, quite a large value. Reduction of the threshold current is one of the main problems in the semiconductor laser technology.

Laser Spectrum

The active region of a laser is a resonator, where only photons of certain wavelength can exist. Those photons must satisfy the standing wave condition

$$L = m\lambda'_m/2 \ ,$$

where L is the distance between the reflecting sides, m is an integer, and λ'_m is the photon wavelength inside the active region. Hence, the laser produces not a single wavelength, but many spectral lines of the wavelength

Fig. 12.10 Schematic illustration of the laser intensity vs. current density relation

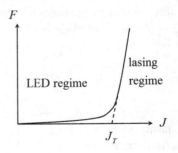

$$\lambda'_m = 2L/m .$$

Each such spectral line is called the mode.

The frequency of the photons that leave the active laser is the same as the frequency of the photons inside the active region. On the other hand, the velocity of light in the semiconductor is

$$c' = c/n_a ,$$

where n_a is the refractive index of the active region. Therefore, the wavelength of the photons emitted by the laser is

$$\lambda_m = c/f = c'n_a/f = n_a\lambda'_m = 2Ln_a/m,$$

and their frequencies are quantized as

$$f_m = c/\lambda_m = cm/(2Ln_a) .$$

The separation between two spectral lines is

$$\Delta\lambda = \lambda_m - \lambda_{m+1} = 2Ln_a \left(\frac{1}{m} - \frac{1}{m+1} \right) = \frac{2Ln_a}{m(m+1)} = \frac{\lambda_m}{m+1} .$$

For large $m \gg 1$, the separation between two modes can be approximated as

$$\Delta\lambda = \frac{\lambda}{m} = \frac{\lambda^2}{2n_a L} .$$

The frequency and energy separations are

$$\Delta f = f_{m+1} - f_m = \frac{c}{2Ln_a} , \quad h\Delta f = \frac{hc}{2Ln_a} = \frac{1.24\,\text{eV} \cdot \mu\text{m}}{2Ln_a} .$$

The energies of those photons range from E_g to $eV = E_{Fe} - E_{Fh} > E_g$. Hence, the number of lines in the spectrum of a laser is $\frac{eV - E_g}{h\Delta f}$. Assuming L to be of the order of $100\,\mu\text{m}$ and n_a of the order of 4, we have the energy separation of the order of $1\,\text{meV}$. For $eV - E_g = 1\,\text{eV}$, this means that the number of spectral lines is of the order of a thousand.

At a first glance, the higher the bias V, the broader the spectrum of the emitted light should be and the more spectral lines it should contain. In reality, quite the opposite happens: increasing the bias results in a decrease of the number of the spectral lines present in the laser light. In fact, for a large enough bias, the number of spectral lines amounts not to a few thousands, but to just a few.

The reason for this paradoxical phenomenon is mode competition. The number of the radiative recombination events that may happen per unit time in the active region is limited by the injection current J; hence, the growth of the laser intensity is not infinite but must saturate at some limit intensity value F_L. Now, suppose that for some reason, one of the modes contains slightly more photons than others. This means that this mode intensity growth rate due to the stimulated emission exceeds that of other modes. Hence, the more photons a particular mode has, the sooner it should reach the limit intensity. Once a particular mode reaches F_L, all other modes cannot be supported because all of the pumping is "consumed" by a particular mode that initially had slightly more energy than the others. As a result, the winner mode gets it all, and a single narrow spectral line is emitted.

12.4 Problems

12.4.1 Solved Problems

Problem 1 The irradiance spectrum of the Sun just outside of the Earth's atmosphere can be modeled as that of a black body with the temperature $T_{Sun} = 5\,778$ K. Assume that each photon of energy $hf > E_g$ is absorbed by a SC and produces just one electron–hole pair, which then contributes the energy E_g to the output power. Based on this information, write down an expression for the limit SC efficiency as a function of the band gap, plot the efficiency vs. the ratio E_g/kT_{Sun}, and find the E_g value at which it is maximal. Which semiconductor has this band gap value? When solving this problem, you may need to integrate a certain function numerically (use Mable, Mathematica, WolframAlpha, or a similar software). You may also need an identity $\int_0^\infty dx \frac{x^3}{e^x - 1} = \frac{\pi^4}{15}$.

Problem 2 Consider a p-i-n structure of a heterojunction LED from Fig. 12.7b. It has the following parameter values at $T = 300$ K: the band gap $E_g = 2.5$ eV; the width of the i-region $d = 0.1\,\mu$m; intrinsic concentration $n_i = 10^7$ cm^{-3}; radiative recombination coefficient $B = 10^{-8}$ cm^3/s; Auger recombination coefficients $C_e = C_h = 10^{-30}$ cm^6/s; external efficiency $\eta_e = 2\%$. The relationship between the excess concentration Δn in the i-region and the applied voltage V follows immediately from the generalization of the mass–action law, $(n_0 + \Delta n)(p_0 + \Delta n) = n_i^2 e^{(E_{Fe} - E_{Fh})/kT} = n_i^2 e^{V/V_{th}}$.

(a) Assuming the i-region to be non-degenerate, find $\Delta n(V)$.
(b) Assuming that only radiative and Auger recombination mechanisms are operative in the i-region with the respective coefficients B and $C_{e,h}$, find the optimal excess concentration Δn at which the recombination efficiency in the i-region, $\eta_{rec} = U_r/(U_r + U_A)$, is the highest. What voltage is needed to maintain this excess concentration at room temperature?
(c) Find the output intensity value obtained at the maximal recombination efficiency.
(d) Find the voltage that needs to be applied to the device to have the output light intensity of 0.1 W/cm^2, which matches the typical intensity of the Sun.

Problem 3 Show that the average time for an atom to remain in the excited state is $1/b_{spon}$, where b_{spon} is Einstein's coefficient of spontaneous emission from Sect. 12.3.1.

12.4.2 Practice Problems

Problem 1 Find the intensity of the sunlight just outside of the Earth's atmosphere. The Sun radiation can be modeled as that of a black body with the temperature $5\,778$ K, the radius of the Sun is $R_S = 696\,340$ km, and the Earth-to-Sun distance is $R_{E-S} = 1.5 \cdot 10^8$ km. You may need an identity $\int_0^\infty dx \frac{x^3}{e^x - 1} = \frac{\pi^4}{15}$.

Problem 2 The total intensity of the monochromatic light incident onto a silicon slab of thickness $d = 100\,\mu$m is $F = 0.01\,$W/cm^2. Its wavelength is $1\,\mu$m. The refractive index of Si at this wavelength is $n_r = 3.572$, and the absorption coefficient $\alpha = 62.94\,$cm^{-1}. Calculate light intensity $F(x)$ and the electron–hole pair generation rate $G(x)$ for two cases: (a) when the rear surface is untreated and (b) when the rear surface is covered by a perfectly reflecting coating. Here, the distance x is measured from the front surface. Perform these calculations without and with taking into account Fresnel reflection by the silicon–air interface. You may need the sum of a geometric series, $1 + x + x^2 + \ldots = 1/(1 - x)$.

12.4.3 Solutions

Problem 1 The spectral irradiance of the Sun is given by (see Appendix A.5)

$$F(f) = A \frac{f^3}{e^{hf/kT_{Sun}} - 1} \,,$$

where the constant $A = \frac{2\pi h}{c^2} \frac{R_S^2}{R_{E-S}^2}$, where $R_S = 696\,340$ km is the radius of the Sun and $R_{E-S} = 1.5 \cdot 10^8$ km is the Earth-to-Sun distance. The ratio of the two radii takes care of the fact that the light intensity decreases with distance r from a point-like source as $1/r^2$.

With the simplifying assumptions made in the problem formulation, the incident and the output power per unit surface area of a SC can be written as

$$P_{in} = \int_0^\infty df\, F(f)\,, \quad P_{out} = E_g \int_{E_g/h}^\infty df\, \frac{F(f)}{hf}\,.$$

In the expression for P_{out}, we took care of the fact that only photons of energy exceeding E_g are absorbed and contribute the energy E_g to the output power; the spectral photon flux density is related to the spectral irradiance by $F(f)/(hf)$. The efficiency is

$$\eta = \frac{P_{out}}{P_{in}} = \frac{\int_{E_g/h}^{\infty} df\, F(f)}{\int_0^{\infty} df\, F(f)}.$$

Using the above expression for $F(f)$ and changing the integration variables in both integrals to $x = hf/kT_{Sun}$, we have

$$\eta = \frac{E_g}{kT_{Sun}} \frac{\int_{E_g/kT_{Sun}}^{\infty} dx\, \frac{x^2}{e^x-1}}{\int_0^{\infty} dx\, \frac{x^3}{e^x-1}} = \frac{15}{\pi^4} \frac{E_g}{kT_{Sun}} \int_{E_g/kT_{Sun}}^{\infty} dx\, \frac{x^2}{e^x - 1}.$$

The efficiency vs. the ratio of the band gap to the thermal energy of the surface of the Sun is shown in Fig. 12.11. This curve has a maximum at $E_g \approx 2.17\,kT_{Sun}$. Given that the thermal energy $kT_{Sun} = (0.02585\,\text{eV}) \cdot 5778/300 = 0.498\,\text{eV}$, the optimal band gap equals 1.08 eV, which is quite close to the band gap of silicon.

This analysis is severely simplified, as it does not account for the radiative and Auger recombination of the photogenerated electron–hole pairs and the optical losses due to the light reflection by the front surface and the escape of some of those photons that got reflected by the rear surface of the SC, as well as a few other, less significant, mechanisms. Nevertheless, the optimal band gap value obtained above is quite close to the truth.

Problem 2 (a) With $n_0 = p_0 = n_i$, we rewrite the modified mass–action law as $(n_i + \Delta n)^2 = n_i^2 e^{V/V_{th}}$, from which the excess concentration follows immediately as

$$\Delta n(V) = n_i (e^{V/(2V_{th})} - 1).$$

(b) The concentration of electrons and holes in the i-region is $n = p = n_i + \Delta n$. Hence, the radiative and Auger net recombination rate (see Sects. 7.3 and 7.4, respectively) recombination efficiencies are expressed in terms of the excess concentration as

$$\eta_{rec} = \frac{B(2n_i \Delta n + \Delta n^2)}{B(2n_i \Delta n + \Delta n^2) + (C_e + C_h)(3n_i^2 \Delta n + n_i \Delta n^2 + \Delta n^3)}$$

$$= \frac{1}{1 + \frac{C_e + C_h}{B} n_i f(x)},$$

where in the second equality, we divided the numerator and the denominator of the first expression by $B(2n_i \Delta n + \Delta n^2)$ and denoted

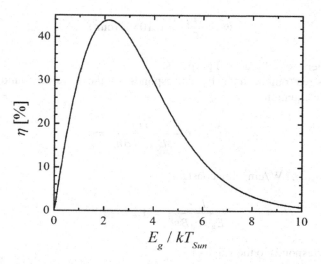

Fig. 12.11 The limit efficiency of a SC as a function of the semiconductor band gap normalized to the thermal energy on the surface of the Sun

$$x = \frac{\Delta n}{n_i}, \quad f(x) = \frac{3x + x^2 + x^3}{2x + x^2}.$$

Clearly, the recombination efficiency is maximal at that x-value at which the function $f(x)$ has a minimum. We find the minimum of $f(x)$ by differentiation at

$$x_0 = \sqrt{5} - 2 \approx 0.2361,$$

corresponding to $\Delta n = x n_i = 0.2361\, n_i$. Based on the result from part (a), we find the voltage needed to be

$$V_0 = 2V_{th} \ln(1 + x) = 2V_{th} \ln(\sqrt{5} - 1) = 11\,\text{mV}.$$

(c) The electroluminescence intensity is

$$I_{EL} = E_g\, U_r(\Delta n) d\eta_e = E_g d\eta_e\, B n_i^2 (2x + x^2).$$

With the $x = 0.2361$, we have

$$I_{EL} = 4.23 \cdot 10^{-20}\,\text{W/cm}^2,$$

corresponding to the photon flux density

$$\Phi = \frac{I_{EL}}{E_g} = 0.105 \ 1/(cm^2 s) \ .$$

That is zero for all practical purposes.

(d) The desired output intensity corresponds to the x-value found from the quadratic equation

$$x^2 + 2x - \frac{I_{EL}}{E_g d\eta_e B n_i^2} = 0 \ .$$

For $I_{EL} = 0.1 \ W/cm^2$, the constant

$$\frac{I_{EL}}{E_g d\eta_e B n_i^2} = 1.248 \cdot 10^{18} \ .$$

This corresponds to the value of x

$$x = -2 + \sqrt{4 + \frac{I_{EL}}{E_g d\eta_e B n_i^2}} = \sqrt{\frac{I_{EL}}{E_g d\eta_e B n_i^2}} = 1.117 \cdot 10^9 \ ,$$

as the parameters of the order of 1 can be neglected for all practical purposes. The voltage necessary to obtain this intensity is

$$V = 2V_{th} \ln x = 2 \cdot 0.02585 \ln 1.117 \cdot 10^9 = 1.08 \ V \ .$$

Problem 3 The probability $p_{exc}(t)$ for an atom to remain in the excited state is related to the concentration of atoms as $n_2(t) = p_{exc}(t)n$. It decays in time as

$$\frac{dp_{exc}(t)}{dt} = -b_{spon} p_{exc}(t) \ .$$

The solution of this equation with the initial condition $p_{exc}(0) = 1$ is $p_{exc}(t) = e^{-b_{spon}t}$. The derivative $-dp_{exc}(t)/dt$ has the physical meaning that the atom will return to the ground state at time between t and $t + dt$. The average time spent in the excited state is, then,

$$t_{exc} = \int_0^\infty dt \ t \left(-\frac{dp_{exc}(t)}{dt} \right) = b_{spon} \int_0^\infty dt \ t \ e^{-b_{spon}t} = \frac{1}{b_{spon}} \ ,$$

as can be obtained by changing the integration variable to $x = b_{spon}t$.

Appendices

A.1 A Crash Course in Complex Numbers

A complex number z is formed by a pair of two real numbers, x and y, as

$$z = x + iy .$$

The number i is called the imaginary unit and is defined as the positive square root of -1:

$$i^2 = -1 , \quad i = \sqrt{-1} .$$

The two components of a complex number z are called the real part and the imaginary part of z:

$$x = \mathrm{Re}z , \quad y = \mathrm{Im}z .$$

The operation of complex conjugation is denoted by a star and is defined by changing the sign in front of the imaginary part:

$$z^* = x - iy .$$

Complex numbers can be added and subtracted. If we have another complex number $w = u + iv$, then

$$z + w = (x + u) + i(u + v) , \quad z - w = (x - u) + i(y - v) .$$

Two complex numbers can be multiplied together:

$$zw = (x + iy)(u + iv) = x(u + iv) + iy(u + iv) = xu + ixv + iyu - yv$$
$$= (xy - yv) + i(xv + yu) .$$

The product of a complex number with its conjugate,

$$zz^* = (x + iy)(x - iy) = x^2 - (iy)^2 = x^2 + y^2$$

is a positive real number. Its positive square root,

$$|z| = \sqrt{zz^*} = \sqrt{x^2 + y^2}$$

is called the modulus of z.

A complex number can also be divided by another complex number:

$$\frac{z}{w} = \frac{zw^*}{ww^*} = \frac{zw^*}{|w|^2} = \frac{(x + iy)(u - iv)}{u^2 + v^2} = \frac{xu + yv}{u^2 + v^2} + i\frac{yu - xv}{u^2 + v^2} .$$

Euler's identity is an extremely useful theorem. Let θ be a real number, and thus $i\theta$ is purely imaginary. Then

$$e^{i\theta} = \cos(\theta) + i\sin(\theta) .$$

To prove this identity, we denote two complex-valued functions of θ:

$$f(\theta) = e^{i\theta} , \quad g(\theta) = \cos(\theta) + i\sin(\theta) .$$

Their logarithms have the same derivatives. Indeed, for the function $f(\theta)$,

$$\frac{d\ln f(\theta)}{d\theta} = \frac{d(i\theta)}{d\theta} = i .$$

As for the function $g(\theta)$, we first observe that

$$\frac{dg(\theta)}{d\theta} = -\sin(\theta) + i\cos(\theta) = i(\cos(\theta) + i\sin(\theta)) = ig(\theta) ,$$

and hence

$$\frac{1}{g(\theta)}\frac{dg(\theta)}{d\theta} = \frac{d\ln g(\theta)}{d\theta} = i .$$

From the identity $\frac{d\ln f(\theta)}{d\theta} = \frac{d\ln g(\theta)}{d\theta}$ it follows that

$$f(\theta) = Cg(\theta) ,$$

where C is some unknown constant. In order to find this constant, we focus on the specific value of $\theta = 0$, for which $f(0) = g(0) = 1$. Hence, $C = 1$, i.e., $f(\theta) = g(\theta)$ for all θ. This completes the proof.

As an illustration of usefulness of Euler's identity, it can be used to derive various trigonometric identities, which one may have a hard time memorizing. Consider the product

$$e^{i\alpha} e^{i\beta} = (\cos\alpha + i\sin\alpha)(\cos\beta + i\sin\beta)$$
$$= \cos\alpha\cos\beta - \sin\alpha\sin\beta + i(\cos\alpha\sin\beta + \sin\alpha\cos\beta) .$$

On the other hand,

$$e^{i\alpha} e^{i\beta} = e^{i(\alpha+\beta)} = \cos(\alpha + \beta) + i\sin(\alpha + \beta) .$$

Two complex numbers are equal if both their real and imaginary parts are equal. We thus conclude that

$$\cos(\alpha + \beta) = \cos\alpha\cos\beta - \sin\alpha\sin\beta ,$$
$$\sin(\alpha + \beta) = \cos\alpha\sin\beta + \sin\alpha\cos\beta .$$

These identities can be proven without complex numbers, but the proof will be much longer.

Note that the trigonometric functions can be expressed in terms of complex exponentials as follows. Write Euler's identity for angles θ and $-\theta$:

$$e^{i\theta} = \cos(\theta) + i\sin(\theta) , \quad e^{-i\theta} = \cos(\theta) - i\sin(\theta) .$$

Adding up these two expressions and dividing the result by 2 gives $\cos(\theta)$, whereas subtracting these expressions from each other and dividing the result by $2i$ gives $\sin(\theta)$:

$$\cos(\theta) = \frac{e^{i\theta} + e^{-i\theta}}{2} , \quad \sin(\theta) = \frac{e^{i\theta} - e^{-i\theta}}{2i} .$$

More importantly, Euler's identity allows one to represent a complex number z in a convenient manner. Namely, a complex number z can be represented as a point in the complex plane whose coordinates are formed by its real and imaginary parts, see Fig. A.1. The distance from that point to the origin is modulus of the complex number z. The angle formed by the line joining the origin with that point is called the argument of z. On the diagram, this angle must be measured in the counterclockwise direction. Geometrically

$$\theta = \arg(z) = \tan^{-1}\frac{y}{x} ; \quad x = |z|\cos\theta ; \quad y = |z|\sin\theta .$$

Fig. A.1 Representation of a complex number $z = x + iy$ as a point in the complex plane

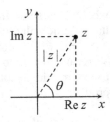

Then, a complex number z can be represented as

$$z = x + iy = |z|(\cos\theta + i\sin\theta) \, .$$

Applying Euler's identity, we obtain the aforementioned representation of a complex number:

$$z = |z|e^{i\theta} \, .$$

A.2   Proof of Bloch's Theorem

When a periodic lattice potential is present, then all measurable physical properties of the electron gas, such as its density, should likewise be periodic. The single-electron wave function $\Psi(\vec{r})$, however, does not have to have the lattice periodicity. But the wave function itself is not physically measurable; it is the probability density that can be measured. Hence, we must have for any translation vector $\vec{d} = i\vec{a} + j\vec{b} + k\vec{c}$

$$|\Psi(\vec{r})|^2 = |\Psi(\vec{r} + \vec{d})|^2 \, .$$

The most general choice for a wave function must with this property is

$$\Psi(\vec{r}) = \frac{1}{\sqrt{V}} e^{i\theta(\vec{r})} u(\vec{r}) \, ,$$

where the factor $\frac{1}{\sqrt{V}}$ is introduced for later convenience, the function $u(\vec{r})$ is periodic,

$$u(\vec{r} + \vec{d}) = u(\vec{r}) \, ,$$

and the unknown function $\theta(\vec{r})$ is not periodic.[1]

---

[1] If $\theta(\vec{r})$ had a periodic and a non-periodic part, i.e., $\theta(\vec{r}) = \theta_{per}(\vec{r}) + \theta_{nonper}(\vec{r})$ with $\theta_{per}(\vec{r} + \vec{d}) = \theta_{per}(\vec{r})$, then we could define our new periodic function $u(\vec{r})$ as a product of the old $u(\vec{r})$ and $e^{i\theta_{per}(\vec{r})}$. Hence, we can explicitly demand that $\theta(\vec{r})$ be not periodic in $\vec{r}$ without loss of generality.

Let us define a new wave function, which differs from the original one $\Psi(\vec{r})$ in that its argument $\vec{r}$ is shifted by the translation vector $\vec{a}$:

$$\Psi_{shft}(\vec{r}) = \Psi(\vec{r}+\vec{a}) = \frac{1}{\sqrt{V}}e^{i\theta(\vec{r}+\vec{a})}u(\vec{r}+\vec{a}) = \frac{1}{\sqrt{V}}e^{i\theta(\vec{r}+\vec{a})}u(\vec{r}) \,,$$

where in the last equality, we used the periodicity property of $u(\vec{r}) = u(\vec{r}+\vec{a})$. Let us define the difference

$$\Delta\theta(\vec{r}) = \theta(\vec{r}+\vec{a}) - \theta(\vec{a}) \,,$$

in terms of which

$$\Psi_{shft}(\vec{r}) = \frac{1}{\sqrt{V}}e^{i(\theta(\vec{r})+\Delta\theta(\vec{r}))}u(\vec{r}) = \frac{1}{\sqrt{V}}e^{i\Delta\theta(\vec{r})}e^{i\theta(\vec{r})}u(\vec{r}) = e^{i\Delta\theta(\vec{r})}\Psi(\vec{r}) \,.$$

Because of the translational symmetry of the potential $U(\vec{r}) = U(\vec{r}+\vec{a})$, shifting the coordinate $\vec{r}$ by a translation vector $\vec{a}$ does not affect the time-independent Schrödinger's equation. This means that $\Psi_{shft}(\vec{r})$ should satisfy the same Schrödinger's equation as $\Psi(\vec{r})$, that is,

$$-\frac{\hbar^2}{2m_e}\vec{\nabla}^2\Psi_{shft}(\vec{r}) + U(\vec{r})\Psi_{shft}(\vec{r}) = E\Psi_{shft}(\vec{r}) \,.$$

Substitution of the relation $\Psi_{shft}(\vec{r}) = e^{i\Delta\theta(\vec{r})}\Psi(\vec{r})$, however, produces two new terms, which are proportional to $\vec{\nabla}\Delta\theta(\vec{r})$ and $\vec{\nabla}^2\Delta\theta(\vec{r})$. Those terms are not present in the original Schrödinger's equation for $\Psi(\vec{r})$. The only way to get rid of those terms is to demand that

$$\Delta\theta(\vec{r}) = \Delta\theta_a$$

is a constant independent of $\vec{r}$.

Hence, shifting the coordinate by one lattice vector $\vec{a}$ results in an increment of the phase by a constant:

$$\theta(\vec{r}+\vec{a}) = \theta(\vec{r}) + \Delta\theta_a \,.$$

Likewise, we could shift the coordinate by either of the remaining two lattice vectors, $\vec{b}$ or $\vec{c}$, and repeated the above argument. This would lead us to the conclusion that

$$\theta(\vec{r}+\vec{b}) = \theta(\vec{r}) + \Delta\theta_b \,, \quad \theta(\vec{r}+\vec{c}) = \theta(\vec{r}) + \Delta\theta_c$$

with constant θ_b and θ_c.

The only function $\theta(\vec{r})$ that has these properties is a linear function:

$$\theta(\vec{r}) = \vec{k} \cdot \vec{r} \, ,$$

up to an immaterial additive constant. Then,

$$\theta_a = \vec{k} \cdot \vec{a} \, , \quad \theta_b = \vec{k} \cdot \vec{b} \, , \quad \theta_c = \vec{k} \cdot \vec{c} \, .$$

We now need to figure out the wave vector $\vec{k}$. We do this by imposing the periodic boundary conditions. Namely, we view the crystal as a large parallelepiped containing N^3 unit cells. It is built on the three translation vectors $N\vec{a}$, $N\vec{b}$, and $N\vec{c}$. We demand that the wave function $\Psi(\vec{r})$ repeat itself every time its argument gets shifted by one of those vectors:

$$\Psi(\vec{r} + N\vec{a}) = \Psi(\vec{r} + N\vec{b}) = \Psi(\vec{r} + N\vec{c}) = \Psi(\vec{r}) \, .$$

This is only possible if $N\vec{k} \cdot \vec{a}$, $N\vec{k} \cdot \vec{b}$, and $N\vec{k} \cdot \vec{c}$ are all integer multiples of 2π. In turn, this implies that the wave vector

$$\vec{k} = \frac{2\pi \vec{m}}{N} \, ,$$

where $\vec{m}$ is a vector whose all three components are integers.

We finally substitute the wave function $\Psi(\vec{r}) = \frac{1}{\sqrt{V}} e^{i\vec{k} \cdot \vec{r}} u(\vec{r})$ into the Schrödinger's equation:

$$-\frac{\hbar^2}{2m_e} \vec{\nabla}^2 \left(e^{i\vec{k} \cdot \vec{r}} u(\vec{r}) \right) + U(\vec{r}) \, e^{i\vec{k} \cdot \vec{r}} u(\vec{r}) = E \, e^{i\vec{k} \cdot \vec{r}} u(\vec{r}) \, .$$

The first term can be transformed to

$$-\frac{\hbar^2}{2m_e} \vec{\nabla}^2 \left(e^{i\vec{k} \cdot \vec{r}} u(\vec{r}) \right) = -\frac{\hbar^2}{2m_e} \left(\vec{\nabla}^2 u(\vec{r}) - 2i\vec{k} \cdot \vec{\nabla} u(\vec{r}) - \vec{k}^2 u(\vec{r}) \right) e^{i\vec{k} \cdot \vec{r}} \, .$$

Recalling (see Sect. 1.8.1) that electron momentum is identified with the operator $\hat{\vec{p}} u(\vec{r}) = -i\hbar \vec{\nabla} u(\vec{r})$, we rewrite this in terms of the momentum as

$$\frac{1}{2m_e} \left(\hat{\vec{p}}^2 u(\vec{r}) - 2\hbar\vec{k} \cdot \hat{\vec{p}} u(\vec{r}) + \hbar^2 \vec{k}^2 u(\vec{r}) \right) e^{i\vec{k} \cdot \vec{r}} = \frac{e^{i\vec{k} \cdot \vec{r}}}{2m_e} \left(\hat{\vec{p}} - \hbar\vec{k} \right)^2 u(\vec{r}) \, .$$

After canceling the factors $e^{i\vec{k} \cdot \vec{r}}$, we arrive at the equation for $u(\vec{r})$:

$$\frac{1}{2m_e} \left(\hat{\vec{p}} - \hbar\vec{k} \right)^2 u(\vec{r}) + U(\vec{r}) u(\vec{r}) = E u(\vec{r}) \, ,$$

where the momentum operator $\hat{\vec{p}} = -i\hbar\vec{\nabla}$, and the square of the operator $\hat{\vec{p}} - \hbar\vec{k}$ simply means applying it twice to the function $u(\vec{r})$. If not for the term $\hbar k$ in the brackets, the equation obtained would look like a time-independent Schrödinger's equation for the wave function $u(\vec{r})$ in the periodic potential $U(\vec{r} + \vec{d}) = U(\vec{r})$ produced by the lattice.

In the absence of the lattice potential, i.e., with $U(\vec{r}) = 0$, the function $u(\vec{r}) = 1$ for all positions $\vec{r}$. Then, $(\hat{\vec{p}} - \hbar\vec{k})u(\vec{r}) = -\hbar\vec{k}$, and $E = (\hbar\vec{k})^2/(2m_e)$, as it should be.

Together with the periodicity requirement, $u(\vec{r} + \vec{d}) = u(\vec{d})$, this equation provides full information from which $u(\vec{r})$ and E can be found. We indicate the $\vec{k}$-dependence of its solution as a subscript in $u_{\vec{k}}(\vec{r})$ and as an argument of the function $E(\vec{k})$. Furthermore, at each given $\vec{k}$, another quantum number n should arise, which labels different solutions of this equation for fixed $\vec{k}$. This quantum number—the band index—is indicated as a subscript both in $u_{\vec{k}}(\vec{r})$ and $E(\vec{k})$. Hence,

$$u(\vec{r}) = u_{n\vec{k}}(\vec{r}) \,, \quad E = E_n(\vec{k}) \,.$$

A.3 Properties of Si, Ge, and GaAs

Unless stated otherwise, the numerical data in this section are taken from the online data base of the Ioffe Institute at http://www.ioffe.ru/SVA/NSM/Semicond. Crystal structure and lattice constant:

$$\text{Si (diamond)} : \quad a = 5.431\,\text{Å} \,;$$

$$\text{Ge (diamond)} : \quad a = 5.658\,\text{Å} \,;$$

$$\text{GaAs (zincblende)} : \quad a = 5.65325\,\text{Å} \,.$$

Bandgap temperature dependence:

$$E_g(T) = E_g(0) - \frac{\alpha T^2}{T + \beta}$$

with the parameters:

$$\text{Si} : \quad E_g(0) = 1.17\,\text{eV} \,; \quad \alpha = 4.73 \cdot 10^{-4}\,\text{eV/K} \,; \quad \beta = 636\,\text{K} \,;$$

$$\text{Ge} : \quad E_g(0) = 0.742\,\text{eV} \,; \quad \alpha = 4.8 \cdot 10^{-4}\,\text{eV/K} \,; \quad \beta = 235\,\text{K} \,;$$

$$\text{GaAs} : \quad E_g(0) = 1.519\,\text{eV} \,; \quad \alpha = 5.405 \cdot 10^{-4}\,\text{eV/K} \,; \quad \beta = 204\,\text{K} \,.$$

Transverse and longitudinal effective mass components of an electron and the number of valleys in the conduction band:

$$Si : m_{e,t}^*/m_e = 0.19 , \quad m_{e,l}^*/m_e = 0.98 ; \quad \nu_C = 6 ;$$

$$Ge : m_{e,t}^*/m_e = 0.0815 , \quad m_{e,l}^*/m_e = 1.59 , \quad \nu_C = 4 ;$$

$$GaAs : m_{e,t}^*/m_e = m_{e,l}^*/m_e = 0.063 , \quad \nu_C = 1 .$$

Heavy and light hole effective masses:

$$Si : m_{hh}^*/m_e = 0.49 , \quad m_{lh}^*/m_e = 0.16 ;$$

$$Ge : m_{hh}^*/m_e = 0.33 , \quad m_{lh}^*/m_e = 0.043 ;$$

$$GaAs : m_{hh}^*/m_e = 0.51 , \quad m_{lh}^*/m_e = 0.082 .$$

Effective densities of states at 300 K:

$$Si : N_C = 3.2 \cdot 10^{19}\,cm^{-3} , \quad N_V = 1.8 \cdot 10^{19}\,cm^{-3} ;$$

$$Ge : N_C = 1.0 \cdot 10^{19}\,cm^{-3} , \quad N_V = 5.0 \cdot 10^{18}\,cm^{-3} ;$$

$$GaAs : N_C = 4.7 \cdot 10^{17}\,cm^{-3} , \quad N_V = 9.0 \cdot 10^{18}\,cm^{-3} .$$

Bandgap and intrinsic concentration at $T = 300\,K$:[2]

$$Si : E_g(300\,K) = 1.12\,eV ; \quad n_i(300\,K) = 8.59 \cdot 10^9\,cm^{-3} ;$$

$$Ge : E_g(300\,K) = 0.661\,eV ; \quad n_i(300\,K) = 1.97 \cdot 10^{13}\,cm^{-3} ;$$

$$GaAs : E_g(300\,K) = 1.424\,eV ; \quad n_i(300\,K) = 2.31 \cdot 10^6\,cm^{-3} .$$

Relative dielectric permittivities:

$$Si : \epsilon = 11.7 ;$$

$$Ge : \epsilon = 16.2 ;$$

$$GaAs : \epsilon = 12.9 .$$

---

[2] Some discrepancy exists in the literature regarding the intrinsic concentration in Si at 300 K, which is $10^{10}\,cm^{-3}$ according to the data base of the Ioffe Institute. For the sake of formal consistency, it is calculated here as $n_i = \sqrt{N_C N_V}\, e^{-E_g/(2kT)}$ rather than taken from this data base. The most recent experimental result in Si is $n_i(300\,K) = 9.65 \cdot 10^9\,cm^{-3}$, see P.P. Altermatt, A. Schenk, F. Geelhaar, and G. Heiser, Journal of Applied Physics **93**, 1598 (2003).

Ionization energies of shallow donor impurities:

$$\text{Si} \; : \; \text{As: } 0.054\,\text{eV} \; ; \; \text{P: } 0.045\,\text{eV} \; ; \; \text{Sb: } 0.043\,\text{eV} \; ;$$

$$\text{Ge} \; : \; \text{As: } 0.014\,\text{eV} \; ; \; \text{P: } 0.013\,\text{eV} \; ; \; \text{Sb: } 0.010\,\text{eV} \; ;$$

$$\text{GaAs} \; : \; \text{Si: } 0.006\,\text{eV} \; ; \; \text{Ge: } 0.006\,\text{eV} \; ; \; \text{Sn: } 0.006\,\text{eV} \; .$$

Ionization energies of shallow acceptor impurities:

$$\text{Si} \; : \; \text{Al: } 0.072\,\text{eV} \; ; \; \text{B: } 0.045\,\text{eV} \; ; \; \text{Ga: } 0.074\,\text{eV}; \; \text{In: } 0.157\,\text{eV} \; ;$$

$$\text{Ge} \; : \; \text{Al: } 0.011\,\text{eV} \; ; \; \text{B: } 0.011\,\text{eV} \; ; \; \text{Ga: } 0.011\,\text{eV}; \; \text{In: } 0.012\,\text{eV} \; ;$$

$$\text{GaAs} \; : \; \text{Si: } 0.1\,\text{eV} \; ; \; \text{Ge: } 0.03\,\text{eV} \; ; \; \text{Sn: } 0.2\,\text{eV}; \; \text{Zn: } 0.025\,\text{eV} \; .$$

Radiative recombination coefficient at $T = 300\,\text{K}$:[3]

$$\text{Si} \; : \; B = 4.7 \cdot 10^{-15}\,\text{cm}^3/\text{s} \; ;$$

$$\text{Ge} \; : \; B = 6.41 \cdot 10^{-14}\,\text{cm}^3/\text{s} \; ;$$

$$\text{GaAs} \; : \; B = 7.2 \cdot 10^{-10}\,\text{cm}^3/\text{s} \; .$$

Auger recombination coefficients at $T = 300\,\text{K}$:

$$\text{Si} \; : \; C_e = 1.1 \cdot 10^{-30}\,\text{cm}^6/\text{s} \; ; \; C_h = 3.0 \cdot 10^{-29}\,\text{cm}^6/\text{s} \; ;$$

$$\text{Ge} \; : \; C_e \sim C_h \sim 10^{-30}\,\text{cm}^6/\text{s} \; ;$$

$$\text{GaAs} \; : \; C_e \sim C_h \sim 10^{-30}\,\text{cm}^6/\text{s} \; .$$

Electron affinities at $T = 300\,\text{K}$:

$$\text{Si} \; : \; e\chi = 4.05\,\text{eV} \; ;$$

$$\text{Ge} \; : \; e\chi = 4.00\,\text{eV} \; ;$$

$$\text{GaAs} \; : \; e\chi = 4.07\,\text{eV} \; .$$

---

[3] The value of B in silicon is taken from T. Trupke, M.A. Green, P. Würfel, P.P. Altermatt, A. Wang, J. Zhao, and R. Corkish, "Temperature dependence of the radiative recombination coefficient of intrinsic crystalline silicon," Journal of Applied Physics, **94**, 4930 (2003).

A.4   Evaluation of Exponential Integrals

Here, we would like to find the value of the integral

$$I_n = \int_0^\infty dx\, x^{n+1/2}\, e^{-x} \text{ for } n = 0, 1, 2, 3, \ldots$$

First, let us focus on the simplest case $n = 0$. We change the integration variables to $y = \sqrt{x}$, giving

$$I_0 = \int_0^\infty dx\, \sqrt{x}\, e^{-x} = 2 \int_0^\infty dy\, y^2\, e^{-y^2} = \int_{-\infty}^\infty dy\, y^2\, e^{-y^2}$$

$$= -\frac{d}{da} \left(\int_{-\infty}^\infty dy\, e^{-ay^2} \right)\Bigg|_{a=1} = -\frac{df(a)}{da}\Bigg|_{a=1}.$$

In the last line, we introduced a parameter a, later to be set to 1. Differentiation with respect to the parameter a turns the more complicated integral of $y^2 e^{-y^2}$ into the one with a simpler integrand e^{-ay^2}. Let us call the function of a that is to be differentiated

$$f(a) = \int_{-\infty}^\infty dy\, e^{-ay^2}$$

and consider the square of this function:

$$f^2(a) = \int_{-\infty}^\infty dy_1 \int_{-\infty}^\infty dy_2\, e^{-a(y_1^2 + y_2^2)}.$$

Transforming to the polar coordinates, $r = \sqrt{y_1^2 + y_2^2}$, $\theta = \cos^{-1} \frac{y_1}{\sqrt{y_1^2 + y_2^2}}$, we must also replace the integration element—the area $dy_1\, dy_2$—with $r\, dr\, d\theta$. Noting that the integrand does not depend on the polar angle, we can write:

$$f^2(a) = 2\pi \int_0^\infty dr\, r\, e^{-ar^2},$$

where integration over the polar angle $\theta = 0 \ldots 2\pi$ has been performed, giving 2π. In other words, the surface area $dy_1\, dy_2$ got replaced by the area $2\pi r\, dr$ of a ring of radius r and thickness dr. The last integral is found by changing the integration variables to $z = r^2$:

$$f^2(a) = \pi \int_0^\infty dz\, e^{-az} = \frac{\pi}{a}.$$

Hence,

$$f(a) = \int_{-\infty}^{\infty} dy\, e^{-ay^2} = \sqrt{\frac{\pi}{a}}\,.$$

Coming back to our original task,

$$I_0 = \int_0^{\infty} dx\, \sqrt{x}\, e^{-x} = 2\int_0^{\infty} dy\, y^2\, e^{-y^2} = -\frac{d\sqrt{\pi/a}}{da}\bigg|_{a=1} = \frac{\sqrt{\pi}}{2}\,.$$

Now, let us evaluate I_n for $n \geq 1$. For this, let us define a function

$$F(a) = \int_0^{\infty} dx\, \sqrt{x}\, e^{-ax}\,,$$

which, upon a change of the integration variable to $y = ax$, becomes

$$F(a) = \frac{I_0}{a^{3/2}}\,.$$

It allows one to obtain I_n by differentiating with respect to a and setting a to one:

$$I_n = (-1)^n \frac{d^n F(a)}{da^n}\bigg|_{a=1}\,,$$

as can be verified by using the integral formula for $F(a)$ and the fact that $\frac{d^n}{da^n} e^{-ax} = (-1)^n x^n e^{-ax}$. In particular, we have

$$I_1 = \frac{3}{2}I_0 = \frac{3\sqrt{\pi}}{4}\,; \quad I_2 = \frac{3\cdot 5}{2^2}I_0 = \frac{15\sqrt{\pi}}{8}\,; \quad I_3 = \frac{3\cdot 5\cdot 7}{2^3}I_0 = \frac{105\sqrt{\pi}}{16}\,; \quad \text{etc.}$$

A.5   Planck's Radiation Law

The quantum state of a photon is characterized by two quantum numbers: its wave vector $\vec{k}$ and its polarization s. The polarization can have two values, $s = +1$ and -1, which correspond to the counterclockwise and clockwise rotation of the electric field vector around the direction $\vec{k}/|\vec{k}|$ of photon propagation. We consider the photons in a cube of side length L and impose the periodic boundary conditions. Hence, the three components of the wave vector have quantized values:

$$k_\alpha = \frac{2\pi}{L}m_\alpha\,, \quad \alpha = x, y, z \ \text{and} \ m_\alpha = \ldots, -2, -1, 0, 1, 2, \ldots\,.$$

The probability to have n photons in the state with particular values of $\vec{k}$ and s is

$$p_n(\vec{k}, s) = \frac{e^{-nhf(\vec{k})/kT}}{Z} \, ,$$

where the frequency of a photon in this state is

$$f(\vec{k}) = \frac{ck}{2\pi} \, , \quad k = |\vec{k}| \, ,$$

c being the speed of light. The constant Z has the value that normalizes the total probability to one:

$$Z = \sum_{n=0}^{\infty} e^{-nhf(\vec{k})/kT} = \frac{1}{1 - e^{-hf(\vec{k})/kT}} \, ,$$

where we used the expression for the sum of a geometric series, $\sum_{n=0}^{\infty} x^n = 1/(1 - x)$ with $x = e^{-hf/kT} < 1$. The average number of photons in the state $(\vec{k}, s)$ is

$$\langle n(\vec{k}, s) \rangle = \sum_{n=0}^{\infty} n p_n(\vec{k}, s) \, .$$

It can be obtained by setting

$$a = \frac{hf(\vec{k})}{kT}$$

and noting that

$$\langle n(\vec{k}, s) \rangle = \frac{\sum_{n=0}^{\infty} n e^{-na}}{\sum_{n=0}^{\infty} e^{-na}} = -\frac{d \ln \sum_{n=0}^{\infty} e^{-na}}{da} = -\frac{d}{da} \ln \frac{1}{1 - e^{-a}} = \frac{1}{e^a - 1} \, .$$

We finally replace the parameter a with the ratio of photon energy to the thermal energy to obtain

$$\langle n(\vec{k}, s) \rangle = \frac{1}{e^{-hf(\vec{k})/kT} - 1} \, .$$

To find the concentration of photons $n_{ph}(f)$ of frequency f, we need to multiply the average number of photons in the $(\vec{k}, s)$-state with the density of photon states, i.e., the number of $(\vec{k}, s)$ states in a unit volume per unit frequency interval. The total number of the photon states with the wave vector smaller than a given value k equals the ratio of the volume of a sphere of radius k to the "volume" $(2\pi/L)^3$ occupied by one photon state; this ratio needs to be multiplied by 2 in order to account for two possible polarizations. Hence,

$$N(k) = 2 \cdot \frac{4\pi k^3}{3} \cdot \frac{1}{(2\pi/L)^3} = \frac{k^3}{3\pi^2} L^3 \ .$$

The total number of photon states of frequency smaller than a given value f is obtained by replacing k with $2\pi f/c$:

$$N(f) = \frac{8\pi f^3}{3c^3} L^3 \ .$$

The photon density of states is obtained as the derivative

$$g(f) = \frac{1}{L^3} \frac{dN(f)}{df} = \frac{8\pi f^2}{c^3} \ .$$

Hence, in thermal equilibrium, the concentration of photons per unit frequency interval is

$$n_{ph}(f) = g(f)\langle n(\vec{k}, s)\rangle = \frac{8\pi f^2}{c^3} \frac{1}{e^{hf/kT} - 1} \ .$$

In the literature, it is more common to work with the spectral energy density, i.e., the energy of photons in a unit volume per unit frequency interval. It is given by the Planck's radiation law:

$$u(f) = hf n_{ph}(f) = \frac{8\pi h f^3}{c^3} \frac{1}{e^{hf/kT} - 1} \ .$$

The spectral irradiance is given by

$$F(f) = \frac{c}{4} u(f) = \frac{2\pi h f^3}{c^2} \frac{1}{e^{hf/kT} - 1} \ .$$

Index

© The Author(s), under exclusive license to Springer Nature Switzerland AG 2022   319
M. Evstigneev, *Introduction to Semiconductor Physics and Devices*,
https://doi.org/10.1007/978-3-031-08458-4

Printed in the United States
by Baker & Taylor Publisher Services